T0191796

Mathématiques et Applications

80

More information about this series at http://www.springer.com/series/2966

Paul Doukhan

Stochastic Models for Time Series

Paul Doukhan
Laboratory of Mathematics
University Cergy-Pontoise
Cergy-Pontoise
France

ISSN 1154-483X ISSN 2198-3275 (electronic)
Mathématiques et Applications
ISBN 978-3-319-76937-0 ISBN 978-3-319-76938-7 (eBook)
https://doi.org/10.1007/978-3-319-76938-7

Library of Congress Control Number: 2018934882

Mathematics Subject Classification (2010): 60G10, 37M10

Printed on acid-free paper

This Springer imprint is published by the registered company Springer International Publishing AG part of Springer Nature
The registered company address is: Gewerbestrasse 11, 6330 Cham, Switzerland

Cet ouvrage est dédié à Sarah et Jean que j'ai participé à faire vivre et qui me le rendent au centuple…

Preface

Time series appear naturally with data sampled in time, but many other physical situations also lead to evolutions indexed by integers (e.g., enumeration of nucleotides on a DNA chain).

We provide some tools for the study of such statistical models. The purpose of this book is introductory, and this is definitely not a systematic study.

This book is divided into three parts, each including four chapters, and three appendices.

Independence and Stationarity

Even though this part addresses mainly items of the independent world, the choice of subjects is biased and motivated by the fact that they easily extend to a dependent setting.

(a) Independence.
This is a main concept in these notes so we include some simple comments concerning independence as a separate chapter. For instance, we mention all the elementary counterexamples invoking independence. Other examples relating orthogonality with independence may be found in Chap. 8 and in Appendix A.3.

(b) Gaussian convergence and moments.
Special emphasis is given on the Lindeberg method which easily extends to a dependent setting. Applications of the central limit theorems are proved in the independent setting. Moment and exponential inequalities related to Gaussian convergence are also derived.

(c) Estimation concepts.
Standard estimations techniques, such as empirical ones, contrasts and non-parametric techniques are introduced. Kernel density estimators are described

in some detail as an application of previous results in view of their extension to time series in a further chapter.

(d) Stationarity.

The notions of stationarity are essential for spectral analysis of time series. Brockwell and Davis (1991) use filtering techniques in order to return to such a simple stationary case. Indeed, this assumption is not naturally observed. Weak stationarity and strong stationarity are considered together with examples.

Second-order weak dependence or long-range dependence is defined according to the convergence of the series of covariances.

Stationarity and an introduction to spectral techniques are then discussed. We provide the spectral representation for both a covariance and the process itself, and we briefly describe some applications of time series.

Models of Time Series

(a) Gaussian chaos.

Due to the CLT, the Gaussian case plays a central role in statistics. The first time series to be considered are Gaussian. We introduce the Gaussian chaos and Hermite polynomials as well as some of their properties. Gaussian processes and the methods of Gaussian chaos are investigated. Namely, Hermite representations and the Mehler formula for functions of Gaussian processes are developed precisely, while the diagram formula for higher-order moments is simply considered. The fractional Brownian motion essential hereafter for the long-range dependent setting is also introduced. The asymptotic theory for Gaussian functionals is also precisely stated. We recall the fourth-moment method based on Malliavin calculus.

(b) Linear models.

From Lindeberg's lemma, the linear case is the second case to consider after the Gaussian one. For example, ARMA processes are weakly dependent processes, and ARFIMA models are long-range dependent. We again refer to Brockwell and Davis (1991) for further information.

(c) Nonlinear models.

This central chapter proposes a wide botany for the models of time series. Nonlinear models are naturally considered as extensions of the previous ones. The elementary ideas of polynomials and chaoses are first considered. We then check an algebraic approach to the models which are explicit solutions of a recursion equation. General contractive iterative systems with (non-explicit) stationary solutions are introduced. Finally, the abstract Bernoulli shifts yield a general and simple overview of those various examples; their correlation properties are explicitly provided. This class of general nonlinear functionals of independent sequences yields a large number of examples.

(d) Association.
Associated processes are then briefly investigated. It was introduced for reliability and for statistical physics. The association property admits a main common point with the Gaussian case: Independence and orthogonality coincide in both cases. This feature is exploited in the following chapter.

Dependences

(a) Ergodic theorem.
As an extension of the strong law of large numbers, the ergodic theorem is the first result proposed in this chapter. In order to find confidence bounds for asymptotic distribution of the mean, one first needs consistency of the empirical mean. Also needed asymptotic expansions are obtained from SRD/LRD properties.
We then make a tour of the tools for the asymptotic theory under long-range or short-range dependence (resp. SRD and LRD).
(b) Long-range dependence.
Under LRD, the more elementary examples are seen to have such asymptotic explicit expansion in distribution up to non-Gaussian limits. Gaussian and subordinated Gaussians are first considered as well as linear LRD models, and a rapid description of nonlinear LRD models is also included.
(c) Short-range dependence.
In the SRD case, we give a rapid idea of techniques. Namely, the standard Bernstein blocks technique is proposed as a way to derive CLTs by using a recent dependent Lindeberg approach.
(d) Moment methods.
The final chapter is devoted to moment and cumulant inequalities developing the more standard spectral ideas of the Chap. 2.
Such inequalities are needed on many occasions, but first in order to derive CLTs, another application is for subsampling. This technique applies to the kernel density estimator.

Appendices

The appendices recall some basic definitions and some R scripts for figures. The reader is also referred to the index for notations which may differ from one author to another.

(A) Probability.
The first appendix recalls essential concepts of probability, including cumulative distribution functions and some Hoeffding's inequalities.
Useful examples of probability distributions are introduced in relation to the dependence conditions. Standard Gaussians, Gaussian vectors, and γ-distributions are then considered.

(B) Convergence and processes.
In the second appendix, some basic concepts of random processes and convergence are recalled. Sufficient conditions for convergence are also briefly discussed. Basic notations of statistics and martingale theory are also provided.

(C) R scripts used for the figures.
The software R is used for figures (Team 2017). This may be useful for a reader who wants to process alternative simulation procedures.

For elementary sets, we shall use the notations $\mathbb{N} = \{0, 1, 2, \ldots\}$ and $\mathbb{Z} = \{0, \pm 1, \pm 2, \ldots\}$, and $\mathbb{R}$ and $\mathbb{C}$ respectively, denote the sets of rational, real, and complex numbers. Other notations may be found in the index.

Applications of those techniques to spectral estimations are developed in an elegant way in Rosenblatt (1985, 1991). Relations with the asymptotic theory for kernel density estimation are also given. The monographs (Azencott and Dacunha-Castelle 1986 and Rosenblatt 1985) also lead to a large amount of additional developments. Functional estimation frameworks are synthetically described in Rosenblatt (1991). The monograph (Doukhan et al. 2002b) provides a wide amount of directions for the study of LRD. The weakly dependent setting is introduced in two papers Doukhan and Louhichi (1999) and Dedecker and Doukhan (2003); a relevant global reference is the monograph (Dedecker et al. 2007).

Paris, France
November 2017

Paul Doukhan
AGM, UMR 8088
University Cergy-Pontoise
Associate member, SAMM, EA 4543
Paris Panthéon-Sorbonne

References

Azencott R, Dacunha-Castelle D (1986) Series of irregular observations: forecasting and model building. In: Applied Probability. Springer-Verlag, New-York, Paris

Brockwell PJ, Davis RA (1991) Time series: theory and methods. Springer-Verlag Series in Statistics, New-York. 2nd edn. Springer-Verlag, New-York

Dedecker J, Doukhan P (2003) A new covariance inequality and applications. Stoch Proc Appl 106:63–80

Dedecker J, Doukhan P, Lang G, León JR, Louhichi S, Prieur C (2007) Weak dependence: with examples and applications. Lecture Notes in Statistics 190, Springer-Verlag, New-York

Doukhan P, Louhichi S (1999) A new weak dependence condition and applications to moment inequalities. Stoch Proc Appl 84:313–342

Doukhan P, Oppenheim G, Taqqu M (2002b) Theory and applications of long-range dependence. Birkhaüser, Boston

Rosenblatt M (1985) Stationary processes and random fields. Birkhäuser, Boston

Rosenblatt M (1991) Stochastic curve estimation, NSF-CBMS regional conference series in probability and statistics, Vol. 3

Team RC (2017) R: a language and environment for statistical computing. R Foundation for Statistical Computing, Vienna, Austria

Acknowledgements

Preliminary versions of those notes[1] were processed for courses at

- IMPA in Rio (Brazil), during summer 2015,
  and before that in
- University of Rio Grande de Sul, Porto Alegre (Brazil),
- Universidad National, Bogotá (Colombia),
- University of Valparaiso (Chile),
- University of Louvain la Neuve (Belgium),
- University Paris 6 (France),
- University of Cergy-Pontoise (France),
- Hong Kong University (Hong-Kong),
- University Nicolas Copernic, Torún (Poland),
- Steklov Institute, Saint Petersburg (Russia),
- Universidad de la Republica, Montevideo (Uruguay),
- Kiev Polytechnic Institute (Ukraine),
- Columbia University, New York (USA), and
- Universidad Central, Caracas (Venezuela).

The constant support of Silvia and Artur Lopes (Porto Alegre) was essential for the redaction of this book.

Adam Jakubowski (Torún), Konstantinos Fokianos (Cyprus), Yvan Nourdin (Luxembourg), and Gabriel Lang (Paris) provided me with many useful suggestions. Also, I wish to thank Natalia Bahamonde (Valparaiso), Jean-Marc Bardet (Paris 1), and Xiaoyin Li (Cleveland) for kindly providing me with helpful comprehensive illustrative figures. The comments of Jean-Luc Prigent (Cergy-Pontoise), Joseph Rynkiewicz (Paris 1), François Roueff (Telecom Paris), and many others were also precious.

[1]Developed within the MME-DII center of excellence (ANR-11-LABEX-0023-01), and with the help of PAI-CONICYT MEC N° 80170072.

I am also extremely indebted to my friend Alain Latour (Grenoble); he provided me with the final version of figures and the corresponding R codes.

Special thanks are due to the SAMM Laboratory and to all its members who supported me over many years and for important collaborations.

An anonymous referee also helped me to organize the material. Mathieu Rosenbaum (Ecole Polytechnique, Paris) and Marc Hoffmann (Paris 9, Dauphine), as well as the staff of Springer, also helped me to provide a more comprehensive final version of the volume, including an important revision of the language.

I am also grateful to University Cergy-Pontoise for its support.

Contents

Part I Independence and Stationarity

1 Independence 3

2 Gaussian Convergence and Inequalities 9
2.1 Gaussian Convergence 9
2.1.1 Central Limit Theorem 12
2.1.2 Empirical Median 13
2.1.3 Gaussian Approximation for Binomials 14
2.2 Quantitative Results 17
2.2.1 Moment Inequalities 17
2.2.2 Exponential Inequalities 22

3 Estimation Concepts 27
3.1 Empirical Estimators 27
3.2 Contrasts 30
3.3 Functional Estimation 31
3.4 Division Trick 41
3.5 A Semi-parametric Test 45

4 Stationarity 49
4.1 Stationarity 49
4.2 Spectral Representation 53
4.3 Range and Spectral Density 59
4.3.1 Limit Variance 63
4.3.2 Cramer–Wold Representation 64
4.4 Spectral Estimation 65
4.4.1 Functional Spectral Estimation 66
4.4.2 Whittle Estimation 67
4.5 Parametric Estimation 67
4.6 Subsampling 69

Part II Models of Time Series

5 Gaussian Chaos . . . 73
5.1 Gaussian Processes . . . 73
5.1.1 Fractional Brownian Motion . . . 74
5.2 Gaussian Chaos . . . 78
5.2.1 Hermite Polynomials . . . 80
5.2.2 Second Order Moments . . . 86
5.2.3 Higher Order Moments . . . 90
5.2.4 Integral Representation of the Brownian Chaos . . . 94
5.2.5 The Fourth Order Moment Method . . . 96

6 Linear Processes . . . 101
6.1 Stationary Linear Models . . . 101
6.2 ARMA(p, q)-Processes . . . 104
6.3 Yule–Walker Equations . . . 107
6.4 ARFIMA$(0, d, 0)$-Processes . . . 108
6.5 ARFIMA(p, d, q)-Processes . . . 112
6.6 Extensions . . . 113

7 Non-linear Processes . . . 115
7.1 Discrete Chaos . . . 115
7.1.1 Volterra Expansions . . . 115
7.1.2 Appell Polynomials . . . 117
7.2 Memory Models . . . 120
7.2.1 Bilinear Models . . . 122
7.2.2 LARCH(∞)-Models . . . 127
7.3 Stable Markov Chains . . . 128
7.3.1 AR-ARCH-Models . . . 131
7.3.2 Moments of ARCH(1)-Models . . . 133
7.3.3 Estimation of LARCH(1)-Models . . . 135
7.3.4 Branching Models . . . 144
7.3.5 Integer Valued Autoregressions . . . 147
7.3.6 Generalized Linear Models . . . 149
7.3.7 Non-linear AR(d)-Models . . . 154
7.4 Bernoulli Schemes . . . 155
7.4.1 Structure and Tools . . . 155
7.4.2 Couplings . . . 162

8 Associated Processes . . . 167
8.1 Association . . . 167
8.2 Associated Processes . . . 169
8.3 Main Inequality . . . 170
8.4 Limit Theory . . . 172

Part III Dependence

9 Dependence 177
9.1 Ergodic Theorem 177
9.2 Range 186

10 Long-Range Dependence 189
10.1 Gaussian Processes 189
10.2 Gaussian Polynomials 191
10.3 Rosenblatt Process 192
10.4 Linear Processes 195
10.5 Functions of Linear Processes 196
10.6 More LRD Models 198
10.6.1 Integer Valued Trawl Models 198
10.6.2 LARCH-Models 201
10.6.3 Randomly Fractional Differences 202
10.6.4 Perturbed Linear Models 203
10.6.5 Non-linear Bernoulli-Shift Models 203

11 Short-Range Dependence 205
11.1 Weak-Dependence 205
11.2 Strong Mixing 206
11.3 Bootstrapping AR(1)-Models 209
11.4 Weak-Dependence Conditions 211
11.5 Proving Limit Theorems 219

12 Moments and Cumulants 225
12.1 Method of Moments 226
12.1.1 Notations 226
12.1.2 Combinatorics of Moments 228
12.2 Dependence and Cumulants 230
12.2.1 More Dependence Coefficients 231
12.2.2 Sums of Cumulants 235
12.2.3 Moments of Sums 236
12.2.4 Rosenthal's Inequality 238
12.3 Dependent Kernel Density Estimation 240

Erratum to: Non-linear Processes E1

Appendix A: Probability and Distributions 247

Appendix B: Convergence and Processes 275

Appendix C: R Scripts Used for the Figures 287

References 301

Index 305

List of Figures

Fig. 2.1	Accuracy of Gaussian approximation for binomials. We represent the renormalized distribution of $\mathcal{B}(n, 3/10)$ with the $\mathcal{N}(0,1)$ density, for $n = 30$, and $n = 100$	15
Fig. 3.1	Proportion of heads among n tosses of a fair coin	28
Fig. 3.2	Empirical cumulative distribution of fuel consumption of 32 cars	29
Fig. 3.3	Sample distribution function and a kernel estimate of the data of Fig. 3.2	33
Fig. 4.1	Annual flow of Nile River at Aswan 1871–1970	50
Fig. 4.2	Correlograms of the annual flow of the Nile River in Aswan 1871–1970. See Fig. 4.1	59
Fig. 5.1	Fractional Brownian motion simulated with $H = 0.30$ and evaluated in 1024 points	76
Fig. 5.2	Differenced time series of Fig. 5.1. This process is a fractional noise	76
Fig. 5.3	Fractional Brownian motion simulated with $H = 0.90$ and evaluated in 1024 points	77
Fig. 5.4	Differenced time series of Fig. 5.1	77
Fig. 5.5	Hermite polynomials	80
Fig. 6.1	Simulated trajectory of an ARMA (1,1) Here, $X_t = 0.6X_{t-1} + \varepsilon_t + 0.7\varepsilon_{t-1}$ with, $\varepsilon_t \sim \mathcal{N}(0,1)$	105
Fig. 6.2	Sample simple and partial correlograms of the series of Fig. 6.1	106
Fig. 6.3	ARFIMA $(0, d, 0)$ trajectories for different values of d	110
Fig. 6.4	Sample correlograms of ARFIMA trajectories for different values of d. See Fig. 6.3	111

Fig. 7.1 Simulated trajectory of an bilinear process and sample autocorrelation function. Here, $X_t = 0.75X_{t-1} + \varepsilon_{t-1} + 0.6X_{t-1}\varepsilon_{t-1}$ with $\varepsilon_t \sim \mathcal{N}(0,1)$ 123
Fig. 7.2 Simulated trajectory of an ARCH(2) process. Here $X_t = \sqrt{\sigma_t^2}\xi_t$ with $\sigma_t^2 = \alpha^2 + \beta^2 X_{t-1}^2 + \gamma^2 X_{t-2}^2$ and $\xi_t \sim \mathcal{N}(0,1)$. We used $\alpha = 0.5, \beta = 0.6$ and $\gamma = 0.7$ 133
Fig. 7.3 Simulated trajectory of an GARCH(1,1). Here, $X_t = \sqrt{\sigma_t^2}\xi_t$ with $\sigma_t^2 = \alpha^2 + \beta^2 X_{t-1}^2 + \gamma^2 \sigma_{t-1}^2$ and $\xi_t \sim \mathcal{N}(0,1)$. We used $\alpha = 0.5, \beta = 0.6$ and $\gamma = 0.7$ 134
Fig. 7.4 NYSE returns. Source: Shumway and Stoffer (2011), p. 7. The data are daily value weighted market returns from February 2, 1984 to December 31, 1991 (2000 trading days). The crash of October 19, 1987 occurs at $t = 938$. 134
Fig. 7.5 Simulated trajectory and simple correlogram of an LARCH (1,1) process. Here $X_t = \varepsilon_t(1 + \beta_1 x_{t-1})$ with $\varepsilon_t \sim \mathcal{B}(0.95)$. We used $\beta_1 = 0.45$. 136
Fig. 7.6 Simulated trajectory and simple correlogram of a switching process. Here, $X_t = \xi_t^{(1)} X_{t-1} + \xi_t^{(2)}$ with $\xi_t^{(1)} \sim \mathcal{B}(0.5)$ and $\xi_t^{(2)} \sim \mathcal{N}(0,1)$. This model switches between a random walk and an iid behaviour 146
Fig. 7.7 Simulated trajectory and simple correlogram of INAR(1). Here, process satisfying $X_t = \alpha \circ X_{t-1} + \zeta_t$ with $\zeta_t \sim \mathcal{P}(2)$ and $\mathcal{B}(0.5)$ thinning operator 148
Fig. 7.8 Simulated trajectory and simple correlogram of INGARCH. Here, $X_t \sim \mathcal{P}(\lambda_t)$ with $\lambda_t = 0.5 + 0.25X_{t-1} + 0.5\lambda_{t-13}$ 151
Fig. 11.1 Asymptotic independence 206
Fig. 11.2 A non-mixing AR(1)-process, and its autocovariances 207
Fig. A.1 Convex function as supremum of affine functions 252
Fig. A.2 Gaussian white noise of variance 1. 266
Fig. A.3 Standard normal density 266
Fig. A.4 Cumulative distribution function of a $\mathcal{N}(10,2)$ 266

Part I
Independence and Stationarity

This part provides basic references to probability theory useful for time series analysis; namely, we provide some details on stochastic independence. Gaussian approximation is then considered in the same spirit of extensions outside of the independence properties. We then recall some concepts of statistics, namely those which extend to time series. The final chapter is dedicated to introduce the basic concept of stationarity of time series.

Chapter 1
Independence

This chapter deals with the standard notion of stochastic independence. This is a crucial concept, since this monograph aims to understand how to weaken it, in order to define asymptotic independence. We discuss in detail the limits of this idea through various examples and counter-examples. Below we denote by $(\Omega, \mathcal{A}, \mathbb{P})$ the underlying probability space and we shall make use of the notations and concepts in Appendix A without additional reference, e.g. examples of distributions are provided in Sect. A.2, and specific notations are given in the Index.

We first recall independence of two events:

Definition 1.1.1 Events $A, B \in \mathcal{A}$ are independent in case

$$\mathbb{P}(A \cap B) = \mathbb{P}(A)\mathbb{P}(B).$$

To define the independence of more than two events it is worse considering the independence of several random variables:

Definition 1.1.2 The random variables $X_1, \ldots, X_n$ (with values for instance in the same topological space E) are independent in case, for any $g_1, \ldots, g_n : E \to \mathbb{R}$ continuous and bounded:

$$\mathbb{E}\Big(g_1(X_1) \times \cdots \times g_n(X_n)\Big) = \mathbb{E}\big(g_1(X_1)\big) \times \cdots \times \mathbb{E}\big(g_n(X_n)\big).$$

Definition 1.1.3 Events $(A_1, \ldots, A_n)$ are independent if the random variables $X_1 = \mathbb{I}_{A_1}, \ldots, X_n = \mathbb{I}_{A_n}$ are independent.

Setting $g_j(x) = (x \vee 0) \wedge 1$ if $j \in E$ and $g_j(x) = 0$ otherwise, derive as an exercise the more usual definition of a finite family of independent events:

P. Doukhan, *Stochastic Models for Time Series*, Mathématiques et Applications 80,
https://doi.org/10.1007/978-3-319-76938-7_1

Proposition 1.1.1 *Events* $(A_1, \ldots, A_n)$ *are independent if and only if, for each* $E \subset \{1, \ldots, n\}$,

$$\mathbb{P}\Big(\bigcap_{i\in E} A_i\Big) = \prod_{i\in E} \mathbb{P}(A_i). \tag{1.1}$$

Remark 1.1.1 Let I be an arbitrary set (finite or infinite). A family $(A_i)_{i\in I}$ is independent if the previous relation (1.1) still holds for each finite subset $E \subset I$.

Definition 1.1.4 The random variables $X_1, \ldots, X_n$ are called pairwise independent if each couple (X_i, X_j) is independent for $i \neq j$, and $1 \leq i, j \leq n$.

In case $E = \mathbb{R}$ and the characteristic functions $\phi_{X_1}, \ldots, \phi_{X_n}$ (see Definition A.2.3 and Lemma 2.15 on p. 15 in van der Vaart (1998)) are analytic around 0, then the previous remarks imply that the independence of$(X_1, \ldots, X_n)$ holds if and only if:

$$\phi_{X_1+\cdots+X_n} = \phi_{X_1} \times \cdots \times \phi_{X_n}.$$

Assume now that X_j admits a density f_j with respect to some measure ν_j on E_j then an independent random vector $(X_1, \ldots, X_n) \in E_1 \times \cdots \times E_n$, then this vector admits the density

$$f(x_1, \ldots, x_n) = f_1(x_1) \cdots f_n(x_n), \qquad \forall (x_1, \ldots, x_n) \in E_1 \times \cdots \times E_n$$

on the product space $E_1 \times \cdots \times E_n$ with respect to the product measure $\nu_1 \times \cdots \times \nu_n$.

If $A_1, \ldots, A_n \in \mathcal{A}$ are events then simple random variables write $X_k = \mathbb{I}_{A_k} \in \{0, 1\}$ and the independence of couples (X_i, X_j) is easily proved to coincide with the independence of couples of events A_i, A_j.

The independence of the family of events $A_1, \ldots, A_n$ is written as:

$$\mathbb{P}\Big(\bigcap_{i\in E} A_i\Big) = \prod_{i\in E} \mathbb{P}(A_i), \qquad \forall E \subset \{1, \ldots, n\}.$$

Example 1.1.1 As a probability space consider a model $(\Omega, \mathcal{A}, \mathbb{P})$ for two (fair) independent dice

$$\Omega = \{1, 2, 3, 4, 5, 6\}^2, \qquad \mathcal{A} = \mathcal{P}(\Omega),$$

and the uniform probability $\mathbb{P}$ on this finite set with 36 elements.

Let A, B be the events that the dice show an even number, then

$$\mathbb{P}(A) = \mathbb{P}(B) = \frac{1}{2}.$$

Then, those events are independent.

Now let C be the event that the sum of the results of both dice is also even then $A \cap B \subset C$ and on the event $A \cap C$ the second dice is necessarily even too, so that $A \cap C \subset B$.

Analogously $B \cap C \subset A$ so that it is easy to check that A, C and B, C are independent pairs of events,

$$\begin{aligned} \mathbb{P}(A \cap B) &= \mathbb{P}(A)\,\mathbb{P}(B), \\ \mathbb{P}(A \cap C) &= \mathbb{P}(A)\,\mathbb{P}(C), \\ \mathbb{P}(B \cap C) &= \mathbb{P}(B)\,\mathbb{P}(C), \end{aligned}$$

(those values all equal $\frac{1}{4} = \frac{1}{2} \cdot \frac{1}{2}$).

On the other hand $A \cap B \cap B = A \cap B$ thus

$$\mathbb{P}(A \cap B \cap C) = \frac{1}{4} \neq \mathbb{P}(A)\,\mathbb{P}(B)\,\mathbb{P}(C) = \frac{1}{8}.$$

Then the triplet of events (A, B, C) is not independent. We have proved that the events A, B, C are pairwise independent but not independent on this probability set with 36 elements equipped with the uniform law.

Another very similar example is as follows.

Example 1.1.2 Consider $\Omega = [0, 1]^2$ with its Borel sigma-field and with $\mathbb{P}$ the uniform distribution. The events $A = [0, \frac{1}{2}] \times [0, 1]$, $B = [0, \frac{1}{2}]^2 \cup [\frac{1}{2}, 1]^2$ and $C = [0, 1] \times [0, \frac{1}{2}]$ admit probability $\frac{1}{2}$. Further $A \cap B = A \cap C = B \cap C = [0, \frac{1}{2}]^2$ has the probability $\frac{1}{4} = \frac{1}{2} \cdot \frac{1}{2}$, thus those events are pairwise independent. They are not independent since $\mathbb{P}(A \cap B \cap C) = \frac{1}{4} \neq \mathbb{P}(A)\mathbb{P}(B)\mathbb{P}(C) = \frac{1}{8}$.

Remark 1.1.2 (*k-wise independence*)

- From the previous example, it is possible to exhibit three pairwise-independent random variables which are not independent, namely $X = \mathbb{I}_A$, $Y = \mathbb{I}_B$ and $Z = \mathbb{I}_C$.
  Pairwise independence should be carefully distinguished from independence.
- For each p, (Derriennic and Klopotowski 2000) exhibit a vector $X = (X_1, \ldots, X_p) \in \mathbb{R}^p$ whose components are not independent but such that any sub-vector with dimension strictly less than p is independent. A concomitant counter-example to the CLT is given in Bradley and Pruss (2009). It is always possible to build iid sequences with a given marginal distribution on $\mathbb{R}$, see Example A.2.3. Hence the above constructions really make sense.
- Let $X_1, \ldots, X_n$ be independent Bernoulli $b(p)$-distributed random variables, then the calculation of generating functions implies that $X_1 + \cdots + X_n \sim B(n, p)$ admits a binomial distribution.

The following result is essential but very simple; it is thus stated as an exercise in this book:

Exercise 1 *Let $X, Y \in \mathbb{R}$ be real valued random variables with $\mathbb{E}X^2 + \mathbb{E}Y^2 < \infty$. If (X, Y) are independent then $Cov(X, Y) = 0$.*

Solution to Exercise 1. In case the variables are bounded, then independence indeed asserts that $\mathbb{E}XY = \mathbb{E}X\mathbb{E}Y$.

The general unbounded case is derived from a truncation by setting

$$X_M = X \vee (-M) \wedge M$$

and use of the Lebesgue dominated convergence theorem with $M \uparrow \infty$.

Exercise 2 *Let $X, R \in \mathbb{R}$ be independent random variables with X symmetric (i.e. $-X$ admits the same distribution as X), $\mathbb{E}X^2 < \infty$ and if moreover $\mathbb{P}(R = \pm 1) = \frac{1}{2}$, then we set $Y = RX$.*

Prove that:

- *$Cov(X, Y) = 0$,*
- *If $|X|$ is not almost surely (a.s.) constant then X, Y are not independent.*

Remark 1.1.3 An important use of this exercise in provided in Exercise 68, much later in those notes.

Solution to Exercise 2. The first equality follows from independence in the case of bounded X and dominated convergence yields the general case as in Exercise 1.

The second result also follows since because $|X|$ is not a.s. constant there is an even function g such that $\mathrm{Var}g(X) \neq 0$, now since $g(X) = g(Y)$, we have: $\mathrm{Cov}(g(X), g(Y)) \neq 0$.

Exercise 3 *If the random variables $X, Y \in \{0, 1\}$ admit only two values and if they satisfy $Cov(X, Y) = 0$, then prove that the pair (X, Y) is independent.*

Hint for Exercise 3. To prove the independence of those random variables one needs to prove the independence of the four following couples of events:

$$(A_a, B_b) \text{ for all } a, b = 0, \text{ or } 1.$$

Here we set $A_a = (X = a)$ and $B_b = (Y = b)$ for $a, b \in \{0, 1\}^2$.

- Relation $\mathrm{Cov}(X, Y) = 0$ infers as the independence of the events A_1, B_1,
- Relation $\mathrm{Cov}(X, 1 - Y) = 0$ infers as the independence of events A_1, B_0,
- Relation $\mathrm{Cov}(1 - X, Y) = 0$ is the independence of A_0, and B_1,
- Relation $\mathrm{Cov}(1 - X, 1 - Y) = 0$ is the independence of A_0, and B_0.

Note that either Gaussian or associated vectors fit the same property: orthogonality implies independence too, see in Appendix A.3, and Chap. 8 respectively.

Exercise 3 above admits tight assumptions as the following exercise also suggests.

Exercise 4 *Exhibit random variables $X \in \{0, \pm 1\}$, and $Y \in \{0, 1\}$ which are not independent, but are orthogonal, in other terms such that $Cov(X, Y) = 0$.*

Solution to Exercise 4. Consider the uniform random variable X on the set $\{-1, 0, 1\}$ and $Y = \mathbb{I}_{\{X=0\}}$, then $\mathbb{E}X = 0$, $\mathrm{Cov}(X, Y) = \mathbb{E}XY = 0$ because their product vanishes $XY = 0$ (a.s.) while these random variables are not independent. Indeed with $f(x) = \mathbb{I}_{\{x=0\}}$ and $g(x) = x$ we derive

$$\mathbb{E}f(X)g(Y) = \mathbb{P}(X = 0) \neq \mathbb{E}f(X)\mathbb{E}g(Y) = \mathbb{P}^2(X = 0).$$

This concludes the proof.

Example 1.1.3 (Bernoulli INARCH(q) models) Set

$$X_k = \mathbb{I}_{\{U_k \le \lambda_k\}},$$

for some iid and uniform sequence (U_k) on $[0, 1]$ and λ_k is a random stationary sequence measurable wrt $X_{k-1}, X_{k-2}, \ldots$ as in Example 7.3.4. If some function $g : \mathbb{R}^q \to \mathbb{R}$ satisfies

$$|g(x') - g(x)| \le \sum_{j=1}^{q} a_j |x'_j - x_j|, \quad \forall x = (x_1, \ldots, x_q), x' = (x'_1, \ldots, x'_q) \in \mathbb{R}^q$$

for coefficients $a_j \ge 0$ with $\alpha = a_1 + \cdots + a_q < 1$, then Theorem 7.3.1 in Sect. 7.3 proves the existence of a stationary sequence (see Definition 4.1.1) with Bernoulli marginals and such that $\lambda_k = g(X_{k-1}, \ldots, X_{k-q})$. If e.g.

$$\lambda_k = d + \sum_{j=1}^{q} a_j X_{k-j}, \qquad a = \sum_{j=1}^{q} a_j < 1$$

and $X_0 \sim b(p)$ with $p = \mathbb{P}(X_0 = 1) = \mathbb{E}X_0 = \mathbb{E}X_0^2$ from stationarity. We derive $p = d + ap$ so that $p = d/(1 - a)$, and the relation $a + d < 1$ implies $p < 1$. Set $r_k = \mathrm{Cov}(X_0, X_k)$ then $r_0 = p(1 - p) \neq 0$.

Consider $g(x_1, \ldots, x_q) = d + ax_q$ for $a > 0$, $d > 0$ and $a + d < 1$, then for each $k \ge 0$,

$$r_k = \mathbb{E}X_0 X_k - p^2 = \mathbb{E}X_0(d + aX_{k-q}) - p^2.$$

We successively derive:

- $$r_q = \mathbb{E}X_0 X_k - p^2 = p(d + a - p) = \frac{ad}{(1 - a)^2}(1 - a - d) \neq 0,$$

- if $0 < k < q$, then

$$\begin{aligned} r_k &= \mathbb{E}X_0(d + aX_{k-q}) - p^2 \\ &= dp - p^2 + a(r_{k-q} + p^2) \\ &= p(d - (1-a)p) + ar_{k-q} \\ &= ar_{k-q}. \end{aligned}$$

  Thus iterating this relation gives $r_k = a^2 r_k$ so that $r_k = 0$.

We thus proved that for each $q \geq 2$ the vector $(X_0, \ldots, X_{q-1})$ associated to this INARCH(q) model is pairwise independent but $(X_0, X_1, \ldots, X_q)$ is not an independent vector.

We conjecture that the vector $(X_0, \ldots, X_{q-1})$ is in fact independent; Remark 1.1.2 provides an example of this situation.

Chapter 2
Gaussian Convergence and Inequalities

This chapter describes a simple Gaussian limit theory; namely we restate simple central limit theorems together with applications and moment/exponential inequalities for partial sums behaving asymptotically as Gaussian random variables. A relevant reference for the whole chapter is Petrov (1975), results without a precise reference should be found in this reference, and the others are in Hall and Heyde (1980). Topics related to empirical processes are covered by van der Vaart and Wellner (1998) and Rosenblatt (1991).

Gaussian behaviours are often observed in the case of time series from the accumulation of small events.

2.1 Gaussian Convergence

It is a standard feature that accumulation of infinitesimal independent random effects are accurately approximated by the Gaussian distribution (see Sect. A.3 for more on Gaussian distributions) as proved in the monograph (Petrov 1975). The best way to make this rigorous is illustrated by the Lindeberg method. Definitions of the convergence in distribution may be found in Appendix B.

Definition 2.1.1 We denote $\mathcal{C}_b^k([u,v])$ the set of k-times differentiable functions on the interval $]u,v[$, such that $f^{(j)}$ can be continuously extended on $[u,v]$ if $j=0,\ldots,k$.

In case the interval of definition is obvious, we simply write $\mathcal{C}_b^k$.

Lemma 2.1.1 (Lindeberg) *Assume that $U_1,\ldots,U_k$ are centred real valued independent random variables such that $\mathbb{E}|U_j|^{2+\epsilon}<\infty$, for some $\epsilon\in[0,1]$.*

Let $V_1,\ldots,V_k$ be independent random variables, independent of the random variables $U_1,\ldots,U_k$ and such that $V_j\sim\mathcal{N}(0,\mathbb{E}U_j^2)$ are centred Gaussians with the same variance as U_j, and $g\in\mathcal{C}_b^3(\mathbb{R})$.

P. Doukhan, *Stochastic Models for Time Series*, Mathématiques et Applications 80,
https://doi.org/10.1007/978-3-319-76938-7_2

Set $U = U_1 + \cdots + U_k$ *and* $V = V_1 + \cdots + V_k$*, then, we obtain the two bounds:*

$$
\begin{aligned}
|\mathbb{E}(g(U) - g(V))| &\le 4 \sum_{i=1}^{k} \mathbb{E}\left(|U_i|^2 \left(\|g''\|_\infty \wedge \left(\|g'''\|_\infty |U_i|\right)\right)\right). \\
&\le 4 \, \|g''\|_\infty^{1-\epsilon} \|g'''\|_\infty^{\epsilon} \sum_{i=1}^{k} \mathbb{E}|U_j|^{2+\epsilon}.
\end{aligned}
$$

Remark 2.1.1 The first bound may involve $\epsilon = 0$ and only square integrable random variables are needed.

The second bound is meaningful in case $\sum_{i=1}^{k} \mathbb{E}|U_j|^{2+\epsilon} < \infty$ and usually needs $\epsilon > 0$.

Indeed for $\epsilon = 0$, $\mathbb{E}|V|^2 = \sum_{i=1}^{k} \mathbb{E}|U_j|^2$ is the limit variance $\sigma^2 > 0$.

Proof of Lemma 2.1.1. Set $Z_j = U_1 + \cdots + U_{j-1} + V_{j+1} + \cdots + V_k$ and $\delta_j = g(Z_j + U_j) - g(Z_j + V_j)$ for $1 \le j \le k$, then

$$
\mathbb{E}(g(U) - g(V)) = \sum_{j=1}^{k} \mathbb{E}\left(g(Z_j + U_j) - g(Z_j + V_j)\right) = \sum_{j=1}^{k} \mathbb{E}\delta_j.
$$

Set for simplicity $\delta = g(z + u) - ug'(z) - \frac{1}{2}u^2 g''(z)$ then Taylor formula with order 2 entails $|\delta| \le \frac{1}{2}u^2 |g''(z) - g''(t)|$ for some $t \in]z, z + u[$.

This implies from either the mean value theorem or from a simple bound that

$$
\begin{aligned}
|\delta| &\le (u^2 \|g''\|_\infty) \wedge \left(\frac{1}{2}|u|^3 \|g'''\|_\infty\right) \\
&= (u^2 \|g''\|_\infty) \left(1 \wedge \left(\frac{1}{2}|u| \frac{\|g'''\|_\infty}{\|g''\|_\infty}\right)\right) \\
&\le u^2 \|g''\|_\infty \left(\frac{1}{2}|u| \frac{\|g'''\|_\infty}{\|g''\|_\infty}\right)^{\epsilon} \\
&= 2^{-\epsilon} |u|^{2+\epsilon} \|g''\|_\infty^{1-\epsilon} \|g'''\|_\infty^{\epsilon}.
\end{aligned}
$$

Apply the above inequality with $z = Z_j$ and $u = U_j$ or V_j. In order to conclude we also note that

$$
\mathbb{E}|V_j|^2 = \mathbb{E}|U_j|^2, \qquad \text{and} \qquad \mathbb{E}|V_j|^3 = \mathbb{E}|Z|^3 \left(\mathbb{E}U_j^2\right)^{3/2},
$$

for a standard Normal random variable $Z \sim \mathcal{N}(0, 1)$. The Hölder inequality (Proposition A.2.2) yields $\left(\mathbb{E}U_j^2\right)^{3/2} \le \mathbb{E}|U_j|^3$.

Integration by parts implies

$$\mathbb{E}|Z|^3 = \frac{4}{\sqrt{2\pi}} < 2, \qquad \text{thus,} \qquad (\mathbb{E}|Z|^3)^{3/2} \sim 2.015 < 3.$$

From the Jensen inequality (Proposition A.2.1) we derive, for $0 < \epsilon \le 1$:

$$\mathbb{E}|V|^{2+\epsilon} = (\mathbb{E}U^2)^{1+\frac{\epsilon}{2}}\mathbb{E}|Z|^{2+\epsilon} \le (\mathbb{E}|Z|^3)^{\frac{3}{2+\epsilon}}\mathbb{E}|U|^{2+\epsilon} < 3\mathbb{E}|U|^{2+\epsilon}.$$

Now

$$\begin{aligned}\mathbb{E}|\delta_j| &\le 2^{-\epsilon}\|g''\|_\infty^{1-\epsilon}\|g'''\|_\infty^{\epsilon}\mathbb{E}\left(|U_j|^{2+\epsilon} + |V_j|^{2+\epsilon}\right) \\ &\le 4\|g''\|_\infty^{1-\epsilon}\|g'''\|_\infty^{\epsilon}\mathbb{E}|U_j|^{2+\epsilon}.\end{aligned}$$

This yields the desired result.

As a simple consequence of this result we derive:

Theorem 2.1.1 (Lindeberg) *For each integer n, let $(\zeta_{n,k})_{k\in\mathbb{Z}}$ be independent sequences of centred random variables. Suppose*

$$\sum_{k=-\infty}^{\infty} \mathbb{E}\zeta_{n,k}^2 \to_{n\to\infty} \sigma^2 > 0,$$

$$\sum_{k=-\infty}^{\infty} \mathbb{E}\zeta_{n,k}^2\, \mathbb{I}_{\{|\zeta_{n,k}|>\epsilon\}} \to_{n\to\infty} 0, \textit{ for each } \epsilon > 0.$$

Then the following convergence in distribution (defined in Appendix B) holds:

$$\sum_{k=-\infty}^{\infty} \zeta_{n,k} \xrightarrow{\mathcal{L}}_{n\to\infty} \mathcal{N}(0,\sigma^2).$$

Proof Use the notation in Lemma 2.1.1. Set $U_k = \zeta_{n,k}\, \mathbb{I}_{\{|\zeta_{n,k}|\le\epsilon\}}$, for a convenient $\epsilon > 0$. From the first inequality in Lemma 2.1.1 we get that

$$\sup_n \sum_{k=-\infty}^{\infty} \mathbb{E}\zeta_{n,k}^2 = C < \infty,$$

and setting

$$\zeta_n = \sum_{k=-\infty}^{\infty} \zeta_{n,k},$$

we derive

$$\sum_{k=-\infty}^{\infty} \mathbb{E}|U_k|^3 \le C \cdot \epsilon.$$

Now from independence,

$$\mathbb{E}\,(\zeta_n - U)^2 \le \sum_{k=-\infty}^{\infty} \mathbb{E}\zeta_{n,k}^2\ \mathbb{I}_{\{|\zeta_{n,k}|>\epsilon\}} = a_n(\epsilon).$$

The triangle inequality implies $\sigma_n^2 = \mathbb{E}U^2 \to_{n\to\infty} \sigma^2$. Those bounds together imply for $Z \sim \mathcal{N}(0, 1)$, a Normal random variable:

$$\begin{aligned}|\mathbb{E}(g(\zeta_n) - g(\sigma Z))| \le |\mathbb{E}(g(\zeta_n) - g(U))| \\ + |\mathbb{E}(g(U) - g(\sigma_n Z))| \\ + |\mathbb{E}(g(\sigma_n Z) - g(\sigma Z))|.\end{aligned}$$

To prove the result use $|\mathbb{E}(g(\sigma_n Z) - g(\sigma Z))| \le \|g'\|_\infty \mathbb{E}|Z| \times |\sigma_n - \sigma|$ and select $\epsilon = \epsilon_n$ conveniently such that $\lim_n (a_n(\epsilon_n) + \epsilon_n) = 0$.

Then the result follows.

In order to prove the power of this result the following subsections derive some other consequences of the Lindeberg lemma, see van der Vaart (1998) for much more. The classical Central Limit Theorem 2.1.2 is a first consequence of this result. Then the asymptotic behaviour of empirical medians are derived in the Proposition 2.1.1 following the proof in van der Vaart (1998). Finally the validity of the Gaussian approximation of binomial distributions is essential for example in order to assert the validity of χ^2-goodness-of-fit tests. To conclude this section a simple dependent version (see Bardet et al. 2006) of the Lindeberg lemma will be developed in Lemma 11.5.1 below.

2.1.1 Central Limit Theorem

Theorem 2.1.2 *The Central Limit Theorem ensures the convergence*

$$\frac{1}{\sqrt{n}}(X_1 + \cdots + X_n) \to^{\mathcal{L}}_{n\to\infty} \mathcal{N}(0, \mathbb{E}X_0^2),$$

for independent identically distributed sequences with finite variance.

Proof This follows from Theorem 2.1.1. Set $\zeta_{n,k} = X_k/\sqrt{n}$. The only point to check is now $\lim_{n\to\infty} \mathbb{E}X_1^2\ \mathbb{I}_{|X_1|\ge\epsilon\sqrt{n}} = 0$, which follows from $\mathbb{E}X_1^2 < \infty$.

Exercise 5 Provide an alternative proof of Theorem 2.1.2 using Lemma 2.1.1.

Set $k = n$ and $U_j = X_j/\sqrt{n}$. To prove Theorem 2.1.2 simply note that the random variable X_0 satisfies the tightness condition

$$\mathbb{E}|X_0|^2 \wedge \left(\frac{|X_0|^3}{\sqrt{n}}\right) \to_{n\to\infty} 0.$$

The result is derived from Exercise 86.

Remark 2.1.2 Bardet and Doukhan (2017) prove the existence of some convex non-decreasing function on $\mathbb{R}^+ \to \mathbb{R}^+$ such that $\mathbb{E}\psi(X_0|) < \infty$ and $\lim_{x\to\infty}\psi(x)/x^2 = \infty$. It also implies that the Orlicz norm of X_0 is finite, the following expression is indeed a norm:

$$\|X\|_\psi = \inf\left\{u > 0\Big/ \mathbb{E}\psi(\frac{|X|}{u}) \le 1\right\}.$$

The above point is useful to derive CLTs as sketched in Remark 8.4.1.

2.1.2 Empirical Median

Here we follow the elegant method in van der Vaart (1998) to derive the asymptotic behaviour of a median. In order to make it easier we assume the following regularity condition.

Definition 2.1.2 An atom of the distribution of a random variable Y is a point such that $\mathbb{P}(Y = a) \neq 0$.

In case such a distribution admits no atom we shall say that it is atomless or continuous, since its cumulative distribution function is then continuous.

Suppose that the number of observations $n = 2N + 1$ is odd; we consider here an independent identically distributed n-sample $Y_1, \ldots, Y_n$ with median M

$$\mathbb{P}(Y_1 < M) \le \frac{1}{2} \le \mathbb{P}(Y_1 > M).$$

To simplify notations and make this median unique, assume this law is continuous. The empirical median of the sample is the value M_n of the order statistic with rank $N + 1$.

Proposition 2.1.1 *Assume that* (X_k) *is an atomless identically distributed and independent sequence. If the cumulative distribution function* F *of* Y_1 *admits a derivative* γ *at point* M *then*

$$\sqrt{n}(M_n - M) \xrightarrow{\mathcal{L}}_{n\to\infty} \mathcal{N}\left(0, \frac{1}{4\gamma^2}\right).$$

Proof Notice that $\mathbb{P}(\sqrt{n}(M_n - M) \le x) = \mathbb{P}(M_n \le M + \frac{x}{\sqrt{n}})$ is the probability that $N+1$ observations Y_i, among the $n = 2N+1$ considered, satisfy $Y_i \le M + x/\sqrt{n}$:

$$\mathbb{P}(\sqrt{n}(M_n - M) \le x) = \mathbb{P}\Big(\sum_{i=1}^{n} \mathbb{I}_{\{Y_i \le M + x/\sqrt{n}\}} \ge N+1\Big).$$

Setting

$$p_n = \mathbb{P}(Y_1 \le M + x/\sqrt{n})$$

and

$$X_{i,n} = \frac{\mathbb{I}_{\{Y_i \le M + x/\sqrt{n}\}} - p_n}{\sqrt{np_n(1-p_n)}},$$

yields

$$\mathbb{P}(\sqrt{n}(M_n - M) \le x) = \mathbb{P}\Big(s_n \le \sum_{i=1}^{n} X_{i,n}\Big), \quad s_n = \frac{N+1-np_n}{\sqrt{np_n(1-p_n)}}.$$

The continuity of the distribution of Y_1 at point M implies $p_n \to \frac{1}{2}$ and its differentiability yields $s_n \to -2x\gamma$. The Lindeberg theorem implies

$$\sum_{i=1}^{n} X_{i,n} \xrightarrow{\mathcal{L}}_{n\to\infty} \mathcal{N}(0,1),$$

allows to conclude.

Remark 2.1.3 If instead of the continuity of X_0's cdf[1] we deal with more general properties then only the regularity around the median is really required.

2.1.3 Gaussian Approximation for Binomials

Theorem 2.1.3 *Let $S_n \sim B(n,p)$ be a binomial random variable (see Example A.2.1). Fix some $\epsilon \in (0,1]$, then using Landau notation:*

$$\sup_{np(1-p)\epsilon > 1} \sup_{u \in \mathbb{R}} \Delta_{n,p}(u) = \mathcal{O}\Big(\frac{1}{(np(1-p))^{\frac{1}{8}}}\Big),$$

with

$$\Delta_{n,p}(u) = \Big|\mathbb{P}\Big(\frac{S_n - np}{\sqrt{np(1-p)}} \le u\Big) - \Phi(u)\Big|.$$

[1] It is the atomless assumption.

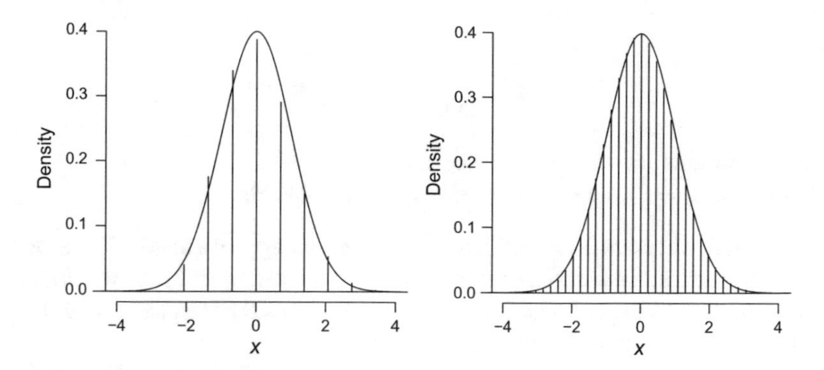

Fig. 2.1 Accuracy of Gaussian approximation for binomials.
We represent the renormalized distribution of $\mathcal{B}(n, 3/10)$ with the $\mathcal{N}(0, 1)$ density, for $n = 30$, and $n = 100$

Proof We shall use Lemma 2.1.1. Rewrite $S_n = b_1 + \cdots + b_n$ with iid $b_1, b_2, \ldots \sim b(p)$. Set

$$X_i = \frac{b_i - p}{\sqrt{np(1-p)}}, \qquad 1 \le i \le n.$$

Then $X_1, \ldots, X_n$ are centred independent identically distributed and

$$\mathbb{E}b_i^3 = \mathbb{E}(b_i - p)^2 = p(1-p).$$

Let $0 < p \le 1 - \epsilon$, $N \sim \mathcal{N}(0, 1)$ and $f \in C_b^3$. We get from Lemma 2.1.1:

$$\begin{aligned} \Delta_n(f) &= \left|\mathbb{E}\Big(f\Big(\frac{S_n - np}{\sqrt{n\theta(1-p)}}\Big) - f(N)\Big)\right| \\ &\le \frac{\|f'''\|_\infty}{2} \sum_{i=1}^n \mathbb{E}|X_i|^3 \\ &\le \frac{4\|f'''\|_\infty}{\epsilon} \frac{1}{\sqrt{np(1-p)}}. \end{aligned}$$

Exercise 6 below is useful. The relation

$$\mathbb{P}(N \in [u, u+\eta]) \le \frac{\eta}{\sqrt{2\pi}}$$

entails

$$\begin{aligned} &\Delta_n(f_{u-\eta,\eta}) + \mathbb{P}(N \in [u, u-\eta]) \\ &\qquad \le \Delta_{n,p}(u) \\ &\qquad\qquad \le \Delta_n(f_{u,\eta}) + \mathbb{P}(N \in [u, u+\eta]). \end{aligned}$$

Thus

$$\Delta_{n,p}(u) \le C\left(\frac{1}{\eta^3\sqrt{np(1-p)}} + \eta\right),$$

for some constant not depending on n, η, ϵ and p.

The choice $\eta = (np(1-p))^{-1/8}$ allows us to conclude.

Remark 2.1.4 Theorem 2.1.3 validates the Gaussian approximation if the product $np(1-p)$ is large; the classical heuristic $np \ge 5$ is used in statistics to validate the Gaussian approximation of binomials. Figure 2.1 reports the effective approximation of renormalized binomials by a Normal distribution.

This result is not optimal and the exponent $\frac{1}{8}$ may be replaced by $\frac{1}{2}$, use (Petrov 1975), theorem 3 on page 111. To this aim assume that $p = p_n$ and $np_n(1-p_n) \to_{n\to\infty} \infty$ then it is possible to choose any sequence $\eta \equiv \eta_n \to_{n\to\infty} 0$ above in order to get the convenient convergence rate.

Exercise 6 For each $\eta > 0$, $u \in \mathbb{R}$ there exists a function $f_{u,\eta} \in \mathcal{C}_b^3$ with

$$\mathbb{I}_{[u+\eta,\infty[} \le \mathbb{I}_{[u,\infty[} \le f_{u,\eta} \quad \text{and} \quad \|f'''_{u,\eta}\|_\infty = \mathcal{O}\left(\eta^{-3}\right).$$

1. Set first $u = 0$, $\eta = 1$. Then we set $g(x) = 0$ if $x \notin]0,1[$ and:

   (a)
   $$g(x) = x^4(1-x)^4, \qquad x \in]0,1[.$$
   Then $g \in \mathcal{C}_b^3$.

   (b)
   $$g(x) = \exp\left(-\frac{1}{x(1-x)}\right), \qquad x \in]0,1[.$$
   Then $g \in \mathcal{C}_b^\infty$. Indeed each of g's derivatives can be written as $g^{(k)}(x) = F(x)g(x)$ for some rational function F with no pole excepted 0 and 1.

   Consider $f(x) = G(x)/G(0)$ where we set

   $$G(x) = \int_x^1 g(s)ds, \quad for\ 0 \le x < 1,$$

   and $f(x) = 0$ for $x \ge 1$ with g as above.
2. General case. With f as before set $f_{u,\eta}(x) = f(u + x/\eta)$:

   $$f_{u,\eta}^{(k)}(x) = \frac{1}{\eta^k}\left(u + \frac{x}{\eta}\right) \le \frac{\|f^{(k)}\|_\infty}{\eta^k}, \quad \text{for } k = 0, 1, 2 \text{ or } 3.$$

   For the second function, the number k may be chosen arbitrarily large. This allows us to conclude.

2.2 Quantitative Results

2.2.1 *Moment Inequalities*

We now derive two important moment inequalities respectively called the Marcinkiewicz–Zygmund and the Rosenthal inequalities, they follow from Hall and Heyde (1980), see respectively Bürkholder and Rosenthal's theorems 2.10 and 2.12.

Alternative proofs of these results will be obtained below.

Lemma 2.2.1 *Let $(X_n)_n$ be a sequence of identically distributed, independent and centred random variables with finite p-th order moment, for some $p \geq 1$.*

Then, there exists a constant $C > 0$ which only depends on p such that the following inequalities hold:

- *Marcinkiewicz–Zygmund inequality:*

$$\mathbb{E}|X_1 + \cdots + X_n|^p \leq Cn^{\frac{p}{2}} EX^p.$$

- *Rosenthal inequality:*

$$\mathbb{E}|X_1 + \cdots + X_n|^p \leq C((n\mathbb{E}X^2)^{\frac{p}{2}} + n\mathbb{E}X_1^p).$$

Remark 2.2.1 (Rosenthal's inequality) For $p = 2$ the Rosenthal inequality is:

$$\mathbb{E}(X_1 + \cdots + X_n)^4 \leq C((n\mathbb{E}X_1^2)^2 + n\mathbb{E}X_1^4).$$

The second inequality also extends to non identically distributed, independent and centred random variables if $p \geq 2$.

There exists a constant C only depending on p, and such that

$$\mathbb{E}|X_1 + \cdots + X_n|^p \leq C\left(\left(\sum_{i=1}^{n} \mathbb{E}X_i^2\right)^{\frac{p}{2}} + \sum_{i=1}^{n} \mathbb{E}|X_i|^p\right).$$

Proof We restrict the proofs to $p \in 2\mathbb{N}^*$ and $p = 4$, respectively. Indeed the technique suitably extends under dependence.

Simple combinatoric arguments yield:

$$\begin{aligned}
\mathbb{E}(X_1 + \cdots + X_n)^{2p} &= \sum_{i_1=1}^{n} \cdots \sum_{i_{2p}=1}^{n} \mathbb{E}X_{i_1} \cdots X_{i_{2p}} \\
&= \sum_{i_1=1}^{n} \cdots \sum_{i_{2p}=1}^{n} T(i_1, \ldots, i_{2p})
\end{aligned}$$

$$\leq \sum_{i_1=1}^{n} \cdots \sum_{i_{2p}=1}^{n} |T(i_1, \ldots, i_{2p})|$$
$$\leq (2p)! \sum_{1 \leq i_1 \leq \cdots \leq i_{2p} \leq n} |T(i_1, \ldots, i_{2p})|.$$

Now from centring conditions we see that terms T vanish except for cases when $i_1 = i_2, \ldots, i_{2p-1} = i_{2p}$, since otherwise an index i would be isolated and the corresponding term vanishes by using independence.

Among $A = \{i_2, i_4, \ldots, i_{2p}\}$ which take precisely n^p values, one needs to make summations according to $\mathrm{Card}(A)$.

If all those indices are equal $T = \mathbb{E}X_0^{2p}$ and there are n such terms, and if they are all different, it is $(\mathbb{E}X_0^2)^p$.

For $p = 2$ we get the second point in this lemma.

For any $p \geq 1$, just use the Hölder inequality (Proposition A.2.2) to derive the first result.

The Rosenthal inequality may be improved:

Exercise 7 *(Rosenthal's inequality with order 4)* For independent and centred random variables with order 4 moments

$$\mathbb{E}|X_1 + \cdots + X_n|^4 = 3\left(\sum_{i=1}^{n} \mathbb{E}X_i^2\right)^2 + \sum_{i=1}^{n} (\mathbb{E}X_i^4 - 3(EX_i^2)^2)$$
$$\leq 3\left(\sum_{i=1}^{n} \mathbb{E}X_i^2\right)^2 + \sum_{i=1}^{n} \mathbb{E}X_i^4.$$

Hint. As above, we write:

$$\mathbb{E}(X_1 + \cdots + X_n)^4 = \sum_{i,j,k,l=1}^{n} \mathbb{E}X_i X_j X_k X_l = M_1 + M_2 + M_3 + M_4.$$

Here

$$M_s = \sum_{\substack{1 \leq i,j,k,l \leq n \\ \mathrm{Card}\{i,j,k,l\} = s}} \mathbb{E}X_i X_j X_k X_l, \qquad s = 1, 2, 3, 4,$$

stands for summations over indices such that s of them are distinct.

This means $M_1 = \sum_{i=1}^{n} \mathbb{E}X_i^4$, and moreover $M_3 = M_4 = 0$ since one index is distinct from all the others in such cases; independence and centring proves that such terms indeed equal 0. Now M_2 deserves a bit more attention and one index i, j, k, or l differs from the others; independence and centring again proves that the contribution of such terms is 0 except in case we have two couples of equal indices.

There exists six choices of such unordered couples:

$$M_2 = 6 \sum_{1 \le i < j \le n} \mathbb{E}X_i^2 \mathbb{E}X_j^2 \le 3 \left(\sum_{i=1}^{n} \mathbb{E}X_i^2 \right)^2.$$

The above bound is the only one which is not an equality so that we have the sharp identity

$$\mathbb{E}(X_1 + \cdots + X_n)^4 = 3 \left(\sum_{i=1}^{n} \mathbb{E}X_i^2 \right)^2 + \sum_{i=1}^{n} \left(\mathbb{E}X_i^4 - 3(EX_i^2)^2 \right). \tag{2.1}$$

This proves the optimality of the constants. The constant 3 is also the fourth order moment of a standard Normal random variable.

As an exercise, we suggest a really simple relation which we were not able to find in the literature.

Exercise 8 For independent and centred random variables with finite third order moments

$$\mathbb{E}(X_1 + \cdots + X_n)^3 = \sum_{i=1}^{n} \mathbb{E}X_i^3. \tag{2.2}$$

Hint. As before

$$\mathbb{E}(X_1 + \cdots + X_n)^3 = \sum_{i=1}^{n} \sum_{j=1}^{n} \sum_{k=1}^{n} \mathbb{E}X_i X_j X_k.$$

To conclude, just note that any non vanishing term in this expansion involves triplets (i, j, k) such that no index is different from the two other indices.

For triplets this simply means $i = j = k$.

Remark 2.2.2 A simple way to check the optimality the above identities (2.1) and (2.2) is to turn to the Gaussian setting, here $X_i \sim \mathcal{N}(0, \sigma_i^2)$ are independent for $i = 1, \ldots, n$.

And

$$S = \sum_{i=1}^{n} \mathbb{E}X_i \sim \mathcal{N}(0, \sigma^2), \qquad \sigma^2 = \sum_{i=1}^{n} \sigma_i^2,$$

then relation (2.1) becomes a tautology

$$\mathbb{E}S^4 = \sigma^4 EN^4 \equiv \mathbb{E}N^4 \sigma^4 + \sum_{i=1}^{n} \sigma_i^4 (\mathbb{E}N^4 - 3),$$

since $\mathbb{E}N^4 = 3$ for a Normal $\mathcal{N}(0, 1)$-r.v. (see Sect. A.3). Relation (2.2) is again the trivial identity $0 = 0$ since $\mathbb{E}N^3 = 0$.

Exercise 9 (The Weierstrass theorem) This result states that a continuous function over the interval is the uniform limit of some sequence of polynomials. Let $g : [0, 1] \to \mathbb{R}$ be a continuous function we define:

$$w(t) = \sup_{|x-y|<t} |g(x) - g(y)|.$$

This expression satisfies $\lim_{t\downarrow 0} w(t) = 0$ since Heine's theorem 2.2.1 (recalled below) entails that the function g is uniformly continuous.

Let $X_{1,x}, X_{2,x}, \ldots$ be iid random variables with marginal Bernoulli $b(x)$ distributions (Bernoulli distributed with the parameter x), we denote

$$S_{n,x} = \frac{1}{n}(X_{1,x} + \cdots + X_{n,x}).$$

Set $g_n(x) = \mathbb{E}g(S_{n,x})$:

1. Prove that g_n is a polynomial with degree n with respect to the variable x.
2. Prove the bound:
$$\operatorname{Var} g(S_{n,x}) = \frac{1}{n} \operatorname{Var} X_{1,x} \le \frac{1}{4n}.$$
3. Apply the Markov inequality to derive:
$$\lim_{n\to\infty} \sup_{0\le x\le 1} |g_n(x) - g(x)| = 0.$$
4. Assume that g is an Hölder function, then there exist constants $c, \gamma > 0$ with
$$|g(x) - g(y)| \le c|x - y|^\gamma, \quad \text{for each } x, y \in [0, 1].$$
Propose explicit convergence decay rates in the Weierstrass approximation theorem.
5. Now use Lemma 2.2.1 for moment inequalities with even order $2m$.
Then
$$\mathbb{E}(S_{n,x} - g_n(x))^{2m} \le cn^{-m},$$
for a constant which does not depend on $x \in [0, 1]$.
6. Use the previous higher order moment inequality to derive alternative convergence rates in the Weierstrass theorem.

Hints.

1. $g_n(x) = \sum_{k=0}^{n} \binom{k}{n} x^k (1-x)^{n-k} g\left(\frac{k}{n}\right).$

2. Prove that $x(1-x) \leq \frac{1}{4}$ if $0 \leq x \leq 1$.
3. Set $t > 0$ arbitrary and $A_{n,p} = (|S_{n,p} - p| > t)$, then:

$$\begin{aligned} g_n(x) - g(x) &= \mathbb{E}(g(S_{n,x}) - g(x)) \\ &= \mathbb{E}(g(S_{n,x}) - g(x))\, \mathbb{I}_{A_{n,x}} + \mathbb{E}(g(S_{n,x}) - g(p))\, \mathbb{I}_{A^c_{n,x}}. \end{aligned}$$

From the Markov inequality and from the above second point

$$\mathbb{P}(A_{n,x}) \leq \frac{1}{4nt^2},$$

a bound of the first term in the previous inequality is

$$|\mathbb{E}(g(S_{n,x}) - g(x))\, \mathbb{I}_{A_{n,x}}| \leq \frac{\|g\|_\infty}{2nt^2},$$

and from definitions the second term is bounded above by $w(t)$.
Let first n tend to infinity, in order to conclude.
4. Here $w(t) \leq ct^\gamma$ and the previous inequality gives

$$\|g_n - g\|_\infty \leq \frac{\|g\|_\infty}{2nt^2} + ct^\gamma.$$

Setting $t^{2+\gamma} = \|g\|_\infty/(2cn)$ provides a rate $n^{-\frac{\gamma}{2+\gamma}}$.
5. From Lemma 2.2.1

$$\mathbb{E}(S_{n,x} - g_n(x))^{2m} \leq c\mathbb{E}X_{1,x}^{2m}\, n^{-m}.$$

6. Now

$$\|g_n - g\|_\infty \leq \frac{c\|g\|_\infty}{2nt^{2m}} + ct^\gamma.$$

Set $t^{2m+\gamma} = C/n^m$ then a rate is $n^{-\frac{m\gamma}{2m+\gamma}}$ is now provided.

Recall that the continuity at point $x_0 \in [0,1]$ and the uniform continuity of $g : [0,1] \to \mathbb{R}$ give respectively

$$\begin{aligned} &\forall \epsilon > 0,\ \exists \eta > 0,\ \forall x \in [0,1]: \quad |x - x_0| < \eta \Longrightarrow |g(x) - g(x_0)| < \epsilon \\ &\forall \epsilon > 0,\ \exists \eta > 0,\ \forall x, y \in [0,1]: |x - y| < \eta \Longrightarrow |g(x) - g(y)| < \epsilon. \end{aligned}$$

In the latter case η does not depend on x_0.

Exercise 10 The function $x \mapsto g(x) = x^2$ is not uniformly continuous over $\mathbb{R}$.

Hint. By contradiction. Set $x = n$ and $y = n + \dfrac{1}{2n}$, then

$$g(y) - g(x) = 1 + \frac{1}{4n^2} \quad \text{does not tend to zero as } n \uparrow \infty.$$

E.g. (Choquet 1973) proves the fundamental and classical result:

Theorem 2.2.1 (Heine) *Let $g : K \to \mathbb{R}$ be a continuous function defined on a compact metric space (K, d) then g is uniformly continuous.*

2.2.2 Exponential Inequalities

Below we develop two exponential inequalities which yields reasonable bounds for the tail of partial sums of independent identically distributed random variables. From the Central Limit Theorem we first check the Gaussian case.

Exercise 11 Let $N \sim \mathcal{N}(0, 1)$ be a standard Normal random variable, then derive:

$$\left(\frac{1}{t} - \frac{1}{t^3}\right)\frac{e^{-\frac{1}{2}t^2}}{\sqrt{2\pi}} \leq \overline{\Phi}(t) = \mathbb{P}(N > t) \leq \frac{1}{t}\frac{e^{-\frac{1}{2}t^2}}{\sqrt{2\pi}}.$$

Hint. Use integration by parts and the Markov inequality.

Analogously we obtain:

Lemma 2.2.2 *(Hoeffding) Let $R_1, \ldots, R_n$ be independent Rademacher random variables (i.e. $\mathbb{P}(R_i = \pm 1) = \frac{1}{2}$).*

For real numbers $a_1, \ldots, a_n$ set

$$\xi = \sum_{i=1}^{n} a_i R_i,$$

and assume that

$$\sum_{i=1}^{n} a_i^2 \leq c.$$

Then:

1. $\mathbb{P}(\xi \geq x) \leq e^{-\frac{x^2}{2c}}$, *for all* $x \geq 0$,
2. $\mathbb{P}(|\xi| \geq x) \leq 2e^{-\frac{x^2}{2c}}$, *for all* $x \geq 0$,
3. $\mathbb{E}e^{\frac{\xi^2}{4c}} \leq 2$.

Proof We first prove that for each $s \in \mathbb{R}$:

$$\mathbb{E}e^{sR_1} \leq e^{s^2/2}. \tag{2.3}$$

The inequality (2.3) is rewritten as

$$\text{ch}\, s = \frac{1}{2}\left(e^s + e^{-s}\right) \le e^{s^2/2}.$$

Indeed the two previous functions may be expanded as analytic functions on the real line $\mathbb{R}$, and:

$$\text{ch}\, s = \sum_{k=0}^{\infty} \frac{s^{2k}}{(2k)!}, \qquad e^{s^2/2} = \sum_{k=0}^{\infty} \frac{s^{2k}}{2^k \cdot k!}.$$

Inequality (2.3) follows from the relation $(2k)! \ge 2^k \cdot k!$ simply restated as

$$(k+1)(k+2)\cdots(k+k) \ge (2\cdot 1)(2\cdot 1)\cdots(2\cdot 1) = 2^k.$$

Markov's inequality now implies

$$\mathbb{P}(\xi \ge x) \le e^{-tx}\mathbb{E}e^{t\xi}, \qquad \forall t \ge 0,$$

because (2.3) entails with independence:

$$\mathbb{E}e^{t\xi} = \prod_{i=1}^{n} \mathbb{E}e^{ta_i R_i} \le e^{ct^2/2}.$$

For $t = x/c$ we derive point *(1)*.

Point *(2)* comes from the observation that ξ is a symmetric random variable and $\mathbb{P}(|\xi| \ge x) = 2\mathbb{P}(\xi \ge x)$ for $x \ge 0$.

Point *(3)* is derived from the following calculations:

$$\begin{aligned}
\mathbb{E}e^{\frac{\xi^2}{4c}} - 1 &= 4c\mathbb{E}\int_0^{\xi^2} \exp\left(\frac{t}{4c}\right) dt \\
&= 4c\mathbb{E}\int_0^{\infty} \mathbb{1}_{\{t\le\xi^2\}} \exp\left(\frac{t}{4c}\right) dt \\
&= 4c\int_0^{\infty} \mathbb{E}\ \mathbb{1}_{\{t\le\xi^2\}} e^{\frac{t}{4c}}\, dt \\
&= 4c\int_0^{\infty} \mathbb{P}(\xi^2 \ge t) e^{\frac{t}{4c}}\, dt \\
&\le 4c\int_0^{\infty} e^{-\frac{t}{4c}}\, dt = 1.
\end{aligned}$$

Here the Fubini–Tonnelli theorem (see e.g. in Doukhan and Sifre 2001) justifies the first inequalities while the last inequality is a consequence of the relation *(2)*.

Remark 2.2.3 Let $R \in [-1,1]$ be a centred random variable, then $\mathbb{E}e^{tR} \le \frac{1}{2}\left(e^t + e^{-t}\right)$, and the Hoeffding Lemma 2.2.2 instantaneously extends to sums $\sum_i a_i R_i$ for R_i with values in $[-1,1]$, centred independent random variables.

Lemma 2.2.3 *(Bennett) Let $Y_1, \ldots, Y_n$ be independent centred random variables with $|Y_i| \le M$ for $1 \le i \le n$, and denote*

$$V = \sum_{i=1}^{n} \mathbb{E}Y_i^2.$$

If $\xi = \sum_{i=1}^{n} Y_i$ then for each $x \ge 0$ the Bennett inequality holds:

$$\mathbb{P}(|\xi| \ge x) \le 2\exp\left(-\frac{x^2}{2V} B\left(\frac{Mx}{V}\right)\right),$$

with $B(t) = \frac{2}{t^2}\left((1+t)\log(1+t) - t\right)$.

The Bernstein inequality also holds:

$$\mathbb{P}(|\xi| \ge x) \le 2\exp\left(-\frac{x^2}{2\left(V + \frac{1}{3}Mx\right)}\right).$$

Proof The proof is again based upon Markov's inequality. We shall make use of the independence of $Y_1, \ldots, Y_n$. We first need to bound above the Laplace transform of Y_i. Using first the facts that $\mathbb{E}Y_i = 0$ and $|\mathbb{E}Y_i^k| \le M^{k-2}\mathbb{E}Y_i^2$ for each $k > 1$ yields:

$$\mathbb{E}e^{tY_i} = \sum_{k=0}^{\infty} \frac{t^k}{k!}\mathbb{E}Y_i^k \le 1 + \mathbb{E}Y_i^2 \sum_{k=2}^{\infty} \frac{t^k}{k!}\mathbb{E}Y_i^k - 1 + \mathbb{E}Y_i^2 g(t) \le \exp\{\mathbb{E}Y_i^2 g(t)\}$$

where we set

$$g(t) = \frac{e^{tM} - 1 - tM}{M^2}.$$

Both from independence and from Markov's inequality we then obtain:

$$\mathbb{P}(\xi \ge x) \le e^{Vg(t) - xt}.$$

Optimizing this bound with respect to V yields $Vg'(t) = x$.

Hence

$$t = \frac{1}{M}\log\left(1 + \frac{xM}{V}\right) > 0,$$

and

$$Vg(t) - xt = \frac{x}{M} - t\Big(\frac{V}{M} + x\Big)$$

yields Bennett's inequality.

The Bernstein inequality follows from the relation

$$(1+t)\log(1+t) - t \geq \frac{t^2}{2(1+\frac{t}{3})}.$$

The latter inequality is rewritten $(1 + \dfrac{t}{3})B(t) \geq 1$.

To prove it one studies the variations of the function

$$t \mapsto f(t) = t^2\Big((1+\frac{t}{3})B(t) - 1\Big).$$

Note that

$$f'(0) = 0, \text{ and } f''(t) = \frac{1}{3}((1+t)\log(1+t) - t) \geq 0,$$

then $f(0) = 0$ and $f'(t) \geq 0$.

Exercise 12 Let $g : \mathbb{R}^+ \to \mathbb{R}^+$ be an a.s. differentiable non-decreasing function, then

$$\mathbb{E}g(|\xi|) = \int g(z)\mathbb{P}_{|\xi|}(dz) = \int g'(z)\mathbb{P}(|\xi| > z)\, dz.$$

and

$$\mathbb{E}|\xi|^p \leq \Big(\frac{3V}{M}\Big)^p + 2p\Big(\frac{4M}{3}\Big)^p \int_{\frac{9V}{4M^2}}^{\infty} x^{p-1}e^{-x}\, dx.$$

Hint. From non-negativity, the Fubini–Tonnelli theorem gives

$$\mathbb{E}g(|\xi|) = \int g(z)\mathbb{P}_{|\xi|}(dz) = \int g'(z)\mathbb{P}(|\xi| > z)\, dz.$$

Set $A = 3V/M$ then from Bernstein's inequality in Lemma 2.2.3 we get

$$\mathbb{E}g(|\xi|) \leq g\Big(\frac{3V}{M}\Big) + 2\int_{\frac{3V}{M}}^{\infty} g'(z)e^{-\frac{3z}{4M}}\, dz,$$

and

$$\mathbb{E}g(|\xi|) \leq g\Big(\frac{3V}{M}\Big) + \frac{8M}{3}\int_{\frac{9V}{4M^2}}^{\infty} g'\left(\frac{4Mx}{3}\right)e^{-x}\, dx,$$

with $x = 3z/4M$.

Hence if $g(x) = |x|^p$ for some $p > 0$,

$$\mathbb{E}g(|\xi|) \leq \left(\frac{3V}{M}\right)^p + 2p\left(\frac{4M}{3}\right)^p \int_{\frac{9V}{4M^2}}^{\infty} x^{p-1}e^{-x}\,dx.$$

This is a more general form of the Rosenthal inequality in Lemma 2.2.1.

Chapter 3
Estimation Concepts

Many statistical procedures are derived from probabilistic inequalities and results; such procedures may need more precise bounds as this is proved in the present chapter for the independent case. Basic notations are those from Appendix B.1. Developments may be found in van der Vaart (1998) and those related with functional estimation may be found in the monograph (Rosenblatt 1991). We begin the chapter with applications of the moment inequalities in Lemma 2.2.1 which are useful for empirical procedures. Then we describe empirical estimators, contrast estimators and non-parametric estimators. The developments do not reflect the relative interest of the topics but are rather considered with respect to possible developments under dependence conditions hereafter.

3.1 Empirical Estimators

The behaviour of empirical means are deduced from the behaviour of partial sums, and below we restate such results in a statistical setting (see Appendix B). The parameter is implicit: this is the distribution of X.

Corollary 3.1.1 *Let $(X_n)_{n\geq 0}$ be an independent and identically distributed sequence. If $\mathbb{E}X_0^4 < \infty$ then,*

$$\overline{X} = \frac{1}{n}(X_1 + \cdots + X_n) \to_{n\to\infty} \mathbb{E}X_0, \qquad a.s.$$

Remark 3.1.1 Note that the ergodic Theorem 9.1.1 proves that the simple assumption $\mathbb{E}|X_0| < \infty$ ensures indeed this SLLN. We give this result as a simple consequence of the previous Marcinkiewicz–Zygmund inequality in Lemma 2.2.1 for clarity of

P. Doukhan, *Stochastic Models for Time Series*, Mathématiques et Applications 80,
https://doi.org/10.1007/978-3-319-76938-7_3

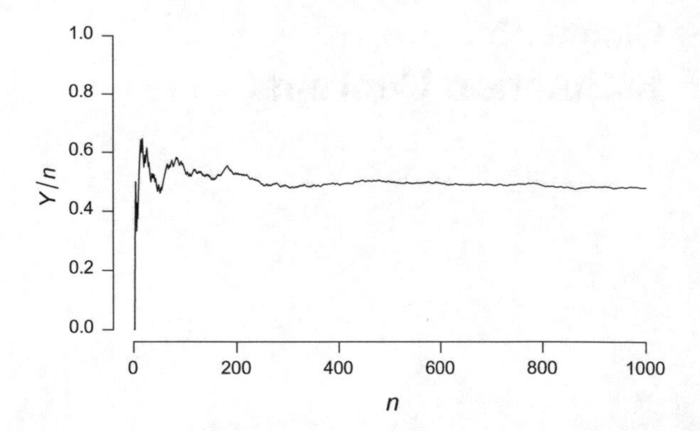

Fig. 3.1 Proportion of heads among n tosses of a fair coin

exposition. Convergence in this LLN is simulated in Fig. 3.1; here the 1000 dots are the values of one trajectory of $(\overline{X}_n)_{1\le n\le 1000}$.

Proof Let $\epsilon > 0$ be arbitrary then Markov's inequality entails

$$\mathbb{P}(|\overline{X}| \ge \epsilon) \le C \cdot \frac{\mathbb{E}X_0^4}{\epsilon^4 n^2}.$$

Hence

$$\sum_{n=1}^{\infty} \mathbb{P}(|\overline{X}_n| \ge \epsilon) < \infty,$$

is a convergent series. The a.s. convergence is a consequence of the Borel–Cantelli lemma B.4.1.

Now when $\mathbb{E}X_0^2 < \infty$, then the Markov inequality yields $\mathbb{L}^2$-convergence of $\overline{X}$, since Var $(\overline{X}) =$ Var $(X_0)/n$; the convergence in probability also holds.

Convergence of the cumulative distribution function for a 1000 sample of binomial distributions with parameter $p = 0.5$ is also illustrated in Fig. 3.2.

This allows us to prove first fundamental statistical result:

Theorem 3.1.1 *Let (Y_n) be a real valued and independent identically distributed sequence such that Y_0 admits the cumulative distribution function $F(y) = \mathbb{P}(Y_0 \le y)$ on $\mathbb{R}$.*

Define the empirical cumulative distribution:

$$F_n(y) = \frac{1}{n}\sum_{j=1}^{n} \mathbb{I}_{\{Y_j \le y\}}.$$

Then $\mathbb{E}F_n(y) = F(y)$, the estimator is said to be unbiased (see Definition B.5.2), and

$$\sup_{y\in\mathbb{R}} |F_n(y) - F(y)| \to_{n\to\infty} 0, \quad a.s.$$

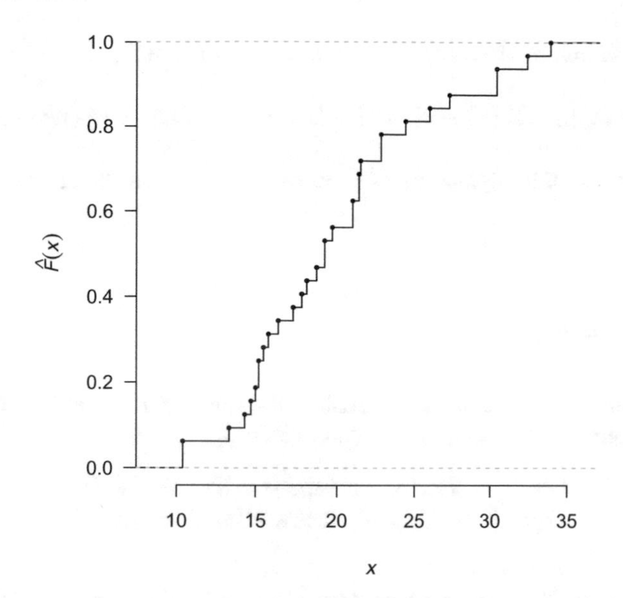

Fig. 3.2 Empirical cumulative distribution of fuel consumption of 32 cars

Remark 3.1.2 Figure 3.2 represents one trajectory of the cumulative distribution function.

The data come from the classical `mtcars` R dataset. According to the documentation, *"the data was extracted from the 1974 Motor Trend US magazine, and comprises fuel consumption in terms of miles per gallon and 10 aspects of car design and performance for 32 cars (1973 and 1974 models)"*.

Proof The previous SLLN (Corollary 3.1.1) implies the convergence.

The uniform convergence follows from the standard variant of Dini's theorem in Exercise 13.

Exercise 13 (*Variant of the Dini theorem*) Assume that a sequence of cdf satisfies $\lim_{n\to\infty} F_n(x) = F(x)$ for each $x \in \mathbb{R}$. If F is a continuous cdf then the convergence is uniform.

Proof Let $\epsilon > 0$ be arbitrary. From the properties of a cdf there exists $A > 0$ such that if $x > A$ then $1 - F(x) < \epsilon/2$ and $x < -A$ implies $F(x) < \epsilon/3$.

From Heine theorem 2.2.1, there exist $x_1 = -A < x_2 < \cdots < x_p = A$ such that if $x \in [x_i, x_{i+1}]$ then

$$F(x_{i+1}) - F(x) < \frac{\epsilon}{3}, \qquad F(x) - F(x_i) < \frac{\epsilon}{3},$$

if $i = 1, \ldots, p-1$. Set $x_0 = -\infty$ and $x_{p+1} = \infty$, and the oscillation of F is less that $\epsilon/3$ over each interval $J_i = [x_i, x_{i+1})$ for each $i = 0, \ldots, p$ (limits are included for each finite extremity).

The relation $\lim_{n\to\infty} F_n(x_i) = F(x_i)$, for $i = 1, \ldots, p$, makes it possible to exhibit N such that, if $n > N$, then $|F_n(x_i) - F(x_i)| < \epsilon/3$.

Each $x \in \mathbb{R}$ belongs to some interval J_i so that in case $i \neq 0$:

$$|F_n(x) - F(x)| \leq |F_n(x) - F_n(x_i)| + |F_n(x_i) - F(x_i)| + |F(x_i) - F(x)| < \epsilon.$$

For $i = 0$ one should replace $x_0 = -\infty$ by $x_1 = -A$ in the above inequality to conclude.

3.2 Contrasts

Assume an independent identically distributed sample with values in a Banach space E admits a marginal distribution in a class $(P_\theta)_{\theta\in\Theta}$.

Definition 3.2.1 Let $X \sim P_{\theta_0}$. A function $\rho : E \times \Theta \to \mathbb{R}$ is a contrast if the expression $\theta \mapsto D(\theta_0, \theta) = \mathbb{E}_{\theta_0}\rho(X, \theta)$ is well defined and if it admits a unique minimum θ_0.

Note that $\rho(X, \theta)$ is an unbiased estimator for the function $g(\theta_0) = D(\theta_0, \theta)$ (for each $\theta \in \Theta$). If we only dispose of a simple realization X of this experiment, then the true parameter θ_0 is estimated by a minimizer $\widehat{\theta}(X)$ of the contrast $\theta \mapsto \rho(X, \theta)$ (we shall assume that such a minimizer exists):

$$\widehat{\theta}(X) = \text{Argmin}_{\theta\in\Theta}\, \rho(X, \theta). \tag{3.1}$$

Assume that $\Theta \subset \mathbb{R}^d$ is open and such that the function $\theta \mapsto \rho(X, \theta)$ is differentiable.

The estimator $\widehat{\theta}(X)$ of the parameter θ_0 satisfies the following condition, usually easier to check than (3.1):

$$\nabla\rho(X, \widehat{\theta}(X)) = 0. \tag{3.2}$$

Example 3.2.1 This situation occurs for example in the cases of:

- Maximum Likelihood Estimator (MLE)

$$\rho(x, \theta) = -\log f_\theta(X)$$

with f_θ the density of P_θ.
Let $X = (X_1, \ldots, X_n)$ be an independent identically distributed sample with marginal densities $p_\theta(x)$.
Then

$$\rho(x, \theta) = -\sum_{k=1}^{n} \log f_\theta(X_k).$$

The point that the above expression is indeed a contrast relies on the following identifiability condition:

$$f_{\theta_1} = f_{\theta_2} \ a.s. \ \Rightarrow \theta_1 = \theta_2.$$

- Least squares estimator (LSE) Assume that $X = G(\theta) + \sigma(\theta)\xi$, and

$$\rho(x, \theta) = \frac{\|X - G(\theta)\|^2}{\sigma^2(\theta)}.$$

If $\xi = (\xi_1, \ldots, \xi_n)$ are independent identically distributed random variables and $G(\theta) = (g(\theta, z_1), \ldots, g(\theta, z_n))$ it is a regression model with a fixed design.

Remark 3.2.1 (*Model selection*) A large part of the modern statistics is based on contrast techniques. Assume that the statistical model itself is unknown but it belongs to a class of models $\mathcal{M}$; precisely each of those models $M \in \mathcal{M}$ is indexed by a parameter set Θ_M and a contrast $(\rho_M(X, \theta))_{\theta \in \Theta_M}$ is given (this is model selection). The price to pay for using the model M is a penalization $p(M)$ which increases with the complexity of the model. One may estimate the model M and the parameter $\theta \in \Theta_M$ as:

$$\text{argmin} \left\{ p(M) + \inf_{\theta \in \Theta_M} \rho_M(X, \theta), \quad M \in \mathcal{M} \right\}.$$

We choose in this book to avoid a precise presentation of those techniques essentially introduced by Pascal Massart, see e.g. Massart (2007). Indeed very intricate concentration inequalities are needed in this fascinating setting. Unfortunately, in the dependent case, no completely satisfactory extension has been developed yet.

3.3 Functional Estimation

We now introduce another standard tool of statistics related to function estimation; Rosenblatt (1991) provides a good presentation of these features. Let $(X_j)_{j \geq 1}$ be an independent identically distributed sequence with a common marginal density f.

In order to fit $f = F'$ the simple plug-in technique consists of deriving an estimator of the cumulative distribution function. This does not work since differentiation is not a continuous function in the Skorohod space $\mathcal{D}[0, 1]$, of right continuous functions with left limits (see Appendix B), moreover F_n's derivative is 0 (a.s.).

Consider a realization of a sample $X_1, \ldots, X_n$, then a reasonable estimator is the histogram; divide the space of values into pieces with a small probability then we may count the proportion of occurrences of X_j's in an interval to fit f by a step function.

Formally this means that

$$\widehat{f}(x) = \frac{1}{n}\sum_{i=1}^{n}\sum_{j=1}^{m} e_{j,m}(x)e_{j,m}(X_i),$$
$$e_{j,m}(x) = \mathbb{1}_{I_{j,m}}(x), \qquad 1 \le j \le m,$$

for a partition $I_{1,m} \bigcup \cdots \bigcup I_{m,m} = \mathbb{R}$.

Remark 3.3.1 A difficulty is that histograms are not smooth even if they estimate possibly smooth densities. More generally $(e_{j,m})_{1\le j\le m}$ may be chosen as an orthonormal system of $\mathbb{L}^2(\mathbb{R})$, such as a wavelet basis; Doukhan (1988) initially introduced a simple linear wavelet estimator. Any orthonormal system $e_{j,m} = e_j$ for $1 \le j \le m$ may also be considered. Note also that

$$\widehat{f}(x) = \sum_{j=1}^{m} \widehat{c}_{j,m} e_{j,m}(x),$$

where we set

$$\widehat{c}_{j,m} = \frac{1}{n}\sum_{i=1}^{n} e_{j,m}(X_i).$$

Note that $\widehat{c}_{j,m}$ is the empirical unbiased estimator of $\mathbb{E}e_{j,m}(X_0)$ (see Definition B.5.2 below).

Such estimators are empirical estimators of the orthogonal projection f_m of f of the vector space spanned by $(e_{j,m})_{1\le j\le m}$; they are known as projection density estimations of f.

In order to make them consistent one needs to choose a sequence of parameters $m = m_n \uparrow \infty$. Such general classes of estimators are reasonable and may be proved to be consistent.

Here we develop an alternative classical smoothing technique, based on kernels. Contrary to the case of projection estimators, an asymptotic expansion of the bias may be exhibited; wavelet estimators (Doukhan 1988 and Doukhan and León 1990 seem to be the first works related to this subject) corrects this real problem of projection estimators.

A simple estimation is introduced through a smoothing argument of F_n. We now introduce kernel estimators of the density:

Definition 3.3.1 Let (X_n) be a real valued and independent identically distributed sequence such that X_0 admits a density f on $\mathbb{R}$.

Assume that $K : \mathbb{R} \to \mathbb{R}$ is a function such that:

$$\int_{\mathbb{R}} (1 + |K(y)|)|K(y)|dy < \infty, \qquad \int_{\mathbb{R}} K(y)dy = 1.$$

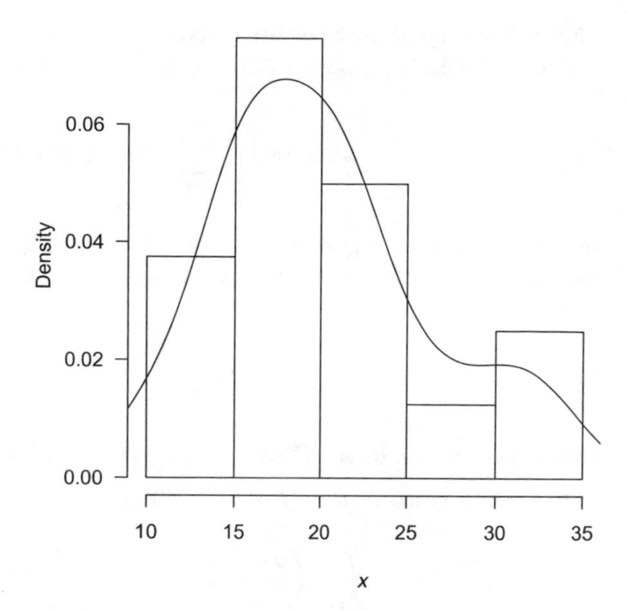

Fig. 3.3 Sample distribution function and a kernel estimate of the data of Fig. 3.2

A kernel estimator of f is defined through a sequence $h = h_n \to_{n\to\infty} 0$ by:

$$\widehat{f}(y) = \frac{1}{nh}\sum_{j=1}^{n} K\left(\frac{X_j - y}{h}\right).$$

Figure 3.3 shows the competitive behaviors of a histogram and a kernel density estimator. Note that dependence of those data is not assumed here.

The first following result allows us to bound the bias of such estimators:

Lemma 3.3.1 *Let g denote a bounded density for some probability distribution with moments up to order $p \in \mathbb{N}^*$, then there exists a polynomial P with degree $\le p$ such that $K = Pg$ is a kernel satisfying*

$$\int_{\mathbb{R}} y^s K(y)\, dy = \begin{cases} 1, & \text{if} \quad j = 0,\ p, \\ 0, & \text{if} \quad 1 \le j < p. \end{cases}$$

Remark 3.3.2 Such functions are called pth order kernels. Note that if $p > 2$ such kernels cannot be non-negative.

For $p = 1$ and g symmetric ($g(-y) = g(y)$) it is simple to see that $P = 1$ satisfies the previous relations but maybe not $\displaystyle\int y^2 g(y)\, dy = 1$, anyway this expression is positive.

Proof It is simple to use the quadratic form associated with the square matrix (with size $(p+1)\times(p+1)$)

$$A = \left(a_{i+j}\right)_{0\le i,j\le n} \quad \text{with} \quad a_k = \int_{\mathbb{R}} y^k g(y)\, dy.$$

This matrix is symmetric positive definite.

Indeed for each $x = (x_0, \dots, x_p)' \in \mathbb{R}^{p+1}$, one obtains:

$$x'Ax = \int_{\mathbb{R}} \Big(\sum_{j=0}^{p} x_j y^j\Big)^2 g(y)\,dy \ge 0.$$

In case the previous expression vanishes, the fact that $g \neq 0$ on a set with positive measure (this set is thus infinite), implies that the polynomial

$$y \mapsto \sum_{j=0}^{p} x_j y^j,$$

vanishes on an infinite set. Hence it admits null coefficients.

The change of variable $u = (v - y)/h$ entails:

$$\mathbb{E}\widehat{f}(y) = \frac{1}{h}\int_{\mathbb{R}} K\left(\frac{v-y}{h}\right) f(v)\,dv = \int_{\mathbb{R}} K(u) f(y - hu)\,du.$$

Together with a simple application of Taylor's formula, left as an exercise, the previous expression yields Proposition 3.3.1:

Proposition 3.3.1 *Assume that* $h \equiv h_n \to_{n\to\infty} 0$. *Let* $[a, b]$ *be a compact interval and* $\epsilon > 0$. *We assume that the function* f *admits* p *continuous and bounded derivatives on* $[a - \epsilon, b + \epsilon]$.
Then if K *is a* p*th order kernel with a compact support:*

$$\lim_{h\downarrow 0} \sup_{y\in[a,b]} h^{-p} \left| \mathbb{E}\widehat{f}(y) - f(y) - \frac{h^p}{p!} f^{(p)}(y) \int u^p K(u)du \right| = 0.$$

Assume now that for some $\rho \in]p, p+1]$ *there exists a constant such that* $|f^{(p)}(x) - f^{(p)}(y)| \le C|x - y|^{\rho-p}$, *for all* $x, y \in [a - \epsilon, b + \epsilon]$.
Then there exists a constant $c > 0$ *with:*

$$\sup_{y\in[a,b]} \left|\mathbb{E}\widehat{f}(y) - f(y)\right| \le ch^{\rho}.$$

Remark 3.3.3 Some details and improvements are needed here.

- Independence of (Y_k) is not necessary but only the fact that Y_k's are identical distributed for $1 \le k \le n$.[1]
- The uniformity over $\mathbb{R}$ may be omitted if K admits a compact support, then

$$\left|\mathbb{E}\widehat{f}(y) - f(y)\right| \le \frac{h^p}{p!} \sup_{u\in y+V} |f^{(p)}(u)| \int |u|^p |K(u)|du.$$

[1] In fact even stationarity may not hold as e.g. if they are subsampled from a stationary process: $Y_k = Z_{j_k}$ for $(Z_j)_{j\in\mathbb{Z}}$ stationary, see Definition 4.1.1.

Here V denotes a neighbourhood of 0 such that $hV \subset \text{Support}(K)$.

- It is possible to describe infinite-order kernels, see Politis (2003).

Use the previous results together with Lindeberg theorem with $k = n$ and

$$U_j = \frac{1}{\sqrt{nh}}\left(K\left(\frac{Y_j - y}{h}\right) - \mathbb{E}K\left(\frac{Y_j - y}{h}\right)\right).$$

Theorem 3.3.1 *Let now $h = h_n \downarrow 0$ as $n \to \infty$.*
Assume that $nh_n \to_{n\to\infty} \infty$:

$$\begin{aligned} nh_n \, Var \, \widehat{f}(y) & \to_{n\to\infty} f(y)\int_{\mathbb{R}} K^2(u)du, \\ \sqrt{nh_n}(\widehat{f}(y) - \mathbb{E}\widehat{f}(y)) & \to^{\mathcal{L}}_{n\to\infty} \mathcal{N}\left(0, f(y)\int_{\mathbb{R}} K^2(u)du\right). \end{aligned}$$

If the conditions in the previous Proposition 3.3.1 and Theorem 3.3.1 hold and if we assume also that $h_n \to_{n\to\infty} 0$, then:

$$\mathbb{E}(\widehat{f}(y) - f(y))^2 \sim_{n\to\infty} \frac{1}{nh_n} f(y)\int_{\mathbb{R}} K^2(u)du + \left(\frac{h_n^p}{p!} f^{(p)}(y)\int u^p K(u)du\right)^2.$$

Convergence in probability holds for such estimators, a CLT is also available. Usual minimax rates (see Tsybakov 2004) of such estimators give $\mathcal{O}(n^{\frac{p}{2p+1}})$: they are obtained by minimizing this expression wrt $h = h_n$ or, equivalently, by equating the squared bias and variance of the estimator.

Moreover, if $nh_n \to \infty$, the use of Rosenthal's moment inequalities from Remark 2.2.1 implies

$$\mathbb{E}|\widehat{f}(y) - \mathbb{E}\widehat{f}(y)|^p \le \frac{C}{(nh)^{\frac{p}{2}}}.$$

This bound together with Markov's inequality and the Borel–Cantelli lemma B.4.1 imply the following result:

Proposition 3.3.2 *Besides the assumptions in Theorem 3.3.1, assume that f is a continuous function around the point y, then*

$$\widehat{f}(y) \to_{n\to\infty} f(y), \quad a.s.$$

as soon as

$$h \equiv h_n \to_{n\to\infty} 0, \qquad \sum_{n=1}^{\infty} \frac{1}{(nh_n)^{\frac{p}{2}}} < \infty.$$

Remark 3.3.4 Sharper results may be developed by using the Bernstein exponential inequality; the present section is only an introduction to statistical applications which will be extended later under dependence.

Rosenthal's inequalities are also proved in Chap. 12 under dependence conditions; the Rosenthal inequalities from Sect. 12.2.4 are then used for dependent kernel estimation in Sect. 12.3.

Exercise 14 Prove the bound:

$$\mathbb{P}(\sqrt{nh_n}|\widehat{f}(x) - \mathbb{E}\widehat{f}(x)| \geq t) \leq 2e^{-\frac{t^2}{cf(x)}}, \tag{3.3}$$

for all $c \leq 2\int_{\mathbb{R}} K^2(u)du$, and for n large enough.

From integration derive that for each $p \geq 1$:

$$\|\widehat{f}(x) - \mathbb{E}\widehat{f}(x)\|_p = \mathcal{O}\left(\frac{1}{\sqrt{nh_n}}\right), \quad \text{if} \quad nh_n \geq 1.$$

Proof The results rely on simple integration tricks. We will use the Bernstein inequality, see Lemma 2.2.3.

Write

$$\sqrt{nh}(\widehat{f}(x) - \mathbb{E}\widehat{f}(x)) = Z_1 + \cdots + Z_n,$$

with

$$Z_j = U_j - \mathbb{E}U_j, \qquad U_j = \frac{1}{\sqrt{nh}}K\left(\frac{Y_j - y}{h}\right).$$

Then the relations

$$\|Z_j\|_\infty \leq \frac{2}{\sqrt{nh}}\|K\|_\infty,$$

and

$$\mathbb{E}Z_j^2 \sim \sqrt{\frac{h}{n}}\, f(y) \int K^2(s)\, ds,$$

complete the proof of the first inequality.

The moment inequality relies on the fact that setting $u = t/\sqrt{f(x)}$ yields:

$$2\int t^p e^{-\frac{t^2}{cf(x)}}\, dt = 2f^{\frac{p+1}{2}}(x) \int u^p e^{-\frac{u^2}{c}}\, du < \infty.$$

This allows us to conclude.

Chaining arguments are detailed below to derive uniform convergence properties.

Exercise 15 (*Uniform convergence*) Consider again the kernel density estimator $\widehat{f}$ based upon a compactly supported and uniformly Lipschitz kernel function.
Assume $\lim_{n\to\infty} \frac{nh_n}{\log n} = \infty$.

- Prove that a.s. uniform convergence holds over any compact interval I, if the density is uniformly continuous.
- As in Proposition 3.3.2 also prove the existence of $C > 0$ such that:

$$\left\| \sup_{x\in I} |\widehat{f}(x) - \mathbb{E}\widehat{f}(x)| \right\|_p \le C\sqrt{\frac{\log n}{nh_n}}.$$

- Assuming that the function f satisfies a ρ-regularity condition uniformly over a compact neighbourhood of I as in Proposition 3.3.1, prove that for a convenient constant $c > 0$

$$\left\| \sup_{x\in I} |\widehat{f}(x) - f(x)| \right\|_p \le C\left(\frac{n}{\log n}\right)^{-\frac{\rho}{2\rho+1}}.$$

Hint. Typically $I = [0, 1]$. Divide $I = [0, 1]$ into m intervals $I_1, \ldots, I_m$ with measure $1/m$. Under ρ-regularity conditions, we obtain:

$$\begin{aligned}\sup_{x\in I} |\widehat{f}(x) - \widehat{f}(x)| &\le \sup_{x\in I} |\widehat{f}(x) - \mathbb{E}\widehat{f}(x)| + \sup_{x\in I} |f(x) - \mathbb{E}\widehat{f}(x)|, \\ &\le \sup_{x\in I} |\widehat{f}(x) - \mathbb{E}\widehat{f}(x)| + Ch_n^\rho.\end{aligned}$$

From uniform continuity of the function and from uniform convergence properties of the bias, only the first member in the RHS of the previous inequality needs consideration.

Then if the chosen kernel is Lispschitz the oscillation of $\widehat{f}$ over each such interval is less than C/mh^2, for some suitable constant.

If $mh^2 > C'$ is large enough for a constant $C' = C'(t, C, f)$, the oscillation of the function $\widehat{f} - \mathbb{E}\widehat{f}$ over each interval I_k will be less that some fixed positive number t.

From the assumption $nh_n/\log n \to \infty$ we also derive that $m = \mathcal{O}(n^s)$ for some $s > 0$.

Choose now some point $x_k \in I_k$ for each $1 \le k \le m$. Then it follows, for each $\epsilon > 0$:

$$\begin{aligned}\mathbb{P}\left(\sup_{x\in[0,1]} |\widehat{f}(x) - f(x)| > 2t\right) &\le m \max_{1\le k\le m} \mathbb{P}(|\widehat{f}(x_k) - f(x_k)| > t), \\ &\le Cn^s \exp\left(-\frac{t^2}{cf(x)}\right).\end{aligned}$$

The more precise calibration of h and m yields a.s. uniform results for the convergence of $\widehat{f}$, improving Proposition 3.3.2.

Integration yields the first moment bound $\mathbb{E}Z = \int_0^\infty \mathbb{P}(Z \geq v)\, dv$ with $v = t^p$ and $Z = \sup_{x\in[0,1]} |\widehat{f}(x) - f(x)|^p$, as in the second part of the proof of Proposition 3.3.2.

Remark that this reinforcement of a.s. convergence is not related to independence, see Exercise 78.

Moreover, the uniform moment bound follows from

$$\begin{aligned}\left\| \sup_{x\in[0,1]} |\widehat{f}(x) - \widehat{f}(x)| \right\|_p &\leq \left\| \sup_{x\in[0,1]} |\widehat{f}(x) - \mathbb{E}\widehat{f}(x)| \right\|_p + \sup_{x\in[0,1]} |f(x) - \mathbb{E}\widehat{f}(x)| \\ &\leq \left\| \sup_{x\in I} |\widehat{f}(x) - \mathbb{E}\widehat{f}(x)| \right\|_p + Ch_n^\rho.\end{aligned}$$

The last bound follows from the choice $h_n = \left(n/\log n\right)^{-\frac{1}{1+2\rho}}$.

Example 3.3.1 Many other functions of interest may be fitted. We rapidly present some of them through kernel estimators.

1. **Non-parametric regression**
   The natural estimator of a mean is the empirical mean, but think now of a centred independent sequence

$$Y_k = r\left(\frac{k}{n}\right) + \xi_k, \qquad k = 1, \ldots, n \tag{3.4}$$

   for some independent identically distributed sequence (ξ_k) and a smooth regression function r.
   A natural estimator would be a local mean

$$\widehat{r}(x) = \frac{\sum_{k=1}^n \mathbb{I}_{\{|x-\frac{k}{n}|<h\}} Y_k}{\sum_{k=1}^n \mathbb{I}_{\{|x-\frac{k}{n}|<h\}}}.$$

   This idea is easily generalized as a kernel regression estimator in the previous fixed regression design:

$$\widehat{r}(x) = \frac{1}{nh} \sum_{k=1}^n Y_k K\left(\frac{\frac{k}{n} - x}{h}\right).$$

   Monographs Priestley and Chao (1972) and Rosenblatt (1991) exhibit the asymptotic properties of such estimators.

2. **Random regression designs**

$$Y_k = r(X_k) + \xi_k, \quad \text{for} \quad k = 1, \ldots, n,$$

where (X_k) is an independent identically distributed sequence.
Here the Nadaraya–Watson estimator gives:

$$\widehat{r}(x) = \begin{cases} \dfrac{\widehat{g}(x)}{\widehat{f}(x)}, & \text{if } \widehat{f}(x) \neq 0, \\ 0, & \text{if } \widehat{f}(x) = 0, \end{cases} \tag{3.5}$$

with

$$\widehat{g}(y) = \frac{1}{nh} \sum_{j=1}^{n} X_j K\left(\frac{X_j - y}{h}\right).$$

The functions f and g estimated are respectively the marginal density and $g = rf$.

3. **Differentiating a density**
For example to estimate f' one may just differentiate f's estimator if K is a smooth function. This makes a change in rates since e.g.

$$\widehat{f}'(y) = -\frac{1}{nh^2} \sum_{j=1}^{n} K'\left(\frac{X_j - y}{h}\right).$$

One may indeed check that each term in the sum above admits a variance equivalent to its second moment.
The usual change in variable $u = y + th$ yields with the dominated convergence theorem:

$$\begin{aligned} \mathbb{E}\left(K'\left(\frac{X_j - y}{h}\right)\right)^2 &= h \int K'^2(t) f(y + th)\, dt \\ &\sim_{h \to 0} hf(y) \int K'^2(t)\, dt. \end{aligned}$$

We differentiate analogously

$$\mathbb{E}K'\left(\frac{X_j - y}{h}\right) = \mathcal{O}(h),$$

and from independence, we obtain with $h \equiv h_n$:

$$\operatorname{Var} \widehat{f}'(y) \sim_{n \to \infty} \frac{1}{nh_n^3} f(y) \int K'^2(t)\, dt.$$

as soon as $\lim_{n\to\infty} h_n = 0$ and $\lim_{n\to\infty} nh_n^3 = \infty$.

The variance of this estimator, $\mathcal{O}\Big(\frac{1}{nh_n^3}\Big)$, admits a different decay rate than $\widehat{f}(y)$, which makes convergence rates pretty distinct.

4. **Differentiating regression functions**
   The same phenomenon occurs for $r' = \dfrac{g'f - gf'}{f^2}$.
   Higher order differentials of f or of r may even be considered.

Such estimators may be analogously controlled. However we shall not develop the theory here and we refer the reader to Rosenblatt (1991, 1985) for several elegant developments.

Exercise 16 (*Regression*) Let $[a, b]$ be a compact interval. Consider the previous fixed design regression setting in Example 3.3.1-1.

Provide bounds for

$$\sup_{x\in[a,b]} \mathbb{E}|\widehat{g}(x) - g(x)|^p.$$

Hint. Proceed as in Exercise 15.

Exercise 17 (*Nadaraya–Watson's estimator*) Let $[a, b]$ be a compact interval. Consider now the previous random regression setting in Example 3.3.1-2. Provide bounds for

$$\sup_{x\in[a,b]} \mathbb{E}|\widehat{g}(x) - g(x)|^p.$$

In order to avoid a division by 0, we now assume that for some $\epsilon > 0$,

$$\inf_{x\in[a-\epsilon,b+\epsilon]} f(x) > 0.$$

Deduce convergence results for the Nadaraya–Watson estimation of a regression function.

Hint. Proceed as in Exercise 15.

Exercise 18 (*Derivative*) Let $[a, b]$ be a compact interval. Consider the estimation of a derivative setting in Example 3.3.1-3.

Provide bounds for

$$\sup_{x\in[a,b]} \mathbb{E}|\widehat{f'}(x) - f'(x)|^p.$$

Hint. Proceed as in Exercise 15 by taking into account the different variance bounds h_n replaced by h_n^3.

3.4 Division Trick

In this section we revisit a result of interest for statistics, as stressed below. This *ratio-trick* was initiated in Colomb (1977) and it was improved in Doukhan and Lang (2009). Due to its importance, we decided to include it in this chapter. It is reformulated in a simplified version below.

Setting $D = \mathbb{E}D_n$, and $N = \mathbb{E}N_n$ where N_n, D_n are random quantities, it is an interesting problem to get evaluations for centred moments of ratios in some special cases, when it may be expected that

$$\left\| \frac{N_n}{D_n} - \frac{N}{D} \right\|_m = \mathcal{O}(\|N_n - N\|_p + \|D_n - D\|_q), \tag{3.6}$$

for convenient values of $p, q \geq m$.

Assume that this ratio appears as a weighted sum where $V_i \geq 0$

$$D_n = a_n \sum_{i=1}^n V_i, \qquad N_n = a_n \sum_{i=1}^n U_i V_i.$$

Maybe more simply, set

$$w_i = \frac{V_i}{\sum_{j=1}^n V_j},$$

then one may rewrite

$$R_n = \frac{N_n}{D_n} = \sum_{i=1}^n w_i U_i, \quad \text{with} \quad \sum_{i=1}^n w_i = 1, \; w_i \geq 0. \tag{3.7}$$

In the general case for the previous relation (3.6) to hold we prove:

Theorem 3.4.1 *Assume that the sequence* (v_n) *is such that* $v_n \downarrow 0$ *(as* $n \uparrow \infty$*) and* $v_n \leq 1$*. Assume also that* $q > m$*. We consider* $a \in]0, 1]$ *with* $q > m(1 + a)$*, and we set* $\frac{1}{p} + \frac{1}{q} = \frac{1}{m}$.

If there exists an absolute constant $M > 0$ *such that:*

$$t > \frac{mq}{q - m(1 + a)} \tag{3.8}$$

$$\max_{1 \leq i \leq n} \|U_i\|_t + n v_n^{at} \leq M \tag{3.9}$$

$$\|D_n - D\|_q + \|N_n - N\|_p \leq v_n \tag{3.10}$$

then the relation (3.6) holds.

Remark 3.4.1 The above result always implies $p, q > m$; possible exponents are $p = q = 2m$ which implies $t > 2m/(1-a)$ (if e.g. $v_n = 1/\sqrt{n}$ this implies $a > 1/(m+1)$).

Now if $a = 1$ the result needs $q > 2m$, and if $a > 0$ is very small this needs high order moments $\max_{1\le i\le n} \|U_i\|_t < \infty$.

If $q > 2m$ then $p < 2m$ (may be close to m), we choose $a = 1$ and $t > qm/(q-2m)$.

If $2m \ge q > m$ the above result needs $p \ge 2m$.

Theorem 3.4.2 *Assume that the sequence (v_n) is such that $v_n \downarrow 0$ as $n \uparrow \infty$, and $v_1 \le 1$. Set $p = m$ and $q > m(1+a)$ for some $0 < a \le 1$.*

Assume also that for some constant $M > 0$, $a_n = \frac{1}{n}$ and $\max_{1\le i\le n} \|V_i\|_\infty < M$; if $q > m(1+a)$, if (3.8), (3.9), and (3.10) hold, then the relation (3.6) holds.

Remark 3.4.2 Here $p = m$ (no further conditions on the convergence rate of N_n besides $\|N_n - N\|_m = \mathcal{O}(v_n)$) and $q > m$ and in this situation we may assume that $v_n = 1/\sqrt{n}$.

If $m < q \le 2m$ then $a < 1$; the result only needs high order moments $\max_{1\le i\le n} \|U_i\|_t < \infty$.

If $q > 2m$ and $a = 1$ then $at > 2$ and, if $\max_{1\le i\le n} \|U_i\|_t$ is bounded, it implies (3.8), since we derive $mq > 2(q-2m)$ from (3.8).

A useful main lemma follows:

Lemma 3.4.1 *For each $z \in \mathbb{R}$, and $0 \le a \le 1$ the following inequality holds:*

$$\left|\frac{1}{1-z} - 1\right| \le |z| + \frac{|z|^{1+a}}{|1-z|}.$$

Proof of Lemma 3.4.1. Begin with the relations

$$\frac{1}{1-z} = 1 + \frac{z}{1-z} = 1 + z + \frac{z^2}{1-z}.$$

Now since $0 \le a \le 1$:

$$\begin{aligned}\left|\frac{1}{1-z} - 1\right| &\le \max\left(|z| + \frac{|z|^2}{|1-z|}, |\frac{|z|}{|1-z|}\right)\\ &\le |z| + |z|\frac{\max(1, |z|)}{|1-z|}\\ &\le |z| + \frac{|z|^{1+a}}{|1-z|}.\end{aligned}$$

The last inequality follows from the elementary inequality (12.18).

Proof of Theorems 3.4.2 *and* 3.4.1. Set $z = (D - D_n)/D$ in the previous lemma.

Then notice with $R = N/D$ that:

$$
\begin{aligned}
|R_n - R| &\le \left| R_n - \frac{N_n}{D} \right| + \frac{1}{D}|N_n - N| \\
&= |N_n| \left| \frac{1}{D_n} - \frac{1}{D} \right| + \frac{1}{D}|N_n - N| \\
&= \frac{|N_n|}{D} \left| \frac{D}{D_n} - 1 \right| + \frac{1}{D}|N_n - N| \\
&= \frac{|N_n|}{D} \left| \frac{1}{1-z} - 1 \right| + \frac{1}{D}|N_n - N| \\
&= \frac{|zN_n|}{D} + |R_n||z|^{1+a} + \frac{1}{D}|N_n - N|.
\end{aligned}
$$

Hence

$$
\|R_n - R\|_m \le A + B + C
$$

with

$$
A = \frac{1}{D^2}\|(D_n - D)N_n\|_m \tag{3.11}
$$

$$
B = \frac{1}{D^{2+a}} \left\| R_n |D_n - D|^{1+a} \right\|_m \tag{3.12}
$$

$$
C = \frac{1}{D}\|N_n - N\|_m \tag{3.13}
$$

Denote generic constants by $c, c', c'', \ldots > 0$. First the bound $C \le cv_n$ follows from the relation $p \ge m$. Now remark that since (3.7) allows to write R_n as a convex combination, a classical idea of Gilles Pisier (see Marcus and Pisier 1981) entails for each $t > 0$,

$$
|R_n|^t \le \max_{1 \le i \le n} |U_i|^t \le \sum_{i=1}^{n} |U_i|^t \le nM^t.
$$

Thus:

$$
\mathbb{E}|R_n|^s \le (\mathbb{E}|R_n|^t)^{\frac{s}{t}} \le (nM^t)^{\frac{s}{t}}, \qquad \text{for } 1 \le s \le t.
$$

Now using Hölder's inequality (Proposition A.2.2) implies

$$
\|YZ\|_m \le \|Y\|_{um}\|Z\|_{vm}, \quad \text{if} \quad \frac{1}{u} + \frac{1}{v} = 1, \tag{3.14}
$$

and with $Y = R_n$, $Z = |D_n - D|^{1+a}$, $um = s$ and $vm = q/(1+a)$:

$$
B \le \frac{1}{D^{2+a}}\|R_n \cdot |D_n - D|^{1+a}\|_m \le c'\|R_n\|_s \|D_n - D\|_q^{1+a} \le c'' n^{\frac{1}{t}} v_n^{1+a}.
$$

Now

$$\frac{m}{s} + \frac{m(1+a)}{q} = 1 \Longrightarrow t > s = \frac{mq}{q - m(1+a)}$$

and $a = \dfrac{qs}{m} - (q+1)$. We need $v_n^a n^{\frac{1}{t}} = \mathcal{O}(1)$, in order that $B \le c' v_n$.

End of the proof of Theorem 3.4.1. Set $\dfrac{1}{p} + \dfrac{1}{q} = \dfrac{1}{m}$, then inequality (3.14) implies:

$$\begin{aligned} A &\le \frac{N}{D^2}\|D_n - D\|_m + \frac{1}{D^2}\|(D_n - D)(N_n - N)\|_m \\ &\le c'''(v_n + \|(D_n - D)(N_n - N)\|_m) \\ &\le c'''(v_n + \|N_n - N\|_p \|D_n - D\|_q) \\ &\le c''''(v_n + v_n^2) \\ &\le 2c'''' v_n, \end{aligned}$$

where the constants $c, c' \dots$ are suitably chosen.

The last inequality follows from $v_n \le 1$.

End of the proof of Theorem 3.4.2. Here the Minkowski triangular inequality implies

$$\|N_n\|_t \le \max_{1 \le i \le n} \|U_i\|_\infty \max_{1 \le i \le n} \|V_i\|_t$$

is bounded, and $A \le c''''' v_n$.

Example 3.4.1 Relations (3.6) are needed in many cases, examples are provided below:

1. **Empirical estimator for non-totally observed data.**
   Here one intends to fit the mean of the incompletely observed iid sequence $(U_t)_{t \ge 0}$. Namely we suppose that this is according to the fact that an independent Bernoulli-distributed sequence $V_t \sim b(p)$ take the value 1. The observed variables are $X_t = U_t V_t$, and their number is $D_n = \sum_{i=1}^{n} V_i$.
   Now with $a_n = n$, $v_n = 1/\sqrt{n}$ we calculate $D = p$ and $N = p \cdot \mathbb{E}V_0$ so that $R = \mathbb{E}V_0$.
2. **Regression with random design.**
   For the previous Nadaraya–Watson estimator (3.5) we may complete Exercise 17. This is indeed important to bound centred moments

$$\|\widehat{r}(x) - r(x)\|_m = \left(\mathbb{E}|\widehat{r}(x) - r(x|^m)\right)^{\frac{1}{m}},$$

as well as uniform moments

$$\left\| \sup_{x \in [a,b]} |\widehat{r}(x) - r(x)| \right\|_m = \left(\mathbb{E} \sup_{x \in [a,b]} |\widehat{r}(x) - r(x)|^m\right)^{\frac{1}{m}}.$$

For clarity we will only address the first question but the other one may be handled analogously to Exercise 17.
In order to use the above results the weights need to be non-negative, and as this was noted above, this condition implies that kernels considered should have regularity less than 2. The regularity 2 is obtained with symmetric kernels.
Assume that the functions admit the regularity $\rho = 2$. Then from the section above, the biases $\widehat{f}(x) - f(x)$ and $\widehat{g}(x) - g(x)$ admit order h^2.
The previous relation does the hard part of the job since with $N_n = \widehat{g}(x)$ and $D_n = \widehat{f}(x)$ and here $a_n = 1$[2] and $v_n = 1/\sqrt{nh}$.
It implies:

$$\begin{aligned}\|\widehat{r}(x) - r(x)\|_m &\le \left\|\widehat{r}(x) - \frac{\widehat{g}(x)}{\widehat{f}(x)}\right\|_m + \left\|\frac{\widehat{g}(x)}{\widehat{f}(x)} - \frac{g(x)}{f(x)}\right\|_m \\ &\le C\left(\frac{1}{\sqrt{nh}} + h^2\right).\end{aligned}$$

With the choice $h = n^{-\frac{1}{5}}$:

$$\|\widehat{r}(x) - r(x)\|_m \le 2Cn^{-\frac{2}{5}}.$$

Exercise 19 (Example 3.4.1-1, continued) Make precise the assumptions in Example 3.4.1-1.

Hint. First conditions (3.8) follow from independence, and $v_n^a n^{\frac{1}{t}} = n^{\frac{1}{t} - \frac{a}{2}}$ is a bounded sequence in case $t \ge \dfrac{2p(s-m)}{(m-p)s + pm}$. This ends the proof.

Exercise 20 (Example 3.4.1-2, continued) Make precise the assumptions in Example 3.4.1-2.

3.5 A Semi-parametric Test

In case the model is indexed by a class of functions but the only parameter of interest is a constant in $\mathbb{R}^d$, the framework is semi-parametric.

An example of such semi-parametric estimation is provided here. Let $w : \mathbb{R} \to \mathbb{R}$ be a weight function such that the following integral converges. We estimate the energy parameter:

$$\theta = \int f^2(x) w(x)\, dx.$$

[2] An alternative choice is $a_n = 1/nh$ and $N_n = nh\widehat{g}(x)$ and $D_n = nh\widehat{f}(x)$.

For this, use a plug-in estimator of the density f, a reasonable estimator is:

$$\widehat{\theta}_n = \int \widehat{f}^2(x)w(x)dx. \tag{3.15}$$

Here $h = h_n \downarrow 0$ will also satisfy additional conditions described later.

Set $\overline{\theta}_n = \int \left(\mathbb{E}\widehat{f}(x)\right)^2 w(x)dx$, then

$$\begin{aligned}
\widehat{\theta}_n - \theta &= (\widehat{\theta}_n - \overline{\theta}_n) + (\overline{\theta}_n - \theta) \\
&= \int \left(\widehat{f}(x) - \mathbb{E}\widehat{f}(x)\right)^2 w(x)\, dx \\
&+ \int \left(\widehat{f}(x) - \mathbb{E}\widehat{f}(x)\right)\left(2\mathbb{E}\widehat{f}(x) \cdot w(x)\right) dx \\
&+ \int \left(\mathbb{E}\widehat{f}^2(x) - f^2(x)\right) w(x)dx \\
&= \int \left(\widehat{f}(x) - \mathbb{E}\widehat{f}(x)\right)\left(2\mathbb{E}\widehat{f}(x) \cdot w(x)\right) dx \\
&+ \mathcal{O}\left(\frac{1}{nh} + h^2\right).
\end{aligned}$$

The Landau expressions $\mathcal{O}$ correspond to bounds obtained in $\mathbb{L}^1$, as well as in probability.

Using the previous bounds in Sect. 3.3, we are in position to derive the following theorem:

Theorem 3.5.1 *Besides the previous assumptions, assume that both the relations $nh_n^2 \to 0$, and $nh_n^4 \to \infty$ hold as $n \to \infty$.*

Then:

$$\sqrt{n}\left(\widehat{\theta}_n - \theta\right) \xrightarrow{\mathcal{L}}_{n\to\infty} \mathcal{N}(0, V), \quad V = 4Var\Big(f(X_1)w(X_1)\Big).$$

Proof Set $v(x) = 2f(x)w(x)$. The above remarks yield the study of the expressions

$$\int \left(\widehat{f}(x) - \mathbb{E}\widehat{f}(x)\right) v(x)\, dx = \frac{1}{n}\sum_{i=1}^{n} (v(X_i) - \mathbb{E}v(X_i) + \Delta_i - \mathbb{E}\Delta_i),$$

the above sums are decomposed as sums of independent random variables with

$$\Delta_i = \int K(s)(v(X_i + sh) - v(X_i))\, ds.$$

The Central Limit Theorem for the iid random variables $v(X_i)$ yields the Gaussian convergence of the expressions:

$$\frac{1}{\sqrt{n}}\sum_{i=1}^{n}(v(X_i)-\mathbb{E}v(X_i)).$$

Now conditions over $h=h_n$ entail that the remainder terms may be neglected.

Indeed, the Lebesgue dominated convergence theorem applies to the sequence $\delta_n=\Delta_1$ (where the dependence with respect to n refers to the decay of $h=h_n$) and the continuity of v implies $\lim_n \mathbb{E}|\delta_n|^2=0$.

Hence

$$\mathbb{E}\left(\frac{1}{\sqrt{n}}\sum_{i=1}^{n}(\Delta_i-\mathbb{E}\Delta_i)\right)^2=\mathbb{E}|\delta_n|^2\to_{n\to\infty}0.$$

This concludes the proof.

Example 3.5.1 (*Some more parameters of interest*) Rosenblatt (1991) suggests additional estimation problems.

Consider e.g.:

- **Fisher information**
  This is the expression
  $$I(f)=\int\frac{f'^2(x)}{f(x)}\,dx.$$
  It may also be estimated under comparable conditions. We leave as an exercise the proof that $f'_{n,h}$ is also a convergent estimator of f' and is asymptotically Gaussian. In this case, one may check that the normalization $\sqrt{nh^3}$ holds.
  The differentiability of the map $(u,v)\mapsto u^2/v$ yields an affine approximation of this non linear functional of the bivariate random process $F_n=\big(f_{n,h},f'_{n,h}\big)$.
- **Regression**
  Using bivariate iid samples (X_n,Y_n) yields estimation of the regression function:
  $$r(x)=\mathbb{E}(Y_0|X_0=x).$$
  We already mentioned that $\widehat{r}=\widehat{g}/\widehat{f}$ with
  $$\widehat{g}(x)=\frac{1}{nh}\sum_{i=1}^{n}Y_iK\left(\frac{X_i-x}{h}\right),$$
  accurately estimates r.
  This is the Nadaraya–Watson estimator, see Rosenblatt (1991).

- **Linearity of regression functions**
  If one wants to test the linearity of r,

$$r'' = \frac{D(f, g, f', g', f'', g'')}{f^3} = 0,$$

  or analogously $D(f, g, f', g', f'', g'') = 0$ where this expression is a polynomial wrt the derivatives of f and g. Since the function D is a polynomial, a Taylor expansion may be derived.
  It is possible to build tests of linearity for r by considering a CLT for the conveniently renormalised expressions:

$$\theta = \int D^2(f(x), g(x), f'(x), g'(x), f''(x), g''(x))w(x)\,dx.$$

- Directly involving the dependence structure, spectral estimation is considered in Sect. 4.4 and multispectral estimation in Rosenblatt (1985), see also Chap. 12.

Exercise 21 Extend ideas in the last item of Example 3.5.1 to propose a goodness-of-fit test for the linearity of a regression function.

Hint. Using notations of Example 3.5.1, a central limit theorem such as Theorem 3.3.1 gives

$$\sqrt{n}(\widehat{\theta} - \theta) \to^{\mathcal{L}}_{n\to\infty} N(0, \sigma^2).$$

Hence under the null hypothesis $\theta = 0$ the above result provides us with a level for the corresponding test of goodness of fit.

The study of the power of this test is a still unsolved and more difficult question.

Chapter 4
Stationarity

Some bases for the theory of time series are given below. The chapter deals with the widely used assumption of stationarity which yields a simpler theory for time series. This concept is widely considered in Rosenblatt (1985) and in Brockwell and Davis (1991). The latter reference is more involved with linear time series.

Time series are sequences $(X_n)_{n\in\mathbb{Z}}$ of random variables defined on a probability space (always denoted by $(\Omega, \mathcal{A}, \mathbb{P})$) and with values in a measurable space $(E, \mathcal{E})$. We assume that sequences of independent random variables can be defined on the same probability space. Another extension is the case of random fields $(X_n)_{n\in\mathbb{Z}^d}$; they are not in our scope.

Nile flooding data, see Fig. 4.1 (ordinates are measured in millions of m^3 of water per day), are classically used as an example of non-linear time series data, see Cobb (1978).[1]

4.1 Stationarity

Definition 4.1.1 (*Strict stationarity*) A random sequence $(X_n)_{n\in\mathbb{Z}}$ is said to be strictly stationary if, for each $k \geq 0$, the distribution of the vector $(X_l, \ldots, X_{l+k})$ does not depend on $l \in \mathbb{Z}$.

Definition 4.1.2 (*Weak stationarity*) A random sequence $(X_n)_{n\in\mathbb{Z}}$ is second order stationary if $\mathbb{E}X_l^2 < \infty$ and if only:

$$\mathbb{E}X_l = \mathbb{E}X_0, \quad \textit{and} \quad \mathrm{Cov}\,(X_l, X_{k+l}) = \mathrm{Cov}\,(X_0, X_k), \quad \text{for each } l, k \in \mathbb{Z}.$$

[1]Nile data may be found on: https:datamarket.com/dataset22w8mean-annual-nile-flow-1871-1970.

P. Doukhan, *Stochastic Models for Time Series*, Mathématiques et Applications 80,
https://doi.org/10.1007/978-3-319-76938-7_4

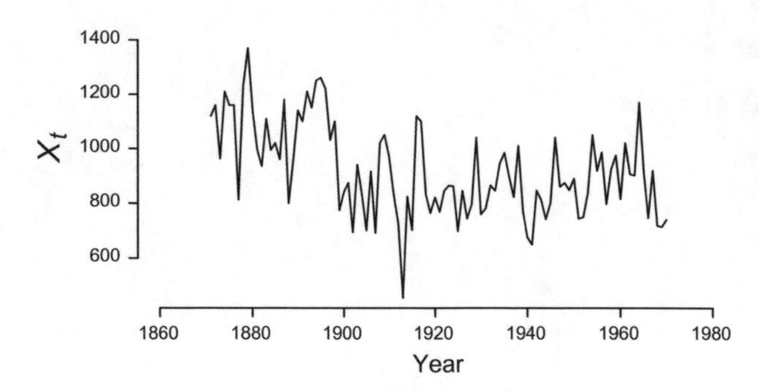

Fig. 4.1 Annual flow of Nile River at Aswan 1871–1970

We shall denote by m the common mean $\mathbb{E}X_0$ of X_n and by $r(k) = \text{Cov}\,(X_0, X_k)$ the covariance of such a process.

In other words $(X_n)_{n\in\mathbb{Z}}$ is strictly stationary if for each $k, l \in \mathbb{N}$ and each continuous and bounded function $h : \mathbb{R}^{k+1} \to \mathbb{R}$:

$$\mathbb{E}h(X_l, \ldots, X_{l+k}) = \mathbb{E}h(X_0, \ldots, X_k).$$

Under second moment assumptions strict stationarity implies second order stationarity (set $k = 1$ and h a second degree polynomial).

Under the Gaussian assumption we will see that both notions coincide. However this is not true in general.

Example 4.1.1 An independent identically distributed sequence is always strictly stationary, however if the variables do not admit finite second order moments, this is an example of a strictly stationary but not a second order stationary sequence.

Exercise 22 Consider a sequence $(\xi_n)_{n\in\mathbb{Z}}$, independent and identically distributed.

1. Assume now that $\mathbb{E}\xi_0 = 0$, then the sequence $X_n = \xi_n\xi_{n-1}$ is centred and orthogonal but not necessarily an independent sequence.
2. There exists a second order stationary sequence which is not strictly stationary.
3. Write now $X_n = \xi_n V_n$ with

$$V_n^2 = c_n\xi_{n-1}^2 + s_n\xi_{n-2}^2.$$

If $a = \mathbb{E}\xi_0^4 < \infty$ then $\mathbb{E}X_n^4$ is not a constant in general.
Again, we obtain a second order stationary sequence which is not strictly stationary.

Hints.

1. Indeed if those variables admit fourth order moments:

$$\begin{aligned}\mathrm{Cov}\,(X_n^2, X_{n-1}^2) &= \mathbb{E}\xi_n^2\xi_{n-1}^4\xi_{n-2}^2 - \mathbb{E}\xi_n^2\xi_{n-1}^2\mathbb{E}\xi_{n-1}^2\xi_{n-2}^2\\ &= \left(\mathbb{E}\xi_0^2\right)^2 \mathrm{Var}\,\xi_0^2.\end{aligned}$$

   does not vanish if ξ_0^2 is not a.s. constant.
2. A modification of the previous example is:

$$X_n = \xi_n\left(\sqrt{1-\frac{1}{n}}\cdot\xi_{n-1} + \frac{1}{\sqrt{n}}\cdot\xi_{n-2}\right).$$

   If $\mathbb{E}\xi_n^2 = 1$ then

$$\mathbb{E}X_nX_m = \begin{cases}0, & n \neq m,\\ 1, & n = m.\end{cases}$$

   Non-stationarity relies on the calculation of $\mathbb{E}(X_nX_{n-1}X_{n-2})$. This expression will be proved to depend on n.
   Write $X_n = \xi_n V_n$ for a sequence with V_n independent of ξ_n. Set similarly as above

$$V_n = c_n\xi_{n-1} + s_n\xi_{n-2}, \quad \text{for constants such that} \quad c_n^2 + s_n^2 = 1.$$

   This sequence is always centred and orthogonal if $\mathbb{E}V_n^2 < \infty$. Also using independence leads to

$$\mathbb{E}X_n^2X_{n-1} = \mathbb{E}V_nV_{n-1}^2\xi_{n-1}^2.$$

   If now the sequence V_n is independent of the sequence ξ_n we consider the similar example

$$V_n = c_n\zeta_{n-1} + s_n\zeta_{n-2},$$

   for a sequence ζ_n independent of ξ_n.
3. Simple calculations follow:

$$\begin{aligned}\mathbb{E}X_n^4 &= a\mathbb{E}V_n^4\\ &= a\mathbb{E}(c_n\xi_{n-1}^2 + s_n\xi_{n-2}^2)^2\\ &= a(a(c_n^2 + s_n^2) + 2s_nc_n)\\ &= a(a + 2s_nc_n).\end{aligned}$$

This exercise provides a family of useful counter-examples.

Remark 4.1.1 Stationarity effects are rather mathematical notions. For any type of financial, the stationarity seems quite problematic. E.g. large peaks appear on

September 11 and many seasonal effects may be seen, corresponding to opening of stock exchange.

The following remark includes the standard non-stationarity useful for statistical uses. Namely we consider "signal + noise" models.

Remark 4.1.2 (*Non-stationarity*) Assume that a process

$$X_t = h_t + Z_t$$

for a stationary and centred (or symmetric if $Z_t \notin \mathbb{L}^1$) sequence is observed at times $t = 1, \ldots, n$.

1. **Periodic signals**. A natural way to fit h_t in case it is T-periodic is to set

$$\widehat{h}_s = \frac{1}{N_s}\sum_{t\in E_{s,n}} X_t, \qquad E_{s,n} = s + (T\cdot\mathbb{Z})\cap[1,n]$$
$$N_s = \text{Card}\,(E_{s,n}), \qquad 1 \le s \le T.$$

2. **Local stationarity**. If $h_t = H(t/n)$ for a smooth function H, then for some bandwidth $b = b_n$ with $b = o(n)$, one sets

$$\widehat{H}(x) = \frac{1}{N_s}\sum_{|nx-k|\le b, t\in[1,n]} X_t \qquad N_s = \sum_{|nx-k|\le b, t\in[1,n]} 1.$$

   Clearly a kernel smoother may also be used here (see Example 3.3.1).
3. If $h_t = p_t + H(t/n)$ then we must normalize the representation with e.g. $p_0 = 0$ and

$$\widehat{p}_s = \frac{1}{N_s}\sum_{t-s\in T\cdot\mathbb{Z}\cap[1,n]} X_t - \widehat{m}, \tag{4.1}$$
$$N_s = \sum_{t-s\in T\cdot\mathbb{Z}\cap[1,n]} 1, \qquad 1 \le s \le T,$$
$$\widehat{m} = \frac{1}{n}\sum_{j=1}^{n} X_j. \tag{4.2}$$

   The above re-centring allows us to assume $p_0 = 0$. In this case we define $\widehat{H}(x)$ as above.
4. **Almost periodicity** (Besicovitch 1954). Let $t \mapsto h_t$ be a function defined on $\mathbb{R}^+ \to \mathbb{R}$, and $\epsilon > 0$, then T is an ϵ-period period of h if $\sup_t |h_{t+T} - h_t| \le \epsilon$; denote $E(h,\epsilon)$ the set of such ϵ-period periods. Then if for each $\epsilon > 0$ there exists some $\ell > 0$ such that $E(h,\epsilon)\cap[a, a+\ell] \ne \emptyset$, the function h is said almost-periodic in the sense of Bohr.

Those functions are the closure in the space of the vector space spanned by exponentials $t \mapsto e^{it\lambda}$ in the Banach space $\mathcal{C}(\mathbb{R}^+)$, $\|\cdot\|_\infty$.
The following mean exists $M(h) = \lim_T \int_t^{T+t} h_s\,ds$ and it does not depend on $t \geq 0$. If now h_t is almost-periodic then (Bohr 1947)'s representation tells us that there exists sequences of numbers a_k, T_k such that

$$h_t = m + \sum_{k=0}^{\infty} p_{t,k}, \qquad p_{t,k} = a_k e^{it/T_k}, \qquad \sum_{k=0}^{\infty} a_k^2 \leq M(h^2),$$

the sequence T_k is assumed to be non-decreasing.
When the frequencies $1/T_k$ are known the above study applies by setting $\widehat{p}_{s,k}$, $\widehat{m}$ (the empirical mean) respectively as in relations (4.1) and (4.2).
If the frequencies are unknown, as in Paraschakisa and Dahlhaus (2012), a minimum contrast estimator based on the periodogram needs to be developed.

Many other simple ways to build non-stationary time series can be considered. As an example for a deterministic sequence ℓ_t the model

$$X_t = \ell_t \cdot Z_t,$$

is simple and its logarithm may be considered as above. Unfortunately a combination of these two difficulties seems really tough, and for the model

$$X_t = h_t + \ell_t \cdot Z_t,$$

this seems reasonable to fit two different T-periodic functions h_t, ℓ_t or even two locally stationary functions (point 2).

The additive part can again be estimated as above, and the multiplicative signal is estimated (through log transforms) with replacing sums by products and $1/n$ by nth order roots.

Other non stationarity situations may be found in Bardet and Doukhan (2017).

4.2 Spectral Representation

It is easy and important to prove the following property of covariances.

Consider $n \in \mathbb{N}^*$. Let $c_l \in \mathbb{C}$ for all $|l| \leq n$, we sett $c = (c_l)_{|l| \leq n}$ and $\Sigma_n = (r_{|i-j|})_{|i|,|j| \leq n}$.

We obtain:

$$c^t \Sigma_n \overline{c} = \sum_{|i| \leq n} \sum_{|j| \leq n} c_i \overline{c_j} r_{|i-j|} = \mathbb{E} \Big| \sum_{|i| \leq n} c_i X_i \Big|^2 \geq 0. \tag{4.3}$$

Theorem 4.2.1 (Herglotz) *If a sequence $(r_n)_{n\in\mathbb{Z}}$ satisfies (4.3) then there exists a non-decreasing function G, unique up to Lebesgue-nullsets, with $G(-\pi) = 0$ and*

$$r_k = \int_{-\pi}^{\pi} e^{ik\lambda} dG(\lambda).$$

Notation. The symbol $dG(\lambda)$ is a Stieljes integral, it may be defined from the measure μ such that:

$$\mu([-\pi, \lambda]) = G(\lambda), \qquad \forall \lambda \in [-\pi, \pi].$$

If $h : [-\pi, \pi] \to \mathbb{R}$ is continuous:

$$\int_{-\pi}^{\pi} h(\lambda)dG(\lambda) = \int_{-\pi}^{\pi} h(\lambda)\mu(d\lambda).$$

Proof of Theorem 4.2.1. Set

$$g_n(\lambda) = \frac{1}{2\pi n} \sum_{s=0}^{n-1} \sum_{t=0}^{n-1} r_{t-s} e^{-i(t-s)\lambda} = \frac{1}{2\pi} \sum_{j=-(n-1)}^{n-1} \left(1 - \frac{|j|}{n}\right) r_j e^{-ij\lambda},$$

and $G_n(\lambda) = \int_{-\pi}^{\lambda} g_n(u)\, du$ then relation (4.3) implies $g_n(u) \geq 0$ hence G_n is continuous, non-decreasing and $G_n(\pi) = r_0$.

From a compactness argument, some subsequence $G_{n'}$ of G_n is convergent[2].

Note that $dG_n(\lambda) = g_n(\lambda)d\lambda$, and then

$$\left(1 - \frac{|k|}{n}\right) r_k = \int_{-\pi}^{\pi} e^{ik\lambda} dG_n(\lambda).$$

Integration by parts yields

$$r_k = (-1)^k r_0 - ik \int_{-\pi}^{\pi} e^{ik\lambda} dG_n(\lambda)\, d\lambda,$$

and implies the uniqueness of G. The existence of G follows from the fact that it is the only possible limit of such a convergent subsequence $G_{n'}$.

Definition 4.2.1 The spectral measure of the second order stationary process $(X_n)_{n\in\mathbb{Z}}$ (defined from G) is such that for each $\lambda \in [-\pi, \pi]$:

$$\mu_X([-\pi, \lambda]) = G(\lambda).$$

---

[2]Use a triangular scheme, by successive extraction of convergent subsequences. Choose a denumerable and dense sequence $(\lambda_k)_k$ in $[-\pi, \pi]$.

Here $\phi_{k+1}(n)$ will denote a subsequence of $\phi_k(n)$ such that $G_{\phi_{k+1}(n)}(\lambda_{k+1})$ converges as $n \to \infty$. Setting $G_{\phi(n)} = G_{\phi_n}(n)$ allows to end the proof.

If G is differentiable, the spectral density of the process $(X_n)_{n\in\mathbb{Z}}$ is the derivative $g = G'$.

Example 4.2.1 (*Spectral measures*)

- For an orthogonal sequence (i.e. $\mathbb{E}X_k X_l = 0$ for $k \neq l$ with $\mathbb{E}X_n = 0$, $\mathbb{E}X_n^2 = 1$), it is clear that $g(\lambda) = \lambda/2\pi$, from integration we derive: $G(\lambda) = 1/2 + \lambda/2\pi$, the measure associated is Lebesgue on $[-\pi, \pi]$.
  In Exercise 22-1 we recall that there exist such non-independent sequences.
- The random phase model admits complex values.
  Given constants $a_1, b_1, \ldots, a_k, b_k \in \mathbb{R}$ and independent uniform random variables $U_1, \ldots, U_k$ on $[-\pi, \pi]$ this model is defined through the relation

$$X_n = \sum_{j=1}^{k} a_j e^{i(nb_j + U_j)}.$$

  Then:

$$\mathrm{Cov}\,(X_s, X_t) = \mathbb{E}X_s\overline{X_t} = r_{s-t} = \sum_{j=1}^{k} |a_j|^2 e^{i(s-t)b_j}.$$

  This model is associated with a stepwise constant function G.
- Let $(\xi_n)_{n\in\mathbb{Z}}$ be a centred and independent identically distributed sequence such that $\mathbb{E}\xi_n^2 = 1$, let $a \in \mathbb{R}$, the moving average model MA(1) is defined as

$$X_n = \xi_n + a\xi_{n-1}.$$

  Here, $r_0 = 1 + a^2$, $r_1 = r_{-1} = a$, and $r_k = 0$ if $k \neq -1, 0, 1$.
  With the proof of the Herglotz theorem we derive

$$\begin{aligned} g(\lambda) &= \frac{1}{2\pi}(r_0 + 2r_1 \cos\lambda) \\ &= \frac{1}{2\pi}\left(1 + a^2 + 2a\cos\lambda\right) \\ &= \frac{1}{2\pi}\left((1 + a\cos\lambda)^2 + a^2\sin^2\lambda\right) \geq 0. \end{aligned}$$

Notation. For a function $g : [-\pi, \pi] \to \mathbb{C}$ denote $g(I) = g(v) - g(u)$ if $I = (u, v)$ is an interval.
If $g : [-\pi, \pi] \to \mathbb{R}$ is non-decreasing, we identify g with the associated non-negative measure.

Definition 4.2.2 (*Random measure*) A random measure is defined with a random function

$$Z : \Omega \times [-\pi, \pi] \to \mathbb{C}, \qquad (\omega, \lambda) \mapsto Z(\omega, \lambda),$$

non-decreasing for each $\omega \in \Omega$, with $\mathbb{E}|Z(\lambda)|^2 < \infty$ and such that there exists a non-decreasing function $H : [-\pi, \pi] \to \mathbb{R}^+$ with,

- $\mathbb{E}Z(\lambda) = 0$ for $\lambda \in [-\pi, \pi]$,
- $\mathbb{E}Z(I)\overline{Z(J)} = H(I \cap J)$ for all the intervals $I, J \subset [-\pi, \pi]$.

Let $g : [-\pi, \pi] \to \mathbb{C}$ be measurable and $\int_{-\pi}^{\pi} |g(\lambda)|^2 dH(\lambda) < \infty$.

Stochastic integrals

$$\int g(\lambda) dZ(\lambda),$$

with respect to a deterministic function may be defined in two steps:

- If g is a step function, $g(\lambda) = g_s$ for $\lambda_{s-1} < \lambda \leq \lambda_s$, $0 < s \leq S$ with $\lambda_0 = -\pi$, $\lambda_S = \pi$, set

$$I(g) = \int g(\lambda) dZ(\lambda) = \sum_{s=1}^{S} g_s Z([\lambda_{s-1}, \lambda_s]).$$

  Notice that

$$\begin{aligned}
\mathbb{E}\,|I(g)|^2 &= \sum_{s,t} g_s\, \overline{g_t}\, \mathbb{E}Z([\lambda_{s-1}, \lambda_s])\, \overline{Z([\lambda_{t-1}, \lambda_t])} \\
&= \sum_{s} |g_s|^2 \mathbb{E}|Z([\lambda_{s-1}, \lambda_s])|^2 \\
&= \sum_{s} |g_s|^2 H([\lambda_{s-1}, \lambda_s]) = \int_{-\pi}^{\pi} g^2(\lambda) dH(\lambda).
\end{aligned}$$

- Otherwise approximate g by a sequence of step functions g_n with

$$\int_{-\pi}^{\pi} |g(\lambda) - g_n(\lambda)|^2 dH(\lambda) \to_{n\to\infty} 0.$$

  The sequence $Y_n = \displaystyle\int g_n(\lambda) dZ(\lambda)$ is such that if $n > m$,

$$\mathbb{E}|Y_n - Y_m|^2 = \int_{-\pi}^{\pi} |g_n(\lambda) - g_m(\lambda)|^2 dH(\lambda) \to_{n\to\infty} 0.$$

  This sequence is proved to be a Cauchy sequence. It converges in $\mathbb{L}^2(\Omega, \mathcal{A}, \mathbb{P})$ and its limit defines the considered integral.

Example 4.2.2 Simple examples are provided from processes with independent increments.

- A natural example of such a random measure is the Brownian measure. Namely, denote $W([a, b]) = W(b) - W(a)$ then this random measure is defined with the Lebesgue measure as a control spectral measure λ.
- Another random measure of interest is the Poisson process on the real line, see Definition A.2.5.
- Compound Poisson processes write through a unit Poisson process P with $X_t = 0$ if $P_t = 0$ and

$$X_t = \sum_{i=1}^{P_t} V_i$$

for some iid sequence (V_i) independent of P_t.
The process satisfies $\mathbb{E}|X_t|^p < \infty$ in case $\mathbb{E}|V_0^p| < \infty$ and if $p = 1$ the process is centred if $\mathbb{E}V_0 = 0$.

Theorem 4.2.2 (Spectral representation of stationary sequences) *Let $(X_n)_{n\in\mathbb{Z}}$ be a centred second order stationary random process then there exists a random spectral measure Z such that*

$$X_n = \int e^{in\lambda} dZ(\lambda),$$

and this random measure is associated with the spectral measure of the process.

Relevant random spectral measures are reported in Example 4.3.1.

Proof The spectral function G of the process X_n is non-decreasing, hence its discontinuities are at most a denumerable set denoted by D_G.[3]

If $I = (a, b)$ is an interval with $a, b \notin D_G$, set

$$Z_n(I) = \frac{1}{2\pi} \sum_{|j|\le n} X_j \int_a^b e^{-iju} du,$$

then the sequence $(Z_n(I))_{n\ge1}$ is Cauchy in $\mathbb{L}^2(\Omega, \mathcal{A}, \mathbb{P})$.

Indeed for $n > m$,

$$\mathbb{E}|Z_n(I) - Z_m(I)|^2 = \frac{1}{4\pi^2} \mathbb{E}\left| \sum_{m<|j|\le n} X_j \int_a^b e^{-iju} du \right|^2 = \int_{-\pi}^{\pi} |h_n(\lambda) - h_m(\lambda)|^2 dG(\lambda).$$

[3]Recall that monotonic functions admit limits on the left and on the right at each point, the non-empty open intervals $(f(x-), f(x+))$ are disjoint in $\mathbb{R}$. Choose a rational number in each of them to conclude.

Denote now by h_n, the truncated Fourier series of the indicator function $\mathbb{I}_I$:

$$h_n(\lambda) = \frac{1}{2\pi} \sum_{|j| \le n} \int_a^b e^{-ij(u-\lambda)} du.$$

Write $Z(I)$ for the limit in $\mathbb{L}^2$ of $Z_n(I)$, then $\mathbb{E}Z(I) = 0$ because $\mathbb{E}X_n = 0$ and with immediate notations

$$\begin{aligned}
\mathbb{E}Z(I)\overline{Z(J)} &= \lim_n \mathbb{E}Z_n(I)\overline{Z_n(J)} \\
&= \lim_n \int_{-\pi}^{\pi} h_{I,n}(\lambda)\overline{h_{J,n}}(\lambda) dG(\lambda) \ = \ G(I \cap J),
\end{aligned}$$

in case the extremities of I, J are not in D_G. The set D_G of continuity points is dense. In case we only consider extremities of this interval in D_G, taking limits also allows us to conclude. In the general case, check that

$$\begin{aligned}
\mathbb{E}X_n\overline{Z_n(I)} &= \frac{1}{2\pi} \sum_{|j| \le n} r_{n-j} \int_a^b e^{iju} du \\
&= \int_{-\pi}^{\pi} \frac{dv}{2\pi} \int_a^b \sum_{|j| \le n} e^{ij(u-v)} dG(u) \\
&= \int_a^b e^{inv} dG(v).
\end{aligned}$$

Hence for step functions f:

$$\mathbb{E}X_n \overline{\int f(\lambda) dZ(\lambda)} = \int_{-\pi}^{\pi} e^{in\lambda} \overline{f(\lambda)} dG(\lambda).$$

This extends to continuous functions f by considering limits.

If now $f(\lambda) = e^{in\lambda}$ then

$$\mathbb{E}\left| X_n - \int e^{in\lambda} dZ(\lambda) \right|^2 = r_0 - 2r_0 + r_0 = 0.$$

Example 4.2.3 Examples of spectral densities may be found in Example 4.2.1. Besides measures with independent increment (Example 4.2.2), some relevant examples are reported in Examples 4.3.1. In Fig. 4.2 we plot empirical autocovariances.

This non-stationary setting is not much addressed but a local empirical covariance around the point $u \in (0, 1)$ may be considered by restricting summations over a set $[nu - m_n, nu + m_n]$ and with the normalization $2m_n$, where $m_n/N \to \infty$, see Dahlhaus (2012).

The following remark is a shy incursion into one of the most accurate proposals for non stationarity.

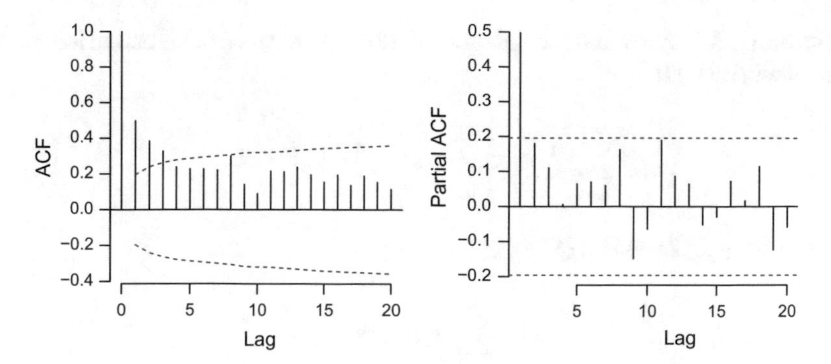

Fig. 4.2 Correlograms of the annual flow of the Nile River in Aswan 1871–1970. See Fig. 4.1

Remark 4.2.1 (*Local stationarity*) Dahlhaus (2012) defined it in the spectral way from a family of spectral densities $f(u,\lambda)$ such that around point $k \sim u\cdot n$ in the sample $\{1,\ldots,n\}$, the spectral density looks like $f(u,\lambda)$.
In other words

$$\mathrm{Cov}\,(X_k, X_{k+\ell}) \sim \int_0^{2\pi} e^{-2i\pi\ell\lambda} f(u,\lambda)\,d\lambda.$$

This is a very geometric idea telling us that some tangent stationary models locally fit such spectral behaviour data with index close to $[u\cdot n]$.

4.3 Range and Spectral Density

Here we denote $(X_n)_{n\in\mathbb{Z}}$ a centred second order stationary process.

Assume that

$$\sum_{k=0}^{\infty} r_k^2 < \infty,$$

then the spectral density

$$g(\lambda) = \frac{1}{2\pi}\sum_{k=-\infty}^{\infty} r_k e^{-ik\lambda},$$

is defined in $\mathbb{L}^2([-\pi,\pi])$. Moreover:

$$r_k = \int_{-\pi}^{\pi} e^{ik\lambda} g(\lambda) d\lambda.$$

Here the spectral measure G of the process is absolutely continuous with derivative $g\in\mathbb{L}^2$.

Definition 4.3.1 A centred second order stationary process (X_n) is called long-range dependent (LRD) if

$$\sum_{k=0}^{\infty} r_k^2 < \infty, \quad \text{and} \quad \sum_{k=0}^{\infty} |r_k| = \infty.$$

It is short-range dependent (SRD), if

$$\sum_{k=0}^{\infty} |r_k| < \infty.$$

In this case, the spectral density g is uniformly continuous and

$$\|g\|_\infty \le \frac{1}{2\pi} \sum_{k=0}^{\infty} |r_k|.$$

Example 4.3.1 Some examples follow:

- If $r_k \sim k^{-\alpha}$ for $\frac{1}{2} < \alpha < 1$ the sequence is LRD and there exists $\beta > 0$ with $g(\lambda) \sim c\lambda^{-\beta}$ as $\lambda \to 0$.
- If the spectral density

$$g(\lambda) = \frac{\sigma^2}{2\pi},$$

is a constant function, then the sequence

$$\xi_n = \int_{-\pi}^{\pi} e^{in\lambda} Z(d\lambda),$$

is a second order white noise with variance σ^2.
This means:

$$\mathbb{E}\xi_n \xi_m = \begin{cases} 0, & \text{if} \quad n \neq m, \\ \sigma^2, & \text{if} \quad n = m. \end{cases}$$

Let W be the Brownian motion (Bm),[4] this is the case if:

$$Z([0, \lambda]) = \frac{\sigma^2}{2\pi} W(\lambda),$$

Here Gaussianness of the white noise also implies its independence and it is an independent identically distributed sequence (a strict white noise).

[4]This process is the centred Gaussian process indexed on $\mathbb{R}^+$ with the covariance $\mathbb{E}W(s)W(t) = s \wedge t$.

If $\lambda \mapsto Z([0, \lambda])$ admits independent increments, the sequence ξ_n is again a strict white noise.

A weak white noise is associated with random spectral measures with orthogonal increments.

- If

$$X_n = \sum_{k=-\infty}^{\infty} c_k \xi_{n-k}, \qquad \sum_{k=-\infty}^{\infty} c_k^2 < \infty,$$

then the spectral density g_X of X gives

$$g_X(\lambda) = \left| \sum_{k=-\infty}^{\infty} c_k e^{-ik\lambda} \right|^2 g_\xi(\lambda).$$

To prove this compute X's covariance.
Moreover

$$Z_X(d\lambda) = \left(\sum_{k=-\infty}^{\infty} c_k e^{ik\lambda} \right) Z_\xi(d\lambda),$$

where Z_ξ denotes the random spectral measure associated with ξ.

E.g. autoregressive models, AR(p), may also be defined for dependent inputs,

$$X_n = \sum_{k=1}^{p} a_k X_{n-k} + \xi_n.$$

In case the sequence (ξ_n) is a white noise with variance 1, they are such that

$$g_X(\lambda) = \frac{1}{2\pi} \left| 1 - \sum_{k=1}^{p} a_k e^{-ik\lambda} \right|^{-2}.$$

Now, the spectral density g_X is continuous if the roots of the polynomial

$$P(z) = z^p - \sum_{k=1}^{p} a_k z^{p-k},$$

are inside the complex unit disk.

Exercise 23 The roots of the polynomial

$$P(z) = z^p - \sum_{k=1}^{p} a_k z^{p-k}$$

are inside the complex unit disk if

$$\sum_{k=1}^{p} |a_k| < 1.$$

Hint. If some z with $|z| \geq 1$ satisfies $P(z) = 0$, then $z^p = \sum_{k=1}^{p} a_k z^{p-k}$.

The triangular inequality implies

$$|z|^p \leq \sum_{k=1}^{p} |a_k||z|^{p-k} \leq |z|^{p-1} \sum_{k=1}^{p} |a_k|.$$

Thus $|z| \geq 1$ implies $\sum_{k=1}^{p} |a_k| \geq 1$.

The previous heredity formulae extend to $\mathbb{L}^2$-stationary sequences ξ_n:

Proposition 4.3.1 *Let (X_n) be a centred second order stationary sequence and c_n be a real sequence:*

$$Y_n = \sum_{k=-\infty}^{\infty} c_k X_{n-k}, \quad h(\lambda) = \sum_{k=-\infty}^{\infty} c_k e^{ik\lambda},$$

with $\sum_{k=-\infty}^{\infty} c_k^2 < \infty$. Then the sequence Y_n is also centred second order stationary sequence and

$$g_Y(\lambda) = |h(\lambda)|^2 g_X(\lambda), \quad Z_Y(d\lambda) = h(\lambda) Z_X(d\lambda).$$

Proof The first claim follows from the bilinearity properties of the covariance:

$$\mathrm{Cov}\,(Y_0, Y_k) = \sum_{m=-\infty}^{\infty} \left(\sum_{j=-\infty}^{\infty} c_j c_{j-m} \right) r_{k+m}.$$

The second claim is just algebra.

4.3.1 Limit Variance

A limit variance is the main difference between expressions the classical central limit theorem under independence and short range dependence. The above definition of the short range of a process is justified as follows in case X_n is centred. Indeed:

$$\mathbb{E}\,|X_1 + \cdots + X_n|^2 = \sum_{s=1}^{n}\sum_{t=1}^{n} \mathbb{E}X_s X_t = \sum_{s=1}^{n}\sum_{t=1}^{n} r_{t-s}$$

Thus:

$$\mathbb{E}\,|X_1 + \cdots + X_n|^2 = \sum_{|k|<n}^{n} (n - |k|) r_k. \tag{4.4}$$

According to the previous section one derives:

Proposition 4.3.2 *If X_n is SRD then*

$$\mathbb{E}\,|X_1 + \cdots + X_n|^2 \sim n g(0).$$

Proof This result is a variant of Cesaro's lemma. It will be enough to prove by using the standard Landau notation[5] that:

$$\sum_{|k|<n} |k| r_k = o(n).$$

For each $\epsilon > 0$ there exists K such that $|r_k| < \epsilon$ for $|k| > K$.

Split the expression

$$\sum_{|k|<n} |k||r_k| \le \sum_{|k|<K} |k||r_k| + \epsilon n.$$

Recall that in case $\mathbb{E}X_0 = 0$, then:

$$g(0) = \sum_{k=-\infty}^{\infty} \mathbb{E}X_0 X_k.$$

[5]Landau notations:
$v_n = o(u_n)$ if $\lim_n (v_n/u_n) = 0$ in case $u_n \neq 0$ for all n,
$v_n = \mathcal{O}(u_n)$ if there exists $C > 0$ such that $|v_n| \le C|u_n|$ for all n.

The previous quantity is of a specific importance. Indeed, according to the independent case a first possibility is to fit each term of the sum above approximated for a convenient sequence $m = m_n$ by

$$\sigma_m^2 = \sum_{k=-m}^{m} \mathbb{E}X_0 X_k.$$

An empirical estimator of this expression gives

$$\widehat{\sigma}_n^2 = \sum_{k=-m}^{m} \frac{1}{n} \sum_{i=1}^{n} \mathbb{E}X_i X_{i+k}. \tag{4.5}$$

or alternatively if one only has a sample $X_1, \ldots, X_n$

$$\widehat{\sigma}_n^2 = \sum_{k=-m_n}^{m_n} \frac{1}{n} \sum_{i=1\vee k}^{n\wedge(n+k)} \mathbb{E}X_i X_{i+k}, \tag{4.6}$$

All the terms in the previous sum do not have the same number of elements.

Namely the k-element of the sum is over $n - |k|$ terms which makes this estimator biased.

A variant of the previous estimator which is unbiased now gives

$$\widehat{\sigma}_n^2 = \sum_{k=-m_n}^{m_n} \frac{1}{n-|k|} \sum_{i=1\vee k}^{n\wedge(n+k)} \mathbb{E}X_i X_{i+k}. \tag{4.7}$$

The previous estimator may also be seen as a non-parametric estimator of the spectral density at the origin which also justifies the introduction of a smoothing parameter even though one only aims at estimating a real parameter.

4.3.2 Cramer–Wold Representation

The second-order stationary processes are represented as infinite order moving average of a weak white noise under a weak assumption. The proof of following results may be found in the volume (Azencott and Dacunha-Castelle 1986):

Theorem 4.3.1 (Cramer–Wold) *Let $(X_n)_{n\in\mathbb{Z}}$ be a second ordered stationary sequence with a differentiable spectral measure G such that $g = G'$ satisfies*

$$\int \log g(x)\, dx > -\infty.$$

Then there exists a unique orthogonal sequence ξ_n second order stationary (weak white noise) with $\mathbb{E}\xi_0^2 = 1$ and a sequence $(c_n)_{n\in\mathbb{N}}$ with

$$\sum_{n=0}^{\infty} c_n^2 < \infty, \qquad c_0 \geq 0$$

such that

$$X_n = \mathbb{E}X_0 + \sum_{k=0}^{\infty} c_k \xi_{n-k}. \tag{4.8}$$

Theorem 4.3.2 (Wold decomposition) *Let $(Z_n)_{n\in\mathbb{Z}}$ be a second order stationary sequence then there exists X_t, V_t with $Z_t = X_t + V_t$ such that (X_t) is as in (4.8) and V_t is measurable wrt to $\sigma(\xi_u / \, u \leq t)$.*

The first part of the representation of Z_t is as before while the second part V_t is something new. That part is called the deterministic part of Z_t because V_t is perfectly predictable based on past observations X_s for $s \leq t$.

A parameter of a main interest for stationary time series is the spectral density.

4.4 Spectral Estimation

This section is a very short survey of the question addressed in several nice volumes: see Azencott and Dacunha-Castelle (1986), and Giraitis et al. (2012) for a complete study of the LRD case (see Sect. 4.3), for parametric setting, see also Brockwell and Davis (1991), and for non-parametric setting see in Rosenblatt (1991).

Our aim is to make explicit how probabilistic limit theory can be used for the development of statistical methods for time series analysis rather than to provide a course of time series analysis since some really good textbooks are already available. The present viewpoint allows us to present many tools usually not considered directly by statisticians.

Definition 4.4.1 For a centred and second order stationary $(X_t)_{t\in\mathbb{Z}}$ define the periodogram:

$$I_n(\lambda) = \frac{1}{2\pi n}\left|\sum_{k=1}^{n} X_k e^{-ik\lambda}\right|^2 = \frac{1}{2\pi}\sum_{|\ell|<n} \widehat{r}_n(\ell) e^{-ik\lambda}.$$

for each $n \geq 1$ and $\lambda \in \mathbb{R}$, where

$$\widehat{r}_n(\ell) = \frac{1}{n}\sum_{k=1\vee(1+\ell)}^{n\wedge(n+\ell)} X_k X_{k+\ell}.$$

Example 4.4.1 An example of classical real data is the annual flow of the river Nile at Aswan 1871–1970 in Figs. 4.1 and 4.2 which show the fitted covariances. A rapid decay of covariances is observed from the covariogram.

Remark 4.4.1 The last sum extends over $(n - |\ell|)$-terms, hence the estimator $\widehat{r}_n(\ell)$ of the covariance $r(\ell) = \mathbb{E}X_0X_\ell$ is biased for $\ell \neq 0$, which means that we do not necessarily have $\mathbb{E}\widehat{r}_n(\ell) = r(\ell)$. Remark that in case $\sum_\ell |r(\ell)| < \infty$ the spectral density of the process f is continuous and that $\mathbb{E}I_n(\lambda) = f(\lambda)$. Unfortunately the variance of this estimator of f does not converge to 0: $I_n(\lambda)$ is not a reasonable estimator of $f(\lambda)$.
The integrated statistics

$$J_n(g) = \int_0^{2\pi} g(\lambda) I_n(\lambda)\, d\lambda,$$

admit smoother behaviours and usually converge to

$$J(g) = \int_0^{2\pi} g(\lambda) f(\lambda)\, d\lambda.$$

They even may be proved to satisfy a central limit theorem.

The previous feature may be used in directions as briefly discussed in the following two subsections.

4.4.1 Functional Spectral Estimation

First, we use a kernel method to consider $g \sim \delta_u$ and for a convenient window width $h = h_n$ and a kernel K we consider the estimator

$$\widehat{f}(\lambda) = I_n \star K_h(\lambda) = \frac{1}{h} \int_0^{2\pi} I_n(\mu) K\left(\frac{\lambda - \mu}{h}\right) d\mu.$$

This allows us to consider reasonable spectral density estimators. If now one replaces the smoothing function $\frac{1}{h}K(\frac{\cdot}{h})$ by the Dirichlet kernel

$$D_m(u) = \sum_{k=-m}^{m} e^{iku} = \frac{\sin\left((2m+1)\frac{u}{2}\right)}{\sin\left(\frac{u}{2}\right)},$$

with order $m = m_n = 1/h_n$ the previous estimators give

$$\widetilde{f}(\lambda) = I_n \star D_{m_n}(\lambda) = \int_0^{2\pi} I_n(\mu) D_{m_n}(\lambda - \mu)\, d\mu.$$

which almost fits the above-mentioned estimator (4.5) of $f(0)$. In fact it can be written in such a way that $\widetilde{f}_n(0)$ is as in (4.6), so contrary to (4.7) this gives a biased estimator:

$$\widetilde{f}(0) = \sum_{k=-m_n}^{m_n} \frac{1}{n} \sum_{i=1\vee k}^{n\wedge(n+k)} \mathbb{E}X_i X_{i+k}.$$

Remark 4.4.2 Asymptotic properties of such estimators may be derived under specific assumptions on the time series. One may prove them by approximating the spectral density by its Fourier expansion. Then standard empirical arguments allow us to derive asymptotic properties of such estimators as for the simple empirical means considered in Sect. 3.1 for independent sequences. Further improvements of inequalities for dependent samples are needed to complete the program. The case of the kernel estimator is in fact analogous since regularity conditions of a spectral density are tightly related to the quality of their approximation by trigonometric polynomials. This point may be proved by using the Jackson polynomials approach, see Lorentz (1966) or Doukhan and Sifre (2001).

4.4.2 *Whittle Estimation*

Assume that the time series is in a parametric set of models; maybe ARMA or others, see hereafter. Then the distribution of the whole process $X = (X_t)_{t\in\mathbb{Z}}$ may depend on a parameter θ, the spectral density which is defined in a family $(f_\theta)_{\theta\in\Theta}$ (for some $\Theta \subset \mathbb{R}^d$) and a suitable estimator, named the Whittle estimator, is the value $\widehat{\theta}$ minimizing the contrast, as defined in Sect. 3.2:

$$\begin{aligned} U_n(\theta) &= \int_0^{2\pi} \left(\log f_\theta(\lambda) + \frac{I_n(\lambda)}{f_\theta(\lambda)} \right) d\lambda, \\ &= \int_0^{2\pi} \log f_\theta(\lambda)\, d\lambda + J_n\left(\frac{1}{f_\theta}\right). \end{aligned}$$

Here again central limit theorems extending these for independent sequences allow us to expand pointwise the previous expression. An additional argument such as for example a uniform result (see e.g. Sect. 3.1) is then necessary so that the Taylor expansion still holds after integration.

4.5 Parametric Estimation

Remark also that parameters based on the spectral density may be estimated from other contrast estimators. Usually there is no close expression for the density $p_\theta(x_1, \ldots, x_n)$ of a sample $(X_1, \ldots, X_n)$ but MLE $\widehat{\theta}$ estimators are defined through the relation:

$$\widehat{\theta} \in \operatorname*{argmax}_{\theta\in\Theta}\; p_\theta(X_1, \ldots, X_n).$$

An interesting special case is that of a homogeneous Markov chain with transitions $P_\theta(x, A) = \mathbb{P}_\theta(X_1 \in A | X_0 = x)$.

If this Markov chain admits a density $\pi_\theta(x, y)$ and an invariant measure with density $\nu_\theta(x)$, then:

$$p_\theta(x_1, \ldots, x_n) = \nu_\theta(x_1)\pi_\theta(x_1, x_2) \cdots \pi_\theta(x_{n-1}, x_n).$$

For instance this applies to the non-linear auto-regressive processes

$$X_t = r_\theta(X_{t-1}) + \xi_t,$$

in case ξ_0 admits a density g_θ, and then

$$\pi_\theta(x, y) = g_\theta(y - r_\theta(x)).$$

Consider now an homogeneous Markov chain, solution of a recursive equation,

$$X_t = \xi_t \sigma_\theta(X_{t-1})$$

with iid centred innovations (ξ_t).

The MLE can be written

$$\pi_\theta(x, y) = \frac{1}{\sigma_\theta(x)} \cdot g\left(\frac{x}{\sigma_\theta(x)}\right),$$

in case ξ_t admits a density g. Instead of considering p_θ it is better to consider the minimization of

$$q_\theta(x_1, \ldots, x_n) = \pi_\theta(x_1, x_2) \cdots \pi_\theta(x_{n-1}, x_n).$$

Usually such maximization problems are numerically unstable; the QMLE is the minimization of the previous expression but with simply $\xi_0 \sim \mathcal{N}(0, 1)$ a Normal distribution. Now the MLE maximizes $\theta \mapsto L_\theta(X_1, \ldots, X_n)$. Even in this simplest case of Gaussian inputs f_θ does not usually admit a closed form. The following expression is simpler to minimize:

$$L_\theta(X_1, \ldots, X_n) = \sum_{t=2}^{n} \frac{X_t^2}{\sigma_\theta^2(X_{t-1})} + \log \sigma_\theta^2(X_{t-1}^2).$$

This estimator is considered in the most general situations in the monograph (Straumann 2005).

Remark 4.5.1 A last related remark is that for Gaussian processes with a fixed variance $\operatorname{Var} X_t = \sigma^2$ the least squares coincide with the MLE because of the quadratic expression of a Gaussian density.

4.6 Subsampling

Besides model-based bootstrap techniques in Sect. 11.3 this section is aimed at explicating the specific features of resampling under dependence.

Namely assume that a limit theorem holds for a sequence

$$t_m(X_1, \ldots, X_m) \xrightarrow{\mathcal{L}}_{m\to\infty} T.$$

It is not unusual that the distribution of T is not accessible. As before a test of goodness-of-fit is based on quantiles of the limiting distribution T. In case one wants more generally to fit the limit distribution of the convergent series of statistics

$$T_m = t_m(X_1, \ldots, X_m), \qquad \text{for some} \qquad m = m_n \ll n.$$

A way to proceed is to consider families of m-samples $(X_{i_1}, \ldots, X_{i_m})$ with $(i_1, \ldots, i_m) \in E_{m,n}$ and $i_1 \le \cdots \le i_m$, then the expression for T_m's distribution is provided from the value of $K(g) = \mathbb{E}g(T_m)$ which is obtained from the empirical method as

$$\widehat{K}_n(g) = \frac{1}{\text{Card } E_{m,n}} \sum_{(i_1,\ldots,i_m)\in E_{m,n}} g\left(t_m(X_{i_1}, \ldots, X_{i_m})\right). \tag{4.9}$$

In order that the distribution of $t_m(X_{i_1}, \ldots, X_{i_m})$ is the same as for T_m it is natural to assume that the distribution of $(X_{i_1}, \ldots, X_{i_m})$, is the same as for $(X_1, \ldots, X_m)$.

For iid samples the set $E_{m,n}$ may admits the huge cardinality

$$\frac{n!}{(n-m)!} \sim n^m.$$

One may select $E_{m,n}$ as the set of all the ordered m-tuples among $\{1, \ldots, n\}$. This is unfortunately a huge sum and it is better to choose randomly among those sets and use the law of large numbers to exhibit a consistent procedure.

Unfortunately not all m-tuples admit the same distribution when independence is omitted. Two choices of sets are considered to support this distributional equality:

$$E_{m,n} = \left\{(i+1, \ldots, i+m)/\ 0 \le i \le n-m\right\},$$

satisfies Card $E_{m,n} = n - m + 1$ and gives overlapping samples,

$$E_{m,n} = \left\{ \left((i-1)m+1, \ldots, im\right) \Big/ \ 1 \le i \le \frac{n}{m} \right\},$$

satisfies Card $E_{m,n} = n/m + 1$ for n a multiple of m and gives non-overlapping samples.

Again asymptotic consistency of such expressions still relies either on moment inequalities, or on exponential inequalities.

Suppose that we have the following schemes:

$$g_{i,m} = g(t_m(X_{i+1}, \ldots, X_{i+m})) \tag{4.10}$$

$$g_{i,m} = g(t_m(X_{(m(i-1)+1}, \ldots, X_{(i+1)m})) \tag{4.11}$$

and the set $E_{m,n}$ is indexed by an integer $i = 1, \ldots, N$ with either $N \sim n - m$ or $N \sim n/m$.

In order to prove the convergence of such expressions, a simple way is to calculate the variance of the expressions and from Cesaro's lemma to derive that

$$\widehat{K}_n(g) \to_{n\to\infty} \mathbb{E}g(T), \qquad \text{in probability.}$$

Equation (4.4) entails

$$\text{Var}\,\widehat{K}_n(g) \le \frac{1}{\text{Card}\,E_{m,n}} \sum_{i \in E_{m,n}} |\text{Cov}\,(g_{0,m}, g_{i,m})|.$$

Usually $g(x) = \mathbb{I}_{(x\le u)}$ so that using Exercise 69 the limits in probability

$$\sup_u |K_{n,m}(u) - \mathbb{P}(T \le u)| \to_{n\to\infty} 0, \quad K_{n,m}(u) = \widehat{K}_n(g), \tag{4.12}$$

holds uniformly with respect to u by using Exercise 13 as in the proof of the Glivenko–Cantelli theorem 3.1.1.

Remark 4.6.1 Such uniform convergences are taken into account to consider non-convergent cases, in Doukhan et al. (2011); we consider extreme value theory.

The divergent statistic sequence is then

$$t_n(x_1, \ldots, x_n) = \max_{1\le i\le n} x_i.$$

Self-normalization of those series then relies on the uniform convergence properties of the sequence $(K_{n,m_n})_{n\ge1}$.

Remark 4.6.2 In order to prove almost-sure convergence of such expressions, higher order moments need to be accurately bounded, as done in Doukhan et al. (2011). Refer to Chap. 12.

Part II
Models of Time Series

This part is of main importance in this volume. The idea is to recall standard techniques and also to introduce new concepts adapted to model time series. In a natural way, the first chapter is restricted to the Gaussian world; Gaussians indeed admit the exceptional feature that all the moments may be explicitly computed.

After this moving averages are the simplest non-Gaussian random processes. Then extensions to nonlinear processes are similar to the Gaussian chaos in the previous moving average setting. Most of the time series models can be written as Bernoulli shifts, and adapted techniques are developed here. In particular, we consider wide classes of memory models, extending on Markov cases.

The final brief chapter is dedicated to association which defines, as Gaussians, a very tiny conic class of time series. It shares the same specific feature of Gaussian processes: independence and orthogonality coincide here too.

Chapter 5
Gaussian Chaos

Gaussian distributions (Appendix A) are natural and play a special role in the field of probability theory since they appear as limit distributions from the CLT (Theorem 2.1.1, Lemma 11.5.1). Gaussian linear spaces admit a simple geometric property:

$\mathbb{L}^2$ *and distributional properties of Gaussian processes are equivalent.*

The Gaussian chaos is the $\mathbb{L}^2$ closure of polynomials of a Gaussian family. For one Normal random variable this chaos admits the Hermite polynomials as a basis.

The organization of the chapter follows. Discretely indexed Gaussian random processes (time series), and Brownian motion, as well as fractional Brownian motion (important for long-range dependence), are first considered. This provides enough tools to study the convergence of functionals of Gaussian processes. The method of moments is briefly reviewed, including the Mehler and the diagram formulae. The final sections introduce the so called fourth-order moment method which proves that in order that a sequence Z_n of random variables belonging to some chaos to converge to the Normal standard distribution, it is enough to prove that only $\lim_n \mathbb{E}Z_n = 0$, $\lim_n \mathbb{E}Z_n^2 = 1$ and $\lim_n \mathbb{E}Z_n^4 = 3$ as proved in Nourdin et al. (2011).

5.1 Gaussian Processes

Definition 5.1.1 A Gaussian process (or a Gaussian family) $Y = (Y_t)_{t\in\mathbb{T}}$ is a collection of random variables defined on the same probability space such that each finite subset defines a Gaussian random vector.

Remark 5.1.1 Alternatively, $Y = (Y_t)_{t\in\mathbb{T}}$ is Gaussian if for $(u_t)_{t\in\mathbb{T}}$, a family of real numbers such that $u_t = 0$ except for finitely many t,

$$\sum_{t\in\mathbb{T}} u_t Y_t$$

is a Gaussian random variable.

P. Doukhan, *Stochastic Models for Time Series*, Mathématiques et Applications 80,
https://doi.org/10.1007/978-3-319-76938-7_5

As an application of Lemma A.4.1:

Proposition 5.1.1 *If a sequence of real numbers* $(r_k)_k$ *satisfies* $r_{-n} = r_n$ *for all* $n \geq 0$ *and*

$$\sum_{i,j=1}^{n} u_i u_j r_{i-j} \geq 0,$$

for all $u_1, \ldots, u_n \in \mathbb{R}$, *then there exists a stationary Gaussian process with covariance* $r_k = \mathbb{E}X_0 X_k$.

Proof From the Lemma A.4.1, for each $d \in \mathbb{N}^*$, the law $\mathcal{N}_d(0, \Sigma_d)$ is well defined with $\Sigma_d = (r_{i-j})_{1\leq i,j\leq d}$.

The Kolmogorov consistency Theorem B.1.1 entails the existence of such a process.

More generally:

Theorem 5.1.1 *Let* $G : \mathbb{T}^2 \to \mathbb{R}$ *be such that the matrix*

$$\big(G(t_i, t_j)\big)_{1\leq i,j\leq n}$$

satisfies (4.3) for all possible choices $t_i \in \mathbb{T}$, *then there exists a Gaussian process with covariance* G.

An example, central for the study of dependence, is described below.

5.1.1 Fractional Brownian Motion

Definition 5.1.2 (Hurst 1951; Dobrushin and Major 1979) The fractional Brownian motion (fBm) with Hurst exponent $H \in (0, 1]$ is a centred Gaussian process $(Z_t)_{t\in\mathbb{R}}$ with covariance $\Gamma_H(s, t) = \mathrm{Cov}\,(Z_s, Z_t)$ defined as

$$\Gamma_H(s, t) = \frac{1}{2}\left(|s|^{2H} + |t|^{2H} - |s - t|^{2H}\right), \quad \forall s, t \in \mathbb{R}. \tag{5.1}$$

Proposition 5.1.2 *The function* Γ_H *in (5.1) for* $s, t \in \mathbb{R}$ *is indeed the covariance of a centred Gaussian process* $(B_H(t))_{t\in[0,1]}$.

Proof See Taqqu in Doukhan et al. (2002b). From Theorem 5.1.1 we need to prove that for all $0 \leq t_1 < \cdots < t_n \leq 1$, and $u_1, \ldots, u_n \in \mathbb{C}$

$$A = \sum_{i,j=1}^{n} \Gamma_H(t_i, t_j) u_i \overline{u_j} \geq 0.$$

- *Step 1*. Set $t_0 = 0$, $u_0 = -\sum_{i=1}^n u_i$ then

$$\sum_{i=1}^n \sum_{j=1}^n |t_i|^{2H} u_i \overline{u_j} = -\sum_{i=0}^n |t_i|^{2H} u_i \overline{u_0} = -\sum_{i=0}^n |t_i - t_0|^{2H} u_i \overline{u_0}.$$

Analogously

$$\sum_{i=1}^n \sum_{j=1}^n |t_j|^{2H} u_i \overline{u_j} = -\sum_{j=0}^n |t_j - t_0|^{2H} u_0 \overline{u_j}$$

hence

$$A = -\sum_{i=0}^n \sum_{j=0}^n |t_i - t_j|^{2H} u_i \overline{u_j}.$$

- *Step 2*. For $\epsilon > 0$ set

$$B_\epsilon = \sum_{i,j=0}^n e^{-\epsilon |t_i - t_j|^{2H}} u_i \overline{u_j}.$$

Then the Taylor formula simply implies

$$B_\epsilon \sim \epsilon A, \qquad \epsilon \downarrow 0.$$

- *Step 3*. For each $\epsilon > 0$ and $H \in (0, 1]$, there exists a real random variable ξ with

$$\phi_\xi(t) = \mathbb{E} e^{it\xi} = e^{-\epsilon |t|^{2H}}$$

(the law is $2H$-stable); this non-trivial point may be derived from Fourier inversion, as in Taqqu (Doukhan et al. 2002b).
Then

$$B_\epsilon = \mathbb{E} \left| \sum_{j=0}^n u_j e^{it_j \xi} \right|^2 \geq 0.$$

This ends the proof.

Remark 5.1.2 The case $H = \frac{1}{2}$ yields the Brownian motion $W = B_{\frac{1}{2}}$ defined on $\mathbb{R}^+$. In this case:

$$\Gamma_{\frac{1}{2}}(s, t) = s \wedge t.$$

Lemma 5.1.1 *Let* $0 \leq h < H$ *then, for almost all* $\omega \in \Omega$, *there exist constants* $c, C > 0$ *with*

$$|B_H(s) - B_H(t)| \leq C |t - s|^h, \qquad \text{if} \quad 0 \leq s, t \leq 1, \ |s - t| < c.$$

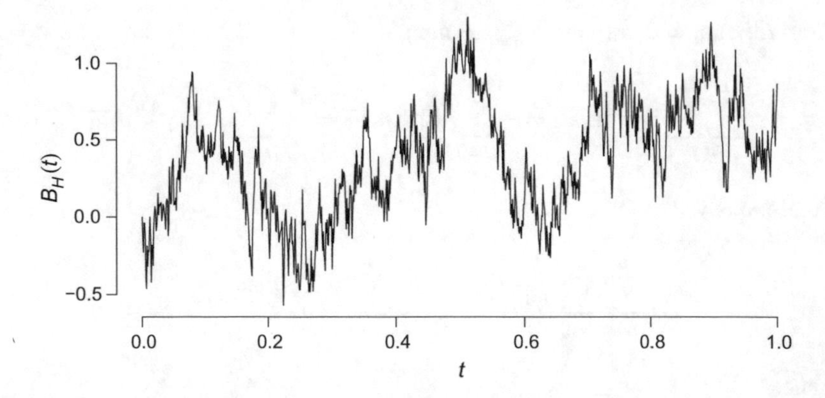

Fig. 5.1 Fractional Brownian motion simulated with $H = 0.30$ and evaluated in 1024 points

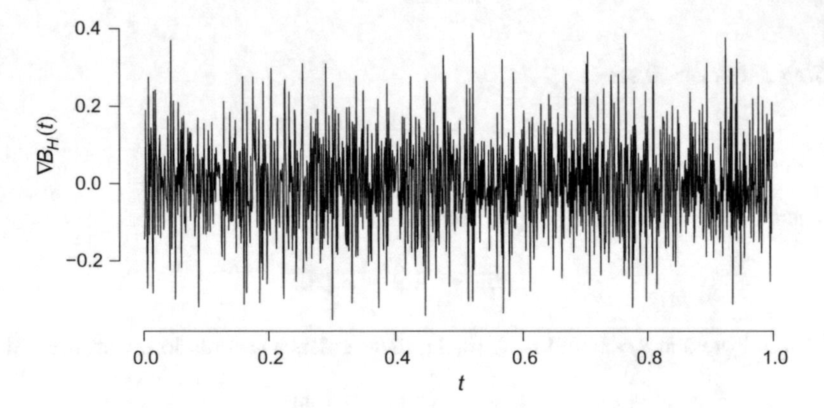

Fig. 5.2 Differenced time series of Fig. 5.1. This process is a fractional noise

Proof Note that:

$$2\mathbb{E}(B_H(s) - B_H(t))^2 = |s|^{2H} + |t|^{2H} - \left(|s|^{2H} + |t|^{2H} - |s-t|^{2H}\right) = |s-t|^{2H}.$$

The result is a consequence of both the first point in the Chentsov Lemma B.2.1, and of the above calculation.

Remark 5.1.3 The regularity properties of the fBm are clear from Figs. 5.1 and 5.3 representing its trajectories respectively for $H = 0.3$ and 0.9. while their differentiates are provided in Figs. 5.2 and 5.4. The larger H is, and the more regular are the trajectories. We use the R package dvfBM, see Coeurjolly (2009).

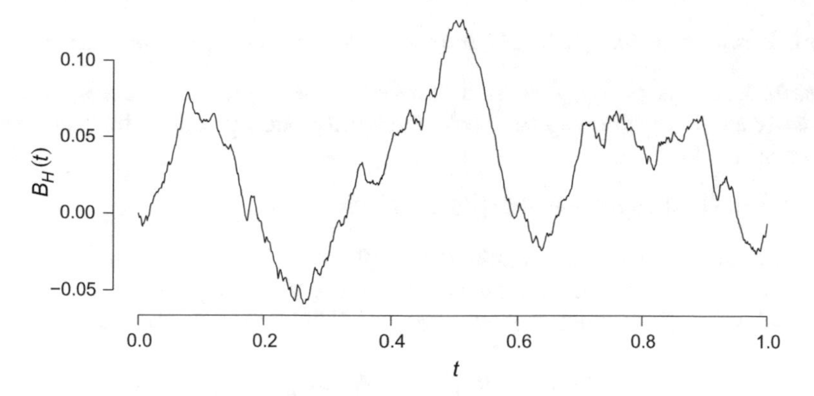

Fig. 5.3 Fractional Brownian motion simulated with $H = 0.90$ and evaluated in 1024 points

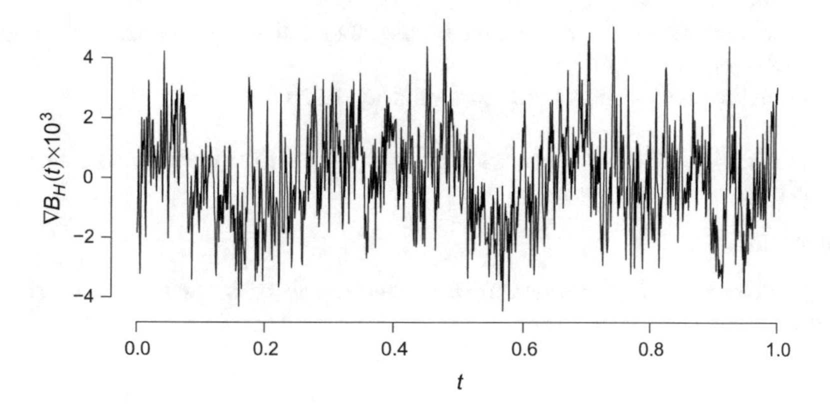

Fig. 5.4 Differenced time series of Fig. 5.3

- Hence for $H = 0.9$, close to 1, the trajectories are rather regular.
- For $H = 0.3$ the trajectories of the fBm are similar to those of a Gaussian noise, in Fig. A.2.

Definition 5.1.3 The process $(Z(t))_{t\in\mathbb{R}^+}$ is H-self-similar if for all $a > 0$

$$(Z(at))_{t\in\mathbb{R}^+} = (a^H Z(t))_{t\in\mathbb{R}^+}, \qquad \text{in distribution.}$$

Figure 5.4 is dedicated to represent the fractional noise.

We leave the following point as exercises for the reader:

Proposition 5.1.3 *Let Z be a random process on $\mathbb{R}^+$. The previous condition of H-self-similarity (Definition 5.1.3) is equivalent to the stationarity of the process*

$$(Y(t))_{t\in\mathbb{R}}, \qquad Y(t) = e^{-tH} Z(e^t), \qquad t \in \mathbb{R}.$$

For this, only check that finite-dimensional repartitions of both processes coincide.

Remark 5.1.4 As $(Y(t))_{t\in\mathbb{R}}$ is easily proved to be a Gaussian process, strict stationarity and $\mathbb{L}^2$-stationarity (or weak stationarity) are equivalent; this clarifies the previous result.

Also it is quite straightforward to prove that:

Exercise 24 1. If Z is self-similar then $Z(0) = 0$.
2. If Z is self-similar and its increments $(Z(t+s) - Z(t))_{t\in\mathbb{R}}$ are stationary for each s then: $\mathbb{E}Z(t) = 0$ if $H \neq 1$ because $\mathbb{E}Z(2t) = 2^H\mathbb{E}Z(t)$ and

$$\mathbb{E}(Z(2t) - Z(t)) = \mathbb{E}(Z(t) - Z(0)) = \mathbb{E}Z(t)$$

implies $(2^H - 2)\mathbb{E}Z(t) = 0$.
3. If increments of Z are stationary we obtain the equality in distribution $\mathcal{L}(Z(-t)) = -\mathcal{L}(Z(t))$[1].
4. From the previous point and self-similarity: $\mathbb{E}Z^2(t) = |t|^{2H}$.
5. $H \le 1$.[2]
6. For $H = 1$, $\mathbb{E}Z(s)Z(t) = \sigma^2 st$ implies $\mathbb{E}(Z(t) - tZ(1))^2 = 0$ and the process is degenerated $Z(t) = tZ(1)$.

We obtain:

Proposition 5.1.4 *B_H is Gaussian centred and H-self-similar with stationary increments.*

5.2 Gaussian Chaos

Linear combinations of Gaussian random variables were investigated above. In order to leave this Gaussian world a first question is as follows:

What are products of Gaussian random variables?

or equivalently

Do Gaussian polynomials admit a specific structure?

Polynomials of Gaussian random variables are needed and in order to consider any asymptotic one needs a closed topological vector space. A simple topology of the Hilbert space $\mathbb{L}^2(\Omega, \mathcal{A}, \mathbb{P})$ of the set of classes[3] of squared integrable random variables may be used. The Gaussian chaos is convenient for deriving expressions of any

[1]It follows from the equality of distributions $Z(0) - Z(-t)$ and $Z(t) - Z(0)$.

[2]Because $\mathbb{E}|Z(2)| = 2^H\mathbb{E}|Z(1)| \le \mathbb{E}|Z(2) - Z(1)| + \mathbb{E}|Z(1)| = 2\mathbb{E}|Z(1)|$, hence $2^H \le 2$.

[3]This means the quotient space of the set of $\mathbb{L}^2$-integrable functions, identified through $\mathbb{P}$-almost sure equality: $f \sim g$ in case $f - g = 0$, $\mathbb{P}$-a.s.

moment expression and yields limit theory in this chaos, through the Mehler formula and the diagram formula respectively. The diagram formula is complicated and we present the so-called fourth order moment method; this is a powerful technique proving Gaussian asymptotic behaviours. Namely any element Z in the Gaussian chaos such that $\mathbb{E}Z = 0$ and $\mathbb{E}Z^2 = 1$ is Gaussian if and only if $\mathbb{E}Z^4 = 3$ (it belongs to the first order chaos). This method needs an integral representation of elements of the chaos which we first explain.

Definition 5.2.1 Let $Y = (Y_t)_{t\in\mathbb{T}}$ be a Gaussian process defined on some probability space $(\Omega, \mathcal{A}, \mathbb{P})$. The Gaussian chaos Chaos(Y) associated with Y is the smallest complete vector sub-space $\mathbb{L}^2(\Omega, \mathcal{A}, \mathbb{P})$ containing Y_t (for all $t \in \mathbb{T}$) as well as the constant 1 and which is stable under products; this is the closure in $\mathbb{L}^2(\Omega, \mathcal{A}, \mathbb{P})$ of the algebra generated by Y.

Remark 5.2.1 Chaos(Y)'s elements are $\mathbb{L}^2$-limits of polynomials:

$$Z = \sum_{d=1}^{D} \sum_{t_1\in T'} \cdots \sum_{t_d\in T'} a^{(d)}_{t_1,\ldots,t_d} Y_{t_1} \cdots Y_{t_d}$$

for some finite subset $T' \subset T$, $d \geq 1$ and $a^{(d)}_{t_1,\ldots,t_d} \in \mathbb{R}$, for $t_1, \ldots, t_d \in \mathbb{R}$. This is a Hilbert space. In order to get easy calculations in this space, a basis is first provided in case $\mathbb{T} = \{t_0\}$ is a singleton. Further subsections allow calculations of second order moments and of higher order moments respectively.

Contrary to the conventions in ergodic theory, chaoses have nothing to do with a erratic behaviour; their origin lies in the tough expression of polynomials with several variables. The annulus of such polynomials of several variables does not share any of the standard properties of spaces of polynomials of one variable, such as **principality** or the **Noether property**, the first of which characterizes ideal sub-rings as generated from products with a fixed polynomial, **principal rings**: this property is essential for factorization.

Example 5.2.1 (Hermite expansions)

- An interesting example of such random variables that concerns the case of singletons $T = \{0\}$ is

$$Z = g(Y_0), \qquad Y_0 \sim \mathcal{N}(0, 1).$$

  If $Z \in L^2$ then we will prove that such expansions exist

$$Z = \sum_{k=0} \frac{g_k}{k!} H_k(Y_0), \quad g_k = \mathbb{E}H_k(N)g(N), \quad N \sim \mathcal{N}(0, 1),$$

  with H_k some polynomial to be defined below, called Hermite polynomials, see Remark 5.2.4.
  Z is also a $\mathbb{L}^2$-limit of polynomials in Y_0.

- A second case is more suitable for time series analysis $T = \mathbb{Z}$ and $(Y_t)_{t\in\mathbb{Z}}$ is a stationary time series with $Y_0 \sim \mathcal{N}(0, 1)$: one may consider partial sum processes

$$Z = g(Y_1) + \cdots + g(Y_n), \qquad \mathbb{E}g^2(Y_0) < \infty.$$

It will be proved that such expressions are again $\mathbb{L}^2$-limits of polynomials; they belong to the chaos.

A difficult question is to determine the asymptotic behaviour of such partial sums. This will be addressed below.

We aim to provide the reader with the tools necessary for Gaussian calculus.

5.2.1 Hermite Polynomials

The Normal density $\varphi(x) = \exp(-x^2/2)/\sqrt{2\pi}$ is described with some details in the Appendix A.3.

Definition 5.2.2 *(Hermite polynomials)* Let $k \geq 0$ be an arbitrary integer. We set

$$H_k(x) = \frac{(-1)^k}{\varphi(x)} \frac{d^k \varphi(x)}{dx^k}.$$

Then H_k is a kth degree polynomial with leading term 1.

Those polynomials are graphically represented in Fig. 5.5. The above degree considerations are easily deduced from the following exercise.

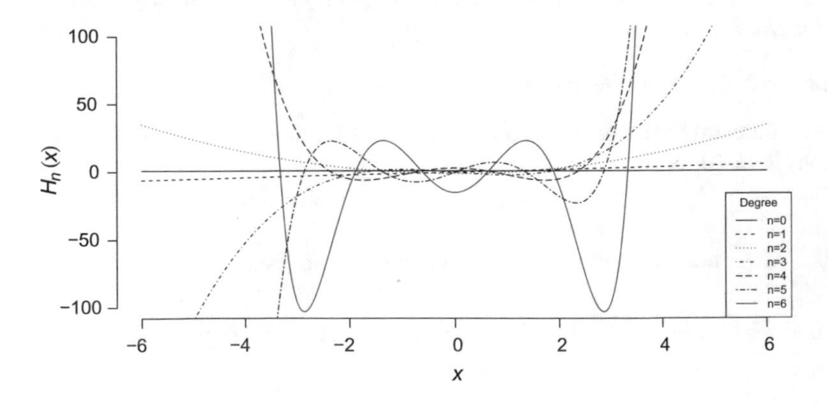

Fig. 5.5 Hermite polynomials

Exercise 25 Prove that for $k \geq 1$:

$$H_{k+1}(x) = x\,H_k(x) - H_k'(x).$$

Proof This follows from the relation

$$H_k(x)\varphi(x) = (-1)^k \varphi^{(k)}(x).$$

By differentiation: $H_k'(x)\varphi(x) + H_k(x)\varphi'(x) = (-1)^k \varphi^{(k+1)}(x)$.

Using $\varphi'(x) = -x\varphi(x)$ we get

$$(H_k'(x) - x\,H_k(x))\varphi(x) = (-1)^k \varphi^{(k+1)}(x).$$

Hence Exercise 25 follows.

Hence $d^\circ H_{k+1} = d^\circ H_k + 1$ admits the same leading coefficient and $H_0(x) = 1$ concludes the above definition.

For example

$$\begin{aligned}
H_0(x) &= 1\\
H_1(x) &= x\\
H_2(x) &= x^2 - 1\\
H_3(x) &= x^3 - 3x\\
H_4(x) &= x^4 - 6x^2 + 3\\
H_5(x) &= x^5 - 10x^3 - 9x.
\end{aligned}$$

Exercise 26 Hermite polynomials $(H_k)_{k \geq 0}$ form an orthogonal system with respect to the Gaussian measure $\varphi(x)\,dx$. Moreover $\| H_k \|_\varphi^2 = k!$, for each $k \in \mathbb{N}$.

Hint. Let $N \sim \mathcal{N}(0, 1)$ be a standard Normal rv. Then k integrations by parts yield for $k \geq l$:

$$\begin{aligned}
\mathbb{E} H_k(N) H_l(N) &= \int_{-\infty}^{\infty} H_k(x) H_l(x) \varphi(x) dx\\
&= (-1)^k \int_{-\infty}^{\infty} \frac{d^k \varphi(x)}{dx^k} H_l(x) dx\\
&= \int_{-\infty}^{\infty} \frac{d^k H_l(x)}{dx^k} \varphi(x) dx.
\end{aligned}$$

The above expression vanishes if $k > l$.

In case $k = l$, this yields $\dfrac{d^k H_k(x)}{dx^k} = k!$.

Hence $\mathbb{E} H_k^2(N) = k!$, which ends the proof.

This system is also total as proved e.g. in Choquet (1973), it means that if $\mathbb{E}|g(N)|^2 < \infty$ for a standard Normal random variable, then $\mathbb{E}g(N)H_k(N) = 0, \forall k \Rightarrow g = 0$. Hence any measurable function g with $\mathbb{E}|g(N)|^2 < \infty$ admits the $\mathbb{L}^2$-representation:

$$\begin{aligned} g(x) &= \sum_{k=0}^{\infty} \frac{g_k}{k!} H_k(x), \\ g_k &= \mathbb{E}g(N)H_k(N), \qquad k = 0, 1, 2, \dots \\ \mathbb{E}|g(N)|^2 &= \sum_{k=0}^{\infty} \frac{|g_k|^2}{k!}. \end{aligned}$$

Definition 5.2.3 Assume that $g \in \mathbb{L}^2(\varphi)$ is not the null-function. Define as before $g_k = \mathbb{E}g(N)H_k(N)$.

The Hermite rank of the function g ($\neq 0$) is the smallest index $k \geq 0$ such that $g_k \neq 0$.

It will be denoted m, or $m(g)$.

Proposition 5.2.1 *This orthonormal basis in* $\mathbb{L}^2(\varphi(x)dx)$ *also satisfies:*

$$\sum_{k=0}^{\infty} \frac{z^k}{k!} H_k(x) = e^{zx - z^2/2}. \tag{5.2}$$

This equality is only an equality in the Hilbert space $\mathbb{L}^2(\varphi(x)dx)$.

The previous series converges (normally) in $\mathbb{L}^2(\varphi(x)dx)$ because:

$$\mathbb{E}\left(\frac{z^k}{k!} H_k(N) \overline{\frac{z^l}{l!} H_l(N)}\right) = \begin{cases} 0, & \text{if } k \neq l \\ \dfrac{|z|^{2k}}{k!}, & \text{if } k = l. \end{cases}$$

We shall need the lemma:

Lemma 5.2.1 $H_k' = kH_{k-1}$.

Proof of Lemma 5.2.1. Since the leading term of H_ℓ is x^ℓ it is simple to check that $d(H_k' - kH_{k-1}) < k - 1$.

The lemma will follow from the relation

$$\int (H_k'(x) - kH_{k-1}(x))H_l(x)\varphi(x)\,dx = 0, \quad \text{for all} \quad l < k.$$

First

$$k \int H_{k-1}(x)H_l(x)\varphi(x)dx = \begin{cases} 0, & \text{if } l < k - 1 \\ k(k-1)! = k!, & \text{if } l = k - 1. \end{cases}$$

An integration by parts implies

$$\begin{aligned}\int H_k'(x)H_l(x)\varphi(x)dx &= (-1)^l \int H_k'(x)\varphi^{(l)}(x)dx \\ &= (-1)^{l+1} \int H_k(x)\varphi^{(l+1)}(x)dx \\ &= \int H_k(x)H_{l+1}(x)\varphi(x)dx.\end{aligned}$$

This expression vanishes if $l < k-1$. If now $l = k-1$ we get the same value, $k!$, as for the other quantity which implies $H_k' = kH_{k-1}$.

Remark 5.2.2 An alternative and more elementary proof of the previous relation begins with the identity $\varphi'(x) = x\varphi(x)$.

From the definition $\varphi^{(k)}(x) = (-1)^k\varphi(x)$ hence the previous expression can be rewritten as

$$H_{k+1}(x) = xH_k(x) - H_k'(x).$$

Derive k times this relation with the Leibniz formula,[4] then

$$\varphi^{(k+1)}(x) = -x\varphi^{(k)}(x) - k\varphi^{(k-1)}(x).$$

We obtain

$$H_{k+1}(x) = xH_k(x) - kH_{k-1}(x).$$

The formula follows from comparing the two previous expressions of H_{k+1}.

Now the function

$$x \mapsto g_z(x) = e^{zx - z^2/2}$$

belongs to $\mathbb{L}^2(\varphi)$, it admits the Hermite expansion

$$g_z = \sum_{k=0}^{\infty} \frac{g_{z,k}}{k!} H_k.$$

[4]If the functions $f, g : \mathbb{R} \to \mathbb{R}$ are differentiable enough then:

$$(fg)^{(n)} = \sum_{k=0}^{n} \binom{n}{k} f^{(k)} g^{(n-k)}.$$

This function indeed satisfies:

$$\begin{aligned}
g_{z,k} &= \mathbb{E} g_z(N) H_k(N) \\
&= \int_{-\infty}^{\infty} H_k(x) e^{zx - z^2/2} \varphi(x) dx = \int_{-\infty}^{\infty} H_k(x) e^{-(z-x)^2/2} \frac{dx}{\sqrt{2\pi}} \\
&= \int_{-\infty}^{\infty} H_k(t+z) \varphi(t) dt = \sum_{l=0}^{k} \frac{z^l}{l!} \int_{-\infty}^{\infty} H_k^{(l)}(t) \varphi(t) dt \\
&= \sum_{l=0}^{k} C_k^l z^l \int_{-\infty}^{\infty} H_{k-l}(t) \varphi(t) dt \; = z^k.
\end{aligned}$$

For the above identities, use the change in variables $t = x - z$, a Taylor expansion, and finally the relation $H_k^{(l)} = \frac{k!}{(k-l)!} H_{k-l}$.

We get the $\mathbb{L}^2$-expansion:

$$\sum_{k=0}^{\infty} \frac{z^k}{k!} H_k(N) = e^{zN - \frac{z^2}{2}}, \quad \text{in } \; \mathbb{L}^2(\Omega, \mathcal{A}, \mathbb{P}). \tag{5.3}$$

Consider $\mathbb{L}^2(\varphi)$ the Hilbert space of such measurable functions g with

$$\int_{\mathbb{R}} g^2(x) \varphi(x) \, dx = \mathbb{E} g^2(N) < \infty.$$

The convergence of (5.3) in the space $\mathbb{L}^2(\varphi)$ for each $z \in \mathbb{C}$, also implies the x-a.s. convergence of the series

$$g(x, z) = \sum_{k=0}^{\infty} \frac{z^k}{k!} H_k(x), \qquad \forall z \in \mathbb{C}.$$

Exercise 27 (*Orthogonal polynomials*) Assume that $-\infty \le a < b \le +\infty$. More generally let $p : (a, b) \to \mathbb{R}^+$ be a measurable function such that

$$\lambda(\{x \in (a, b) / \; p(x) = 0\}) = 0.$$

with λ, the Lebesgue measure on $\mathbb{R}$.

Set also

$$(f, g) \mapsto (f, g)_p = \int_a^b f(x) g(x) \, p(x) \, dx,$$

the scalar product on the pre-Hilbert space $\mathbb{L}^2(p)$ of classes (wrt to a.s. equality) of integrable measurable functions with

$$\int_a^b f^2(x)\, p(x)dx < \infty.$$

1. *Schmidt orthogonalization.* Consider $0 < N \leq \infty$. Suppose that the polynomials $x \mapsto 1, \ldots, x^N$ satisfy this relation. Analogously to Hermite polynomials, define recursively a sequence of orthogonal polynomials such that $P_0 = 1$, and such that $P_n(x) - x^n$ is orthogonal to $1, x, \ldots, x^{n-1}$, for each $0 \leq n \leq N \leq +\infty$.
2. *Recurrence relation.* There exist sequences $a_n \in \mathbb{R}$, $b_n > 0$ such that

$$P_n(x) = (x + a_n)P_{n-1}(x) + b_n P_{n-2}(x), \quad \forall x \in (a, b),\ 2 \leq n \leq N.$$

3. *Roots of orthogonal polynomials.* In case (a, b) is any closed, open or semi-open interval of $\mathbb{R}$, then each orthogonal polynomial admits n distinct roots.
   From now on we consider examples with $N = \infty$.
4. If $(a, b) = [-1, 1]$ and $p(x) = (1 - x)^u(1 + x)^v$ we get Jacobi's polynomials for $u, v > -1$. In case $u = v = 1$ one obtains Legendre's polynomials and $u = v = \frac{1}{2}$ yields Tchebichev's polynomials. Prove that (P_n) is a complete system.
5. If $(a, b) = [0, +\infty)$ and $p(x) = e^{-x}$ we get the Legendre polynomials. Analogously to the Hermite case, prove that

$$P_n(x) = \frac{e^x}{n!}\frac{d^n}{dx^n}\left(x^n e^{-x}\right).$$

Proofs.

1. We use the Schmidt orthogonalization technique; this is a recursive technique. Assume that $P_0, \ldots, P_n$ are orthogonal and

$$d^\circ P_k = k, \quad d^\circ(P_k(x) - x^k) < k, \qquad 0 \leq k \leq n$$

   have been constructed, then set $P_{n+1}(x) = x^{n+1} + a_0 + \cdots + a_n x^n$ such that $(P_{n+1}, P_k) = 0$ $(0 \leq k \leq n)$ and thus relations $a_k(P_k, P_k) + (P_k, P_{n+1}) = 0$ determine the coefficients of this expansion of P_{n+1}.
   The system of orthogonal polynomials is a Hilbert basis and we need to prove:

$$\forall f \in \mathbb{L}^2(p): \quad \left\{\forall n \geq 0,\ (f, P_n)_p = 0\right\} \Rightarrow f = 0.$$

2. As the degree of $P_n(x) - xP_{n-1}(x)$ is $< n - 1$, one may write its expansion

$$P_n(x) - xP_{n-1}(x) = c_0 P_0(x) + \cdots + c_{n-2}P_{n-2}(x).$$

Now

$$(xP_{n-1}, P_k)_p = (P_{n-1}, xP_k)_p = c_k(P_k, P_k)_p \geq 0.$$

For $k < n-2$ this entails $c_k = 0$ and if $k+1 = n-1$ this is > 0.

3. Let $x_1 < \cdots < x_k$ be the real roots of P_n with a change of sign.
Set

$$Q(x) = (x - x_1) \cdots (x - x_k)$$

then $P_n(x)Q(x)p(x) > 0$ (a.s.), this excludes the relation $(P_n, Q)_p = 0$ which holds by construction in case $k < n$.

4. Properties of these polynomials may be found in Szegö (1959) or in Sansone (1959), and the Weierstrass theorem (see Exercise 9 for a first approach, and e.g. Doukhan and Sifre 2001 for more comments) entails that these systems are complete.
5. Prove that the leading coefficient of RHS is 1 and that the corresponding system is orthogonal. To this aim again use integrations by parts and due to the fact that integrated terms all vanish we get for $n > k$:

$$(P_n, P_k)_p = \frac{1}{n!}\int_0^\infty P_k(x)\frac{d^n}{dx^n}(x^n e^{-x})dx = \frac{(-1)^n}{n!}\int_0^\infty P_k^{(n)}(x)e^{-x}dx.$$

This ends the proof.

5.2.2 Second Order Moments

The following results allows us to better understand the Euclidean structure of the chaos.

Lemma 5.2.2 (Mehler formula) *Let $Y = (Y_1, Y_2)$ be a Gaussian random vector with law*

$$\mathcal{N}_2\left(0, \begin{pmatrix} 1 & r \\ r & 1 \end{pmatrix}\right),$$

then

$$Cov\ (H_k(Y_1), H_l(Y_2)) = \begin{cases} 0, & \text{if } k \neq l, \\ k!r^k, & \text{if } k = l. \end{cases}$$

Remark 5.2.3 The main Lemma 5.2.2 allows to control the second order structure of elements in the chaos.

Consider a closed Gaussian space V_1 spanned by the Gaussian process $(X_t)_{t\in\mathbb{T}}$.

- Mehler's formula proves the orthogonality of the various chaoses V_k linearly generated by $(H_k(X_t))_{t\in\mathbb{T}}$, for $k \geq 1$.
- Each chaos V_k admits a geometry described by Mehler formula.

Proof If $t_1, t_2 \in \mathbb{R}$ set

$$\sigma^2 = \text{Var}\,(t_1Y_1 + t_2Y_2) = t_1^2 + t_2^2 + 2rt_1t_2,$$

then $t_1Y_1 + t_2Y_2 \sim \sigma N$. The relation (A.6) implies:

$$\mathbb{E}\exp\left(t_1Y_1 + t_2Y_2 - \frac{1}{2}(t_1^2 + t_2^2)\right) = e^{rt_1t_2}.$$

From the $\mathbb{L}^2$-identity (5.3) we may exchange integrals and sums from dominated convergence

$$\begin{aligned}\mathbb{E}\exp\left(t_1Y_1 + t_2Y_2 - \frac{1}{2}(t_1^2 + t_2^2)\right) &= e^{rt_1t_2} \\ &= \sum_{k,l=0}^{\infty} \frac{t_1^k}{k!}\frac{t_2^l}{l!}\mathbb{E}H_k(Y_1)H_l(Y_2).\end{aligned}$$

Identifying the previous expansion with respect to powers of t_1 and t_2 yields the conclusion since $\mathbb{E}H_k(Y_1) \neq 0$ only for the case $k = 0$.

Remark 5.2.4 Let $g : \mathbb{R} \to \mathbb{C}$ be measurable and $\mathbb{E}|g(N)|^2 < \infty$ in the setting of Example 5.2.1.

Then

$$g = \sum_{k=0}^{\infty} \frac{g_k}{k!}H_k, \qquad g_k = \mathbb{E}H_k(N)\overline{g(N)}.$$

Now

$$\begin{aligned}\mathbb{E}g(Y_1)\overline{g(Y_2)} &= \sum_{k=0}^{\infty} \frac{|g_k|^2}{k!}r^k, \\ \text{Cov}\,(g(Y_1), g(Y_2)) &= \sum_{k=1}^{\infty} \frac{|g_k|^2}{k!}r^k.\end{aligned}$$

Below we consider a stationary Gaussian process $(Y_n)_{n\in\mathbb{Z}}$ such that $\mathbb{E}Y_0 = 0$, $\text{Var}\, Y_0 = 1$, then $r_n = \mathbb{E}Y_0Y_n$.

Assume also that $\mathbb{E}g(Y_0) = 0$, which means that the Hermite rank satisfies $m(g) \geq 1$.

Then:

$$\begin{aligned}\mathbb{E}\left|\sum_{j=1}^{n} g(Y_j)\right|^2 &= \sum_{s=1}^{n}\sum_{t=1}^{n} \mathbb{E}g(Y_s)\overline{g(Y_t)} \\ &= n\sum_{|l|<n}^{n}\left(1 - \frac{|l|}{n}\right)\mathbb{E}g(Y_0)\overline{g(Y_l)}\end{aligned}$$

$$
\begin{aligned}
&= n \sum_{|l|<n}^{n} \left(1 - \frac{|l|}{n}\right) \sum_{k=m(g)}^{\infty} \frac{|g_k|^2}{k!} r_l^k \\
&= n \sum_{k=m(g)}^{\infty} \frac{|g_k|^2}{k!} \sum_{|l|<n}^{n} \left(1 - \frac{|l|}{n}\right) r_l^k. \qquad (5.4)
\end{aligned}
$$

In case $\sum_{l=-\infty}^{\infty} |r_l| < \infty$, each series $R_k = \sum_{l=-\infty}^{\infty} r_l^k$ converges (for $k \geq 1$) because $|r_l|^k \leq r_0^{k-1} |r_l| = |r_l|$ and

$$
\mathbb{E} \left| \sum_{j=1}^{n} g(Y_j) \right|^2 \sim n \sum_{k=m(g)}^{\infty} \frac{R_k |g_k|^2}{k!} = \mathcal{O}(n),
$$

if only

$$
S = \sum_{l=-\infty}^{\infty} |r_l|^{m(g)} < \infty.
$$

The Hermite rank in Definition 5.2.3 is written $m(g)$. The previous claim still holds true; indeed all series R_k are then convergent for $k \geq m(g)$.

The Cauchy–Schwarz inequality implies $|r(\ell)| \leq 1 = \mathbb{E}Y_0^2$. Thus

$$
|r(\ell)|^k \leq |r(\ell)|^{m(g)}, \qquad \text{if} \quad k \geq m(g).
$$

Moreover $|R_k| \leq S$ which proves that the previous expansion (5.4) is indeed convergent.

Exercise 28 The empirical cumulative distribution is of a main interest for statistics:

$$
F_n(x) = \frac{1}{n} \sum_{k=1}^{n} \mathbb{I}_{\{Y_k \leq x\}}.
$$

$F_n(x)$ is an unbiased estimator of the cumulative function.

Prove that:

$$
\text{Var } F_n(x) = \frac{1}{n} \sum_{k=m(g)}^{\infty} \frac{|\varphi^{(k-1)}(x)|^2}{k!} \sum_{|l|<n}^{n} \left(1 - \frac{|l|}{n}\right) r_l^k.
$$

This expression is

$$
\text{Var } F_n(x) = \mathcal{O}\left(\frac{1}{n}\right), \text{ as } n \to \infty, \qquad \text{if} \qquad \sum_{l=-\infty}^{\infty} |r_l| < \infty.
$$

If now

$$\sum_{l=-\infty}^{\infty} |r_l| = \infty$$

then

$$\text{Var } F_n(x) = \mathcal{O}\left(\frac{1}{n}\sum_{|l|<n} |r_l|\right) \gg \frac{1}{n}.$$

However, Cesaro's lemma proves that this expression converges to 0 if the sequence r_l converges to 0.

Proof $F_n(x)$ is unbiased, since a simple calculation yields:

$$\mathbb{E}F_n(x) = F(x).$$

The expression for its variance relies on the previous identity written for the function

$$u \mapsto g(u) = \mathbb{I}_{\{u \le x\}}.$$

Here again with N a standard Normal rv:

$$\begin{aligned} g_k &= \mathbb{E}H_k(N)\ \mathbb{I}_{\{N \le x\}} \\ &= \int_{-\infty}^{x} H_k(u)\varphi(u)du \\ &= (-1)^k \int_{-\infty}^{x} \varphi^{(k)}(u)du \\ &= \begin{cases} \Phi(x), & \text{(a primitive of } \varphi) \text{ for } k = 0 \\ -\varphi(x)H_{k-1}(x), & \text{if } k \ne 0. \end{cases} \end{aligned}$$

Hence

$$\text{Var } F_n(x) = \frac{1}{n} \sum_{k=m(g)}^{\infty} \frac{|\varphi^{(k-1)}(x)|^2}{k!} \sum_{|l|<n}^{n} \left(1 - \frac{|l|}{n}\right) r_l^k.$$

If now $\sum_{l=-\infty}^{\infty} |r_l| = \infty$, then its order of magnitude is

$$\text{Var } F_n(x) = \mathcal{O}\left(\frac{1}{n}\sum_{|l|<n}\left(1 - \frac{|l|}{n}\right)|r_l|\right) = \mathcal{O}\left(\frac{1}{n}\sum_{|l|<n}|r_l|\right),$$

admits a rate of growth larger than $\frac{1}{n}$. From Cesaro's lemma this expression converges to 0 if the sequence r_l converges to 0.

Again we assume that the polynomials form a complete system in Chaos(X). Mehler's formula in Lemma 5.2.2 allows us to decouple chaoses of different orders.

We consider a Gaussian process $X = (X_t)_{t\in\mathbb{T}}$ and we denote by Chaos(X), the corresponding chaos.

5.2.3 *Higher Order Moments*

The technique used to derive Mehler's formula suggests an extension for an arbitrary number of factors $H_{l_j}(Y_j)$. Thus let $Y = (Y_1, \ldots, Y_p) \sim \mathcal{N}_p(0, R)$ for a symmetric matrix

$$R = (r_{i,j})_{1\le i,j\le p}$$

with diagonal entries $r_{i,i} = 1$.

Hence $r_{i,j} = \mathrm{Cov}\,(Y_i, Y_j)$. If $(t_1, \ldots, t_p) \in \mathbb{R}^p$ we derive

$$\mathrm{Var}\left(\sum_{j=1}^{p} t_j Y_j\right) = \sum_{j=1}^{p} t_j^2 + 2\rho, \qquad \rho = \sum_{1\le i<j\le p} r_{i,j} t_i t_j.$$

Relation (A.6) proves

$$e^{\rho} = \mathbb{E}\exp\left(\sum_{j=1}^{p}\left(t_j Y_j - \frac{t_j^2}{2}\right)\right).$$

As in the proof of Mehler's formula (Lemma 5.2.2), the idea is to identify the coefficient of these expansions. If the expansion (5.3) was also valid in $\mathbb{L}^p$, then it would be possible to write:

$$\exp\left(\sum_{1\le i<j\le p} r_{i,j} t_i t_j\right) = \mathbb{E}\underbrace{\sum_{l_1=0}^{\infty}\cdots\sum_{l_p=0}^{\infty}}_{p \text{ sums}} \frac{t_1^{l_1}}{l_1!}\cdots\frac{t_p^{l_p}}{l_p!}\,\mathbb{E}\left(\prod_{j=1}^{p} H_{l_j}(Y_j)\right).$$

An argument allowing the inversion of sums and integrals would provide the identification of such moments.

Unfortunately, such convergences are not accessible and to derive expressions of the moments we will use an alternative argument from Slepian (1972).

The characteristic function of the random vector $Y = (Y_1, \dots, Y_k)$ can be written

$$\phi_Y(s) = e^{-\frac{1}{2}s^t \Sigma s},$$

if this is a centred Gaussian vector and its covariance Σ.

Then an alternative representation of its density function follows from Fourier inversion. Assuming Σ to be invertible will imply the convergence of the following integrals:

$$f(y, \Sigma) = \frac{1}{(2\pi)^{\frac{k}{2}}} \int_{-\infty}^{\infty} \cdots \int_{-\infty}^{\infty} e^{is^t y} e^{-\frac{1}{2}s^t \Sigma s} ds.$$

If $\Sigma = (r_{i,j})_{1 \le i,j \le k}$ with $r_{i,i} = 1$ we thus get the heat equation from differentiations:

Exercise 29 (*Heat equation*)

$$\frac{\partial f(y, \Sigma)}{\partial r_{i,j}} = \frac{\partial^2 f(y, \Sigma)}{\partial y_i \partial y_j}$$

if $i \neq j$. The function $f(y, \Sigma)$ is analytic wrt the multidimensional variable Σ.

Hint. Apply Lebesgue's dominated convergence.
This will allow the expansion below. Let $n = (n_{i,j})_{1 \le i < j \le k}$ be such that $n_{i,j} \in \mathbb{N}$ for each couple $1 \le i < j \le k$.

We denote

$$r^n = \prod_{i<j} r_{i,j}^{n_{i,j}}, \qquad n! = \prod_{i<j} n_{i,j}!$$

Also set

$$n_{i,j} = n_{j,i}, \quad \text{if } i > j, \quad \text{and} \quad s_{n,i} = \sum_{j \neq i} n_{i,j}.$$

Then, with

$$f(y, I_k) = \prod_{i=1}^{k} \varphi(y_i),$$

we get

$$\begin{aligned} f(y, \Sigma) &= \sum_{n=(n_{i,j})} \frac{r^n}{n!} \frac{\partial^{\{\sum_{i<j} n_{i,j}\}} f(y, I_k)}{\prod_{i<j} \partial r_{i,j}^{n_{i,j}}} \\ &= \sum_{n=(n_{i,j})} \frac{r^n}{n!} \frac{\partial^{s_{n,i}} f(y, I_k)}{\prod_{i<j} \partial y_i^{n_{i,j}} \partial y_j^{n_{i,j}}} \\ &= \sum_{n=(n_{i,j})} \frac{r^n}{n!} \prod_{i=1}^{k} \frac{\partial^{s_{n,i}} \varphi(y_i)}{\partial y_i^{s_{n,i}}} \end{aligned} \tag{5.5}$$

$$= \sum_{n=(n_{i,j})} \frac{r^n}{n!} \prod_{i=1}^{k} \varphi^{(s_{n,i})}(y_i)$$

Thus,

$$f(y, \Sigma) = \sum_{n=(n_{i,j})} \frac{r^n}{n!} \prod_{i=1}^{k} H_{s_{n,i}}(y_i) \cdot \phi(y) \tag{5.6}$$

where

$$\phi(y) = \prod_{i=1}^{k} \varphi(y_i)$$

denotes the density function of a random vector $\mathcal{N}_k(0, I_k)$ and the previous sums extend to all integer multi-indices $n = (n_{i,j})_{1 \le i < j \le k}$.

Indeed $s_{n,i}$ is the number of appearances of y_i in the second identity.

Relation (5.6) thus implies

$$\mathbb{E} \prod_{i=1}^{k} H_{s_i}(Y_i) = \sum_{n} \frac{r^n}{n!} \prod_{i=1}^{k} \int_{-\infty}^{\infty} H_{s_{n,i}}(y_i) H_{s_i}(y_i) \varphi(y_i) dy_i,$$

and orthogonality of the Hermite polynomials implies:

Proposition 5.2.2 (Diagram formula) *For $k \ge 2$:*

$$\mathbb{E} \prod_{i=1}^{k} H_{s_i}(Y_i) = s_1! \cdots s_k! \sum_{n \in N(s_1, \ldots, s_k)} \frac{r^n}{n!},$$

for sums extended to such multi-indices $n = (n_{i,j})$ with

$$s_{n,i} = s_i, \qquad \textit{if} \quad 1 \le i \le k.$$

The $n_{i,j}$'s correspond to partitions of the array such that

$$\begin{array}{lll} x_1 & \ldots\ x_1 & \textit{appears } s_1 \textit{ times} \\ x_2 & \ldots\ x_2 & \textit{appears } s_2 \textit{ times} \\ \ldots & \ldots\ldots & \qquad \ldots \\ x_k & \ldots\ x_k & \textit{appears } s_k \textit{ times.} \end{array}$$

Precisely the first line of the arrays may be divided into $(k-1)$ parts with respective sizes $n_{1,2}, \ldots, n_{1,k}$.

The number of such multi-indices is also the number of arrays satisfying the constraints $s_{n,i} = s_i$.

Exercise 30 Prove again Mehler's formula in Lemma 5.2.2.

Hint. If $k = 2$ the sum in n is a simple sum on the set of integers $\mathbb{N}$ because $i < j$ implies $i = 1$ and $j = 2$.

Thus the summation

$$\sum_{n \in N(s_1, s_2)}$$

corresponds to the value $n_{1,2} = s_1 = s_2$: this is again Mehler's formula.
Based on Melher's formula, Lemma 5.2.2 for the case $p = 2$ and on the diagram formula for larger values of p, the following decomposition may be derived:

Theorem 5.2.1 (Chaotic decomposition) *Let $X = (X_t)_{t\in\mathbb{T}}$ be a Gaussian process, the decomposition of the chaos $Chaos(X)$ generated by X is orthogonal,*

$$Chaos(X) = \bigoplus_{k=0}^{\infty} \mathcal{H}_k(X).$$

Here $\mathcal{H}_k(X)$ is the subspace of $\mathbb{L}^2(\Omega, \mathcal{A}, \mathbb{P})$ spanned by

$$\prod_{j=1}^{p} H_{k_j}(X_{t_j}), \qquad k_1 + \cdots + k_p = k,\ t_1, \ldots, t_p \in \mathbb{T}.$$

Remark 5.2.5 Various applications of the diagram formula to time series are known. Breuer and Major (1983) prove that if a stationary Gaussian process satisfies $Y_0 \sim \mathcal{N}(0, 1)$,

$$S_n = \frac{1}{\sqrt{n}} \sum_{k=1}^{n} g(Y_k) \xrightarrow{\mathcal{L}} \mathcal{N}(0, \sigma^2),$$

if

$$\sum_{k=-\infty}^{\infty} |r_k|^m < \infty, \qquad r(k) = \mathbb{E}Y_0 Y_k,$$

and $m = m(g)$ denotes the Hermite rank of g. The convergence of moments of S_n to the Gaussian ones is proved with the diagram formula.

Another application is the Arcones inequality for vector valued processes, see Taqqu in Doukhan et al. (2002b). This inequality is extended in Soulier (2001) and further by Bardet and Surgailis. Other developments are also reported in Rosenblatt (1985).

The fourth order moment approach yields an impressive simplification of the calculations.

The two following subsections introduce the technique.

5.2.4 Integral Representation of the Brownian Chaos

Consider a square integrable function $f : \mathbb{R}^+ \to \mathbb{R}$. Wiener integrals are simple to define[5]

$$I_1(f) = \int_0^\infty f(t)dW(t)$$

as centred Gaussian random variables, in the corresponding Gaussian closed space generated by the Brownian process $(W(s))_{s\geq 0}$.

With

$$\|f\|_2 = \left(\int_0^\infty f^2(t)dt\right)^{\frac{1}{2}}$$

the application

$$f \mapsto I_1(f), \qquad \mathbb{L}^2(\mathbb{R}^+) \to \mathbb{L}^2(\Omega, \mathcal{A}, \mathbb{P})$$

is an isometry.

A first simple extension is to define stochastic integrals on the real line. Consider two independent Brownian motions W_- and W_+.

A way to define the Brownian motion on the line is to set $W(t) = W_+(t)$ if $t \geq 0$ and $W(t) = W_-(-t)$ if $t < 0$. Wiener integral is straightforwardly extended on $(-\infty, \infty)$.

There exist two different ways to define

$$I_k(h) = \int_{-\infty}^\infty \cdots \int_{-\infty}^\infty h(t_1, \ldots, t_k)\, dW(t_1) \cdots dW(t_k).$$

We denote by $\mathcal{H}_k$ the set of symmetric functions $h \in \mathbb{L}^2(\mathbb{R}^k)$, i.e. such that for any arbitrary bijection (permutation) $\pi : \{1, \ldots, k\} \to \{1, \ldots, k\}$:

$$h(t_{\pi(1)}, \ldots, t_{\pi(k)}) = h(t_1, \ldots, t_k).$$

We use the symmetrized version of a function $h \in \mathbb{L}^2(\mathbb{R}^k)$ by setting:

$$\mathrm{Sym}(h)(t_1, \ldots, t_k) = \frac{1}{k!} \sum_\pi h(t_{\pi(1)}, \ldots, t_{\pi(k)}).$$

[5]Define it first for step functions and notice that for such functions $f \mapsto I_1(f)$ is an isometry on this dense subspace in order to prove the same for the application defined on $\mathbb{L}^2(\mathbb{R}^+) \to \mathbb{L}^2(\Omega, \mathcal{A}, \mathbb{P})$,

$$\|f\|_2 = \left(\int_0^\infty f^2(t)dt\right)^{\frac{1}{2}} = \left(\mathbb{E} I_1(f)^2\right)^{\frac{1}{2}} = \|I_1(f)\|_{\mathbb{L}^2(\Omega,\mathcal{A},\mathbb{P})}.$$

This is a standard trick to extend it by using a density argument in $\mathbb{L}^2(\mathbb{R})$.

These spaces are naturally equipped with their natural Hilbert norms

$$\|h\|^2_{\mathcal{H}_k} = \int_{\mathbb{R}^k} h^2(t_1, \dots, t_k)\, dt_1 \cdots dt_k,$$

and the triangle inequality justifies the above symmetrization

$$\|\mathrm{Sym}(h)\|_{\mathcal{H}_k} \le \|h\|_{\mathcal{H}_k}.$$

We refer the reader to Major (1981) for precise statements. Questions of convergence are extremely specific and technically difficult in this framework as noticed in a following chapter concerned with dependence.

- A first way is to simply set it by recursion but in this case the stochastic integrals to be considered are anticipative.
- An alternative way to proceed is to consider integrals over sets

$$\{(t_1, \dots, t_k) \in \mathbb{R}^k /\ t_1 \le \cdots \le t_k\},$$

then if the function is invariant through permutations one defines

$$I_k(h) = k! \int_{-\infty}^{\infty} \int_{-\infty}^{t_1} \cdots \int_{-\infty}^{t_{k-1}} h(t_1, \dots, t_k)\, dW(t_1) \cdots dW(t_k).$$

- Assume now that h is a symmetric function with

$$h(\pm t_1, \dots, \pm t_k) = h(t_1, \dots, t_k).$$

An alternative construction in Major (1981) is based again on an approximation by step functions. First if $A_1, \dots, A_k \subset \mathbb{R}^+$ are closed intervals, set $\Delta_j = A_j \cup (-A_j)$ and $\Delta = \Delta_1 \times \cdots \times \Delta_k$. Then define

$$I_k(\mathbb{I}_\Delta) = L_1 \times \cdots \times L_k, \quad \text{with} \quad L_j = W_+(A_j) - W_-(A_j).$$

If $A_1, \dots, A_k$ are pairwise disjoint then these random variables are independent. This definition is extended by linearity to functions constant on such intervals Δ. A uniform continuity argument is thus used to define such multiple integrals for $h \in \mathcal{H}_k$. Namely this integral is an isometry over simple functions; it thus extends to the closure $\mathcal{H}_k$ of this set.

Exercise 31 Prove that:

$$H_k(I_1(f)) = I_k(f^{\otimes k}),$$

with I_k the kth Ito–Wiener integral and

$$f^{\otimes k}(t_1, \ldots, t_k) = f(t_1) \cdots f(t_k).$$

For example this formula is just the Ito formula for $k = 2$.

Hint. As for the construction of multiple Ito integrals, first proceed with simple indicator functions, and then extend it linearly to piecewise constant functions. We conclude with the previous extension argument.

5.2.5 The Fourth Order Moment Method

Peccati and coauthors, see e.g. Nourdin et al. (2011) recently documented important developments.[6]

The fourth order method is a nice alternative to the diagram formula. In order to simplify expressions we consider the chaos generated by $\{W(t)/\ t \geq 0\}$.

From now on we restrict to functions on the interval [0, 1] and we keep using the same notations as above.

For $f \in \mathcal{H}_k$ and $g \in \mathcal{H}_m$, for $1 \leq p \leq k \wedge m$, define with Nourdin et al. (2011) the expression:

$$\begin{aligned} f \otimes_p g(t_1, \ldots, t_{m+k-2p}) &= \int_{\mathbb{R}^p} f(t_1, \ldots, t_{k-p}, s_1, \ldots, s_p) \\ &\quad \times g(t_{k-p+1}, \ldots, t_{k+m-2p}, s_1, \ldots, s_p)\, ds_1 \cdots ds_p. \end{aligned}$$

For example if $m = 0$ or k we have respectively:

$$f \otimes_0 g = f \otimes g, \qquad f \otimes_k g = \int_{\mathbb{R}^k} f(s)g(s)\, ds.$$

Ito's formula is a way to represent product of elements in the k-th and in the mth order chaos in the chaos with order $k + m$. It can be written in this case as the following formula and the other two formulae are also useful:

$$I_k(f)I_m(g) = \sum_{p=0}^{k \wedge m} p! \binom{k}{p} \binom{m}{p} I_{k+m-2p}(f \otimes_p g).$$

[6]Many thanks to Ivan Nourdin for his friendly help for his redaction of this section.

and

$$\frac{(k+m)!}{k!m!}\|\mathrm{Sym}(f\otimes g)\|^2_{\mathcal{H}_{k+m}} = \|f\|^2_{\mathcal{H}_k}\|g\|^2_{\mathcal{H}_m} + \sum_{q=1}^{k\wedge m}\binom{k}{q}\binom{m}{q}\|f\otimes_q g)\|^2_{\mathcal{H}_{k+m-2q}}.$$

Now the fourth order moments may also be calculated:

$$\mathbb{E}I_k^4(f) = 3k!^2\|f\|^4_{\mathcal{H}_k} + \frac{3}{k}\sum_{p=1}^{k-1} p\cdot p!\binom{k}{p}^4 (2(k-p))!\,\|\mathrm{Sym}(f\otimes_p f)\|^2_{\mathcal{H}_{2(k-p)}}.$$

In particular, observe from the above representation that

$$\lim_{n\to\infty}\left(\mathbb{E}I_k^4(f_n) - 3(\mathbb{E}I_k^2(f_n))^2\right) = 0,$$

is equivalent to

$$\lim_{n\to\infty}\|\mathrm{Sym}(f_n\otimes_p f_n)\|^2_{\mathcal{H}_{2(k-p)}} = 0, \qquad \forall p\in\{1,\ldots,k-1\}.$$

We now present the deep rigidity result from the **Nualart–Peccati–Tudor** theory.

Theorem 5.2.2 *Assume that a sequence $f_n\in\mathcal{H}_k$ satisfies*

$$\lim_n \|f_n\|_{\mathcal{H}_k} = 1,$$

then

$$I_k(f_n)\xrightarrow{\mathcal{L}}_{n\to\infty}\mathcal{N}(0,1) \iff \lim_n \mathbb{E}I_k^4(f_n) = 3.$$

Remark 5.2.6

- Essentially a sequence of standard random variables ($\mathbb{E}Z_n = 0$, $\mathbb{E}Z_n^2 = 1$) in the k-order chaos converges to a Gaussian rv if and only if

$$\lim_n \mathbb{E}Z_n^4 = 3.$$

- More simply a standard rv Z in the kth chaos is Normal if and only if $\mathbb{E}Z^4 = 3$.

Proof (thanks to Ivan Nourdin). In fact this will be enough to prove the result if, only

$$\mathbb{E}I_k^2(f_n) = 1, \qquad \text{and} \qquad \lim_n \mathbb{E}I_k^4(f_n) = 3.$$

In order to prove the result, two additional tools will be needed:

1. For each function $\psi : \mathbb{R} \to \mathbb{R}$ in $\mathcal{C}_b^1$,

$$\mathbb{E} I_k(f)\psi(I_k(f)) = kE\psi'(f) \int_{-\infty}^{\infty} I_{k-1}^2(f(\cdot, t))\, dt.$$

2.

$$\begin{aligned}\operatorname{Var} \int_{-\infty}^{\infty} I_{k-1}^2(f(\cdot, t))\, dt \\ &= \frac{1}{k^4} \sum_{p=1}^{k-1} p(p!)^2 \binom{k}{p}^4 (2(k-p))!\, \|\mathrm{Sym}(f \otimes_p f\|_{\mathcal{H}_{2(k-p)}}^2\end{aligned}$$

This entails in particular

$$\begin{aligned}\lim_{n\to\infty} \operatorname{Var} \int_{-\infty}^{\infty} I_{k-1}^2(f_n(\cdot, t))\, dt = 0 \\ \iff \lim_{n\to\infty} \|\mathrm{Sym}(f_n \otimes_p f_n)\|_{\mathcal{H}_{2(k-p)}}^2 = 0,\ (1 \le p < k) \\ \iff \lim_{n\to\infty} \mathbb{E} I_k^4(f_n) = 3.\end{aligned}$$

Now set

$$\psi_n(t) = e^{\frac{t^2}{2}}\, \mathbb{E}(\exp(itI_k(f_n))).$$

Then

$$\begin{aligned}\psi_n'(t) &= t\psi_n(t) + ie^{\frac{t^2}{2}} \mathbb{E}(I_k(f_n) \exp(itI_k(f_n)) \\ &= te^{\frac{t^2}{2}} \mathbb{E}\left(1 - \int_{-\infty}^{\infty} I_{k-1}^2 f_n(\cdot, t)\, dt\right) I_k(f_n) \exp(itI_k(f_n)),\end{aligned}$$

and

$$|\psi_n'(t)| \le te^{\frac{t^2}{2}} \mathbb{E}\left|1 - \int_{-\infty}^{\infty} I_{k-1}^2 f_n(\cdot, t)\, dt\right|.$$

Note that:

$$\mathbb{E} \int_{-\infty}^{\infty} I_{k-1}^2 f_n(\cdot, t) dt = 1,$$

then from the Cauchy–Schwarz inequality we need to control the variance of

$$\int_{-\infty}^{\infty} I_{k-1}^2(f_n(\cdot, t))\, dt$$

which tends to 0 from the above equivalence. Thus we have proved that the sequence of the characteristic functions of $I_k(f_n)$ converge to that of a standard Gaussian. See Nourdin et al. (2011) for more details.

Chapter 6
Linear Processes

We consider stationary sequences generated through independent identically distributed $(\xi_n)_{n\in\mathbb{Z}}$. A reference is Brockwell and Davis (1991). Such models are natural in signal theory since they appear through linear filtering of a white noise. The usual setting is that $(\xi_n)_{n\in\mathbb{Z}}$ is only a $\mathbb{L}^2$-stationary white noise sequence and not an independent identically distributed sequence.

6.1 Stationary Linear Models

Definition 6.1.1 Let $(c_n)_{n\in\mathbb{Z}}$ a sequence of real numbers, and $(\xi_n)_{n\in\mathbb{Z}}$ be an iid sequence. When it makes sense, define stationary linear processes as:

$$X_n = \sum_{k=-\infty}^{\infty} c_k \xi_{n-k}. \tag{6.1}$$

Lemma 6.1.1 *The relation*

$$\sum_{k=-\infty}^{\infty} |c_k|^{m\wedge 1} < \infty$$

implies that the previous series converge if $\mathbb{E}|\xi_0|^m < \infty$ *for some* $m > 0$*, then this series converges in probability.*

If $\mathbb{E}\xi_0^2 < \infty$ $(m = 2)$ *and* $\mathbb{E}\xi_0 = 0$*, then a weaker condition holds for the stationarity and the existence of* (6.1) *in* $\mathbb{L}^2$

$$\sum_{k=-\infty}^{\infty} |c_k|^2 < \infty.$$

P. Doukhan, *Stochastic Models for Time Series*, Mathématiques et Applications 80,
https://doi.org/10.1007/978-3-319-76938-7_6

Proof From Markov's inequality we derive, if $m \le 1$:

$$\mathbb{P}\left(\sum_{k=-\infty}^{\infty} |c_{n-k}||\xi_k| > A\right) \le \frac{1}{A^m}\mathbb{E}\left(\sum_{k=-\infty}^{\infty} |c_k||\xi_{n-k}|\right)^m \le \frac{1}{A^m}\mathbb{E}|\xi_0|^m \sum_{k=-\infty}^{\infty} |c_k|^m.$$

Use Exercise 32-1) to get the last inequality. Now to prove the convergence, consider an arbitrary $A > 0$ and restrict the above sums to $k \ge K(A)$ to derive Cauchy convergence criteria. Since (c_k) is a convergent series, there exists $K(A)$ such that if $k > K(A)$ is large enough then the RHS of the previous relation is arbitrarily small as desired whatever A is chosen.

The case $m > 1$ is analogue and follows with Minkowski inequality.

Exercise 32 Let $a, b \ge 1$:

1. Prove the relation $(a+b)^m \le a^m + b^m$, if $0 \le m \le 1$.
2. Prove the relation $(a+b)^m \le 2^{m-1}(a^m + b^m)$, if $m \ge 1$.

Hints.

1. $m \le 1$. Divide both members by a^m if $a \ne 0$ and set $t = b/a$. Then we need to prove that $g(t) = (1+t)^m - t^m - 1 \le 0$ for $m \le 1$ and $t = b/a$. Here $g(0) = 0$. One remarks that $g'(t) = m((1+t)^{m-1} - t^{m-1}) < 0$ thus $g(t) < 0$ for $t > 0$.
2. $m \ge 1$. The function $h(x) = x^m$ is convex in case $m \ge 1$, indeed it is easy to check that $h''(x) = m(m-1)x^{m-2} \ge 0$. The inequality now follows with the convexity inequality with equal weights

$$h\left(\frac{a+b}{2}\right) \le \frac{1}{2}\Big(h(a) + h(b)\Big).$$

The sequence (ξ_n) considered is zero-mean in case $m \ge 1$ and we assume that this is an independent sequence in order to derive strict stationarity assumptions.

Definition 6.1.2 If $c_k = 0$ for $k < 0$ then the stationary process (6.1) is said to be *causal*.

Assume here that (ξ_n) is a $\mathbb{L}^2$-white noise. This process admits the covariance:

$$r_k = \mathrm{Cov}\,(X_0, X_k) = \sum_{l=-\infty}^{\infty} c_l c_{l+k} = c \star \widetilde{c}_k, \tag{6.2}$$

denoting $\widetilde{c} = (\widetilde{c}_k)_{k\in\mathbb{Z}}$ with $\widetilde{c}_k = c_{-k}$.

Remark that by completing infinite series simply yields:

$$\sum_{k=-\infty}^{\infty} |r_k| \le \left(\sum_{k=-\infty}^{\infty} |c_k|\right)^2,$$

this series converges in case

$$\sum_{k=-\infty}^{\infty} |c_k| < \infty.$$

We thus obtain:

Proposition 6.1.1 *Let (X_t) be a linear process defined from (6.1) (with iid inputs ξ_n) then the above series converge a.s., this process is stationary and in $\mathbb{L}^m$ in case, either*

$$\mathbb{E}|\xi_0|^m < \infty, \qquad \sum_{k=-\infty}^{\infty} |c_k|^m < \infty, \qquad 0 < m \le 1,$$

or it is causal and,

$$\mathbb{E}|\xi_0|^2 < \infty, \qquad \sum_{k=0}^{\infty} |c_k|^2 < \infty, \qquad m = 2.$$

In the latter case the covariance of the process can be written as in (6.2). The series of covariances converges if

$$\sum_{k=0}^{\infty} |c_k| < \infty.$$

Definition 6.1.3 The backward or shift operator B is defined for sequences $x = (x_n)_{n\in\mathbb{Z}}$ by the relation:

$$x = (x_n)_{n\in\mathbb{Z}} \mapsto Bx, \qquad (Bx)_n = x_{n-1}, \quad n \in \mathbb{Z}.$$

Remark 6.1.1 The convention is to write $Bx = (Bx_n)_{n\in\mathbb{Z}}$, or equivalently $Bx_n = x_{n-1}$, e.g. for any discrete time stochastic process we set:

$$BX_t = X_{t-1}, \qquad t \in \mathbb{Z}.$$

In the econometric literature this operator is also denoted by L, the lag-operator.

Using the backward operator B the previous causal models also can be written

$$X = g(B)\xi, \qquad \text{with} \qquad g(z) = \sum_{k=0}^{\infty} c_k z^k, \quad \text{in case} \quad |z| < 1.$$

We now briefly describe some very simple models of constant use in statistics.

Clearly this chapter has no statistical ambition but we shall simply rephrase some currently used models.

Remark 6.1.2 (Centring) In case $(X_t)_{t\in\mathbb{Z}}$ is not a centred process, given a sample $X_1, \ldots, X_n$, the parameter $m = \mathbb{E}X_0$ may be estimated empirically by

$$\widehat{m} = \frac{1}{n}\sum_{k=1}^{n} X_k,$$

the estimation is consistent from the ergodic theorem (Corollary 9.1.3) in case the process $(X_t)_{t\in\mathbb{Z}}$ is ergodic.

Remark 6.1.3 (Local means) Assume now that the process is observed on the period $\{1, \ldots, n\}$ and there exists a continuous function and a centred stationary linear process such that

$$X_t = m\left(\frac{t}{n}\right) + Y_t, \qquad t = 1, 2, \ldots, n.$$

In this case a local mean may be used; the function m is fitted by

$$\widehat{m}(x) = \frac{1}{2k_n + 1}\sum_{k=-k_n}^{k_n} X_{[nx]+k},$$

and for k_n such that $\lim_{n\to\infty}(k_n/n) = 0$ and $\lim_{n\to\infty} k_n = \infty$ this estimation is consistent.

Smoothing techniques, analogously to (3.4), may also be used; regular and more accurate estimators of the function m may thus be deduced by using kernel functions of higher order.

6.2 ARMA(p, q)-Processes

Auto-regressive moving average processes (ARMA) are stationary solutions of the equation

$$X_t - \sum_{j=1}^{p} a_j X_{t-j} = \xi_t - \sum_{k=1}^{q} b_k \xi_{t-k}. \tag{6.3}$$

The above equation is formally written

$$\alpha(B)X_t = \beta(B)\xi_t$$

for polynomials defined as:

$$\alpha(z) = 1 - \sum_{j=1}^{p} a_j z^j, \quad \beta(z) = 1 - \sum_{j=1}^{q} b_j z^j. \tag{6.4}$$

Proposition 6.2.1 (ARMA-Processes) *The recursion (6.3) admits a stationary solution in $\mathbb{L}^p$, in case the inputs satisfy $\xi_j \in \mathbb{L}^p$ for some $p > 0$ and the roots $r_1, \ldots, r_p$ of the polynomial α are such that*

$$|r_1| > 1, \ldots, |r_p| > 1.$$

If moreover $p \geq 2$ then the covariance of this stationary process satisfies:

$$\forall k \in \mathbb{Z} : \quad |r_k| \leq c\rho^{|k|}, \quad \text{for } 0 \leq \rho < 1, c > 0.$$

Remark 6.2.1 The Exercise 23 shows that the condition

$$|a_1| + \cdots + |a_p| < 1$$

implies that the roots of the polynomial α are outside the unit disk.

Trajectories of these ARMA models are reported in Fig. 6.1; here both coefficients equal 0.2 and inputs are standard Gaussian (Fig. 6.2).

ARMA(1,1)-Processes admit quite erratic trajectories as may be seen in the first graphic of the Fig. 6.1. The second graphic proves that they also admit covariances with extremely fast decay rates.

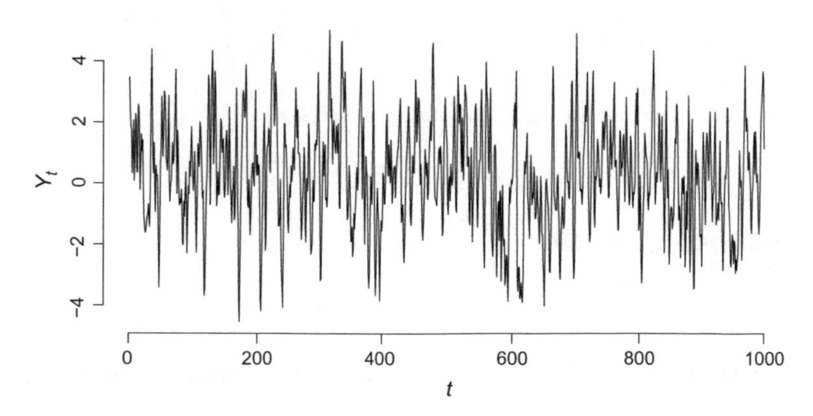

Fig. 6.1 Simulated trajectory of an ARMA (1,1). Here, $X_t = 0.6X_{t-1} + \varepsilon_t + 0.7\varepsilon_{t-1}$ with, $\varepsilon_t \sim \mathcal{N}(0, 1)$

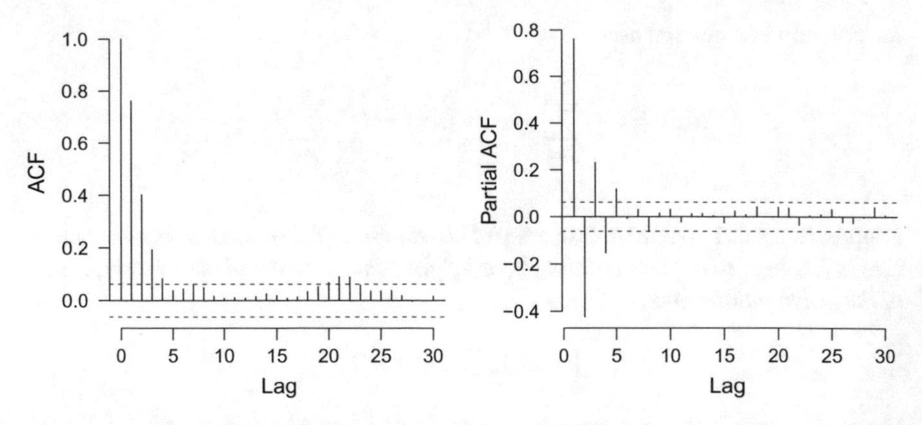

Fig. 6.2 Sample simple and partial correlograms of the series of Fig. 6.1

Sketch of the proof. A solution of (6.3) is written:

$$X_t = \sum_{j=0}^{\infty} c_j \xi_{t-j}$$

where the c_j's are defined from

$$\sum_{j=0}^{\infty} c_j z^j = \frac{\beta(z)}{\alpha(z)},$$

with:

$$\alpha(z) = 1 - a_1 z - \cdots - a_p z^p = \left(1 - \frac{z}{r_1}\right) \cdots \left(1 - \frac{z}{r_p}\right).$$

If the roots $r_1, \ldots, r_p$ of the polynomial α are such that

$$|r_1| > 1, \ldots, |r_p| > 1$$

then the function $1/\alpha$ is analytic if

$$|z| < \min\{|r_1|, \ldots, |r_p|\}$$

and thus on a neighbourhood of the closed complex unit disk. For example

$$\left(1 - \frac{z}{r_1}\right)^{-1} = \sum_{l=0}^{\infty} \frac{z}{r_1^l}.$$

Moreover the analyticity of the function β/α on some disk $D(0, 1+\epsilon)$ implies $|c_k| \le Ce^{-\gamma k}$.

Remark 6.2.2 We use the classical decomposition of rational functions. Let κ be the maximal order of roots r with $r = \min_j |r_j|$ then one may precisely prove that $|c_k| \le Ck^{\kappa-1}r^{-k}$, for some constant $C > 0$ if $k \ne 0$.

6.3 Yule–Walker Equations

This section provides a brief approach to Yule–Walker equations yielding parametric estimation for ARMA models, we refer the reader to the textbook (Brockwell and Davis 1991). Those equations are based on causality:

for each $t \in \mathbb{Z}$ and $p \ge 0$, the innovation is independent of $(X_{t-1}, \ldots, X_{t-p})$.

This condition means that the history of X_t is meaningful and that $X_t = g(X_{t-1}, \ldots, X_{t-p}, \xi_t)$ can be written explicitly as a function of the (finite) past and of some innovation; this is a natural condition for processes indexed by time.

For simplicity we restrict to AR(p) models where $(\xi_t)_{t\in\mathbb{Z}}$ denotes an iid sequence centred and with $\sigma^2 = \mathbb{E}\xi_0^2$, as before

$$X_t = a_1X_{t-1} + \cdots + a_pX_{t-p} + \xi_t. \tag{6.5}$$

We again assume that

$$\alpha(z) = 1 - a_1z - \cdots - a_pz^p = \prod_{j=1}^{p}\left(1 - \frac{z}{r_j}\right),$$

admits roots such that $|r_j| > 1$ for $j = 1, \ldots, p$. Then we just proved that a MA(∞)-expansion indeed holds:

$$X_t = \sum_{j=0}^{\infty} c_j\xi_{t-j}.$$

Parameters of interest in this model are $\theta = (a, \sigma^2)$ with $a^t = (a_1, \ldots, a_p)$. In case the inputs are iid Gaussian $\mathcal{N}(0, \sigma^2)$ these are the only parameters.

We aim at estimating these parameters.

Multiply Eq. (6.5) by X_{t-j} for $0 \le j \le p$ then taking expectations entails

$$R_pa = \mathbf{r}_p, \quad \sigma^2 = r_0 - a'\mathbf{r}_p,$$

with

$$R_p = (r_{i-j})_{1\le i,j\le p}, \quad \mathbf{r}_p = (r_0, \ldots, r_p)'.$$

Plugging-in estimators $\widehat{r}_j$ of covariances r_j as in (9.2) provides us with empirical estimators of the parameters.

It is easy to define $\widehat{R}_p = (\widehat{r}_{i-j})_{1\le i,j\le p}$ and $\widehat{\mathbf{r}}_p$ and thus

$$\widehat{R}_p\widehat{a} = \widehat{\mathbf{r}}_p, \quad \widehat{\sigma}^2 = \widehat{r}_0 - \widehat{a}'\widehat{\mathbf{r}}_p.$$

Remark 6.3.1 (ARMA-Models) The above equations extend to ARMA-Models, see Brockwell and Davis (1991), Chap. 8.

Remark 6.3.2 (Non-Linear models) Extensions to the case of weak-white noise are used; for example non-linear models such as ARCH-models are such white noises and a linear process with such input may also be considered. In the following chapter we describe some elementary versions of this idea.

Remark 6.3.3 (Durbin–Levinson algorithm) From such estimation a plug-in one-step-ahead prediction of the process can be written:

$$\widehat{X}_t = \widehat{a}_1 X_{t-1} + \cdots + \widehat{a}_p X_{t-p},$$

once the parameters have been estimated from the data $X_0, \ldots, X_{t-1}$.

Two-steps ahead predictions are similar by replacing now X_t by $\widehat{X}_t$ in the previous relation and:

$$\widehat{X}_{t+1} = \widehat{a}_1\widehat{X}_t + \widehat{a}_2 X_{t-1} + \cdots + \widehat{a}_p X_{t-p+1}.$$

Now we may replace the covariances by their empirical counterparts, see Brockwell and Davis (1991) and Sect. 8.2.

6.4 ARFIMA(0, *d*, 0)-Processes

Set $\Delta = I - B$ with B the backward operator. The operator Δ allows us to rewrite the previous models but it also helps to define some new models.

We aim at solving the formal equation

$$\Delta^d X_t = \xi_t.$$

- In case $d = 1$ the equation is $X_t - X_{t-1} = \xi_t$ thus

$$X_t = X_0 + \xi_1 + \cdots + \xi_t, \qquad \forall t \ge 1,$$

  which is a random walk if $X_0 = 0$.
- If $d = 2$ the relation is

$$\Delta^2 X_t = \Delta(\Delta X_t) = \xi_t,$$

which leads to a recursive definition with initial condition 0 for the solution of the equation

$$\Delta^d X_t = \xi_t, \quad \text{for} \quad d \in \mathbb{N}.$$

- If $d \in -\mathbb{N}$ the relation is

$$X_t = \Delta^{-d}\xi_t = \sum_{j=0}^{-d} \binom{-d}{j} \xi_{t-j}.$$

- More generally, for $d > -1$, we do not necessarily assume that $d \in \mathbb{N}$, the relation $X_t = (I - B)^{-d}\xi_t$ is interpreted as an expansion for $|z| < 1$ of the function $g(z) = (1 - z)^{-d} = \sum_{j=0}^{\infty} b_j z^j$.

Exercise 33 Prove that if $d > -1$, the coefficients of g's expansion are:

$$b_j = \frac{\Gamma(j + d)}{\Gamma(j + 1)\Gamma(d)} = \frac{1}{\Gamma(d)} \prod_{k=1}^{j} \frac{k - 1 + d}{k}. \tag{6.6}$$

Hint. The analyticity of g over the disk $D(0, 1)$ follows from the representation

$$g(z) = \exp(-d \ln(1 - z)).$$

Now $g(0) = 1 = b_0$ and $(1 - z)g'(z) = dg(z)$ for $|z| < 1$, thus

$$\sum_{j>0} jb_j(1 - z)z^{j-1} = d + d \sum_{j>0} b_j z^j.$$

This relation can be rewritten as

$$d + \sum_{j>0} (d + j)b_j z^j = \sum_{j>0} jb_j z^{j-1} = b_1 + \sum_{k>0} (k + 1)b_{k+1} z^k.$$

The last identity follows with $j = k + 1$.

Thus analytic continuation theorem entails $b_1 = d$, which also may be derived from the relation $g'(0) = d$, and

$$b_{j+1} = \frac{d + j}{1 + j} b_j, \quad \text{thus} \quad b_k = \frac{(d + k - 1) \cdots (d + 1)d}{k(k - 1) \cdots 2 \cdot 1}.$$

The conclusion follows.

E.g. Feller (1968) proves the useful standard Stirling formula:

$$n! \sim \sqrt{2\pi n} \left(\frac{n}{e}\right)^n, \qquad n \to \infty.$$

This implies

$$b_j \sim \frac{1}{\Gamma(d)} j^{d-1}, \quad \text{as} \quad j \to \infty.$$

For $-\frac{1}{2} < d < \frac{1}{2}$, this define the operators $\Delta^{\pm d}$.

To define Δ^d out of this range, use relations $\Delta^{d+1} = \Delta\Delta^d$ and

$$\Delta^{d-1} X_t = \xi_t \Rightarrow \Delta^d X_t = \Delta\xi_t = \xi_t - \xi_{t-1}.$$

The evolution of trajectories of ARFIMA(0, d, 0) is reported in Fig. 6.3 (we use the R package dvfBM, see Coeurjolly 2009).

Clearly the smallest values of $d = 0.01$ yields a white noise behaviour and the trajectories look more and more regular as $d < 0.5$ becomes larger.

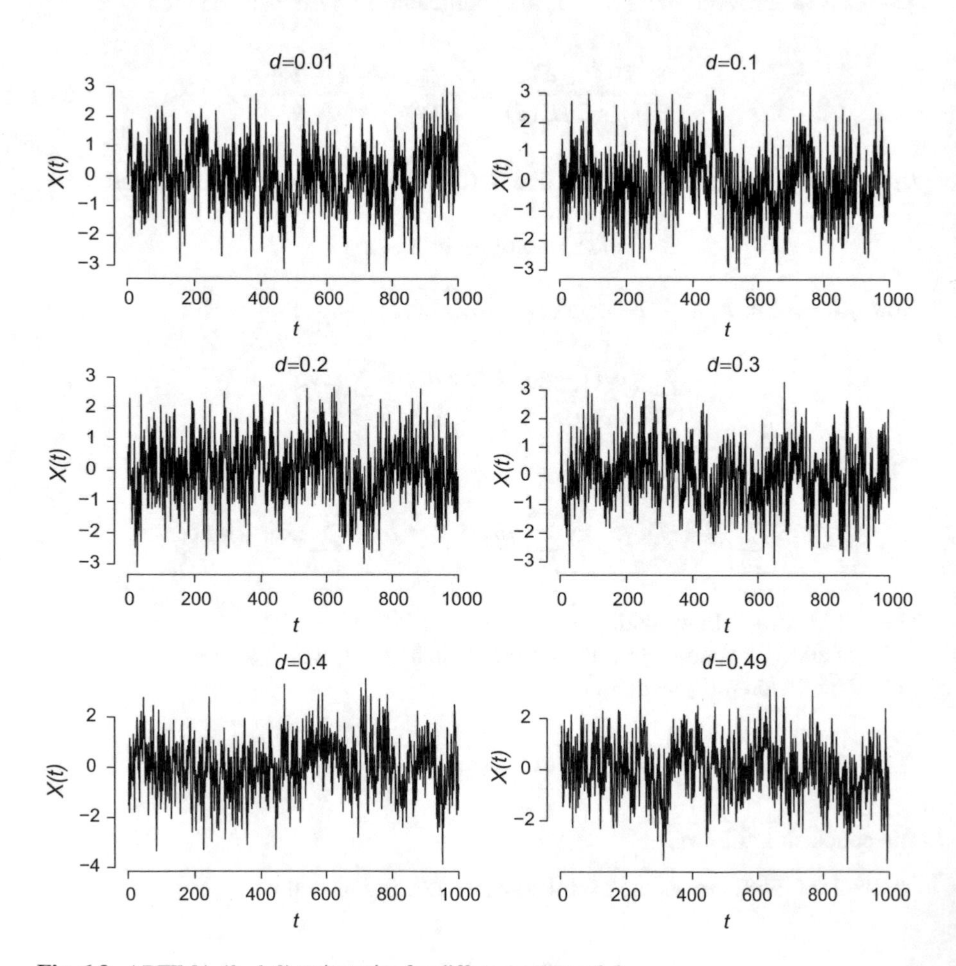

Fig. 6.3 ARFIMA (0, d, 0) trajectories for different values of d

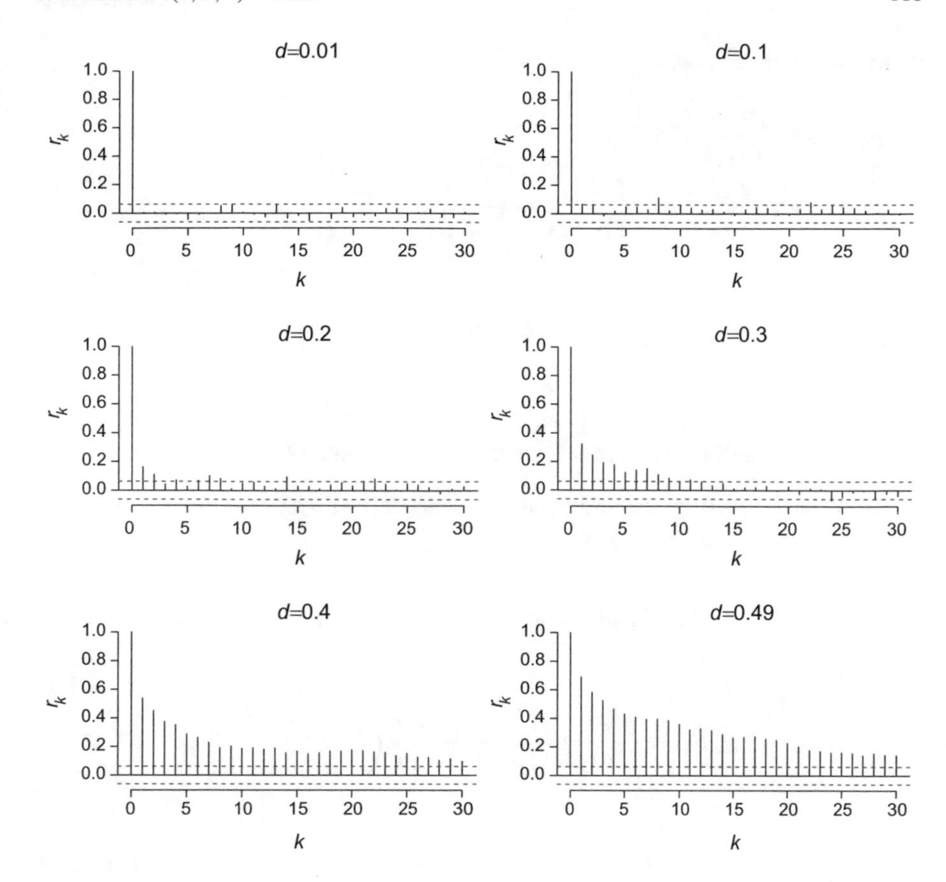

Fig. 6.4 Sample correlograms of ARFIMA trajectories for different values of d. See Fig. 6.3

It is clear from Fig. 6.4 that the corresponding covariogram (listing covariance estimators for such models) confirms the impression provided by trajectories of such ARFIMA models. The covariances are those of white noise again for $d = 0.01$ and then seem more and more cyclical for larger values of d.

Proposition 6.4.1 *Assume that $d < \frac{1}{2}$, and ξ_j is a centred iid sequence in $\mathbb{L}^2$, then* ARFIMA(0, d, 0) *are linear causal processes in $\mathbb{L}^2$.*

The coefficients in (6.6) satisfy:

$$\sum_{j=0}^{\infty} b_j^2 < \infty,$$

and the series

$$X_t = \sum_{j=0}^{\infty} b_j \xi_{t-j} \text{ converge in } \mathbb{L}^2.$$

Moreover, with $\sigma^2 = \mathbb{E}\xi_0^2$*:*

$$r(0) = \sigma^2 \frac{\Gamma(1-2d)}{\Gamma^2(1-d)}$$
$$r(k) = r(0)\frac{\Gamma(k+d)\Gamma(1-d)}{\Gamma(k-d+1)\Gamma(d)} \sim \sigma^2 \frac{\Gamma(1-2d)}{\Gamma(d)\Gamma(1-d)} |k|^{2d-1}, \quad \textit{as } |k| \to \infty.$$

Hence

$$\sum_k |r(k)| = \infty \iff d \in \left(0, \frac{1}{2}\right).$$

Remark 6.4.1 The Hurst coefficient $H = d + \frac{1}{2}$ is used as a parameter for these models. It was introduced to model river flooding in Hurst (1951).

Let Z be the random spectral measure associated with an iid white noise ξ_t such that $\mathbb{E}\xi_0 = 0$, and $\mathbb{E}\xi_0^2 = \sigma^2$. Then

$$X_t = \int_{-\pi}^{\pi} e^{it\lambda}(1 - e^{-i\lambda})^{-d} Z(d\lambda)$$

and

$$g_X(\lambda) = \frac{\sigma^2}{2\pi} \frac{1}{\left|1 - e^{-i\lambda}\right|^{2d}} = \frac{\sigma^2}{2\pi}\left(4\sin^2\frac{\lambda}{2}\right)^{-2d}.$$

Remark 6.4.2 (Simulation) Such integral representations are used to simulate these models. For the case of Gaussian inputs the previous spectral process is Gaussian with independent increments which makes the previous simulation trick possible by providing independent random variables with a given distribution, see Remark A.2.8.

This idea extends to each process with independent increments such as the Poisson unit process.

The other possibility to simulate such time series is to truncate the corresponding series. However the simulations may be inefficient in this case, and alternatives may be preferred, see Doukhan et al. (2002b).

6.5 ARFIMA(p, d, q)-Processes

The models ARFIMA(p, d, q) fit the equation

$$\alpha(B)(I - B)^d X_t = \beta(B)\xi_t.$$

α, β are again polynomials with respective degree p, q and with constant coefficient equal to 1, as in (6.4).

If $d < \frac{1}{2}$ the process is causal and well defined in case the roots of α are not inside the unit disk. It is invertible if $d > -\frac{1}{2}$ and the roots of α are out of the unit disk.

Indeed in this case $\xi_t = \gamma(B)X_t$ admits an MA(∞) representation, for a function γ analytic on the unit disk

$$D(0,1) = \left\{z \in \mathbb{C}/\ |z| < 1\right\}.$$

Let again Z denote the random spectral measure associated with the white noise (ξ_t) then

$$X_t = \int_{-\pi}^{\pi} e^{it\lambda}\left(1 - e^{-i\lambda}\right)^{-d} \frac{\beta(e^{i\lambda})}{\alpha(e^{i\lambda})} Z(d\lambda).$$

Hence

$$g_X(\lambda) = \frac{\sigma^2}{2\pi}\left|1 - e^{-i\lambda}\right|^{-2d}\left|\frac{\beta(e^{i\lambda})}{\alpha(e^{i\lambda})}\right|^2.$$

6.6 Extensions

For any meromorphic function $\gamma : \mathbb{C} \to \mathbb{C}$ without singularities on $\mathcal{D}(0,1)$ with finitely many singularities on the unit circle, we define analogously to Sect. 6.5, a process

$$X_t = \gamma(B)\xi_t.$$

In case $1/\gamma$ satisfies the same assumptions, then the process is reversible (zeros replacing singularities).

The singularities different from 1 on the unit circle are called periodic long-range singularities.

Now let $(c_{k,\ell})_{k,\ell \in \mathbb{Z}}$ be a sequence of real numbers, analogously to (6.1) we may define non stationary linear processes from the relation

$$X_k = \sum_{\ell=-\infty}^{\infty} c_{k,\ell}\xi_{k-\ell}.$$

The existence of such models is proposed as an exercise.

Example 6.6.1 (Non-stationarity) Cases of interest correspond to trends, local stationarity, and periods:

- $c_{k,\ell} = c_\ell + C_k \mathbb{I}_{\{\ell=k\}}$ correspond to trends.
- $c_{k,\ell} = c_\ell\left(\frac{k}{n}\right)$ for a family $c_\ell(\cdot)$ of smooth functions. This class of locally stationary models observed over epochs $1, \ldots, n$ was introduced in Dahlhaus (2012). There exist in fact two ways to define this notion and the other spectral way is sketched in Remark 4.2.1. We prefer the present state domain presentation for possible non-linear extensions which usually are less related with the spectrum.

- $c_{k,\ell} = c_\ell\left(e^{2i\pi\frac{k}{T}}\right)$ corresponds to T-periodic random processes.

All such specific behaviours may be combined in order to provide non-stationary behaviours more adequate to data sets.

Besides periods on also may think of exogenous data interfering with the phenomenon of interest. The point is that such data should also admit some dependence structure.

A simple example is the temperature at some place, indeed hourly and seasonal periodicity appear together with the global warming phenomenon. An example of exogenous data is nebulosity. Analogue modelling considerations may be drawn for electricity consumption and for only retail data.

Chapter 7
Non-linear Processes

This chapter aims at describing stationary sequences generated from independent identically distributed samples $(\xi_n)_{n\in\mathbb{Z}}$. Most of the material in this chapter is specific to this monograph so that we do not provide a global reference. However Rosenblatt (1985) performs an excellent approach to modelling. Generalized linear models are presented in Kedem and Fokianos (2002). The Markov case has drawn much attention, see Duflo (1996), and for example Douc et al. (2015) for the estimation of such Markov models. Many statistical models will be proved in this way. The organization follows the order from natural extensions of linearity to more general settings. From linear processes it is natural to build polynomial models or their limits. Then we consider more general Bernoulli shift models to define recurrence equations besides the standard Markov setting.

7.1 Discrete Chaos

This section introduces some basic tools for algebraic extensions of linear to polynomial models.

7.1.1 Volterra Expansions

Definition 7.1.1 Set $X_n^{(0)} = c^{(0)}$ for some constant and consider arrays $(c_j^{(k)})_{j\in\mathbb{Z}^k}$ of constants and a sequence of arrays of independent identically distributed random variables

$$\left(\left(\xi_n^{(k,j)}\right)_{1\le j\le k}\right)_{n\in\mathbb{Z}}.$$

P. Doukhan, *Stochastic Models for Time Series*, Mathématiques et Applications 80,
https://doi.org/10.1007/978-3-319-76938-7_7

If the following series converge in $\mathbb{L}^p$, for some $p \geq 1$, set:

$$X_n^{(k)} = \sum_{j_1 < j_2 < \cdots < j_k} c_{j_1,\ldots,j_k}^{(k)} \xi_{n-j_1}^{(k,1)} \cdots \xi_{n-j_k}^{(k,k)}.$$

Define now a Volterra process as a process such that the following Volterra expansion holds:

$$X_n = \sum_{k=0}^{\infty} X_n^{(k)}.$$

Remark 7.1.1 According to Chap. 5 devoted to the Gaussian case, such stationary models can also be written in the chaotic form generated from

$$\left((\xi_n^{(k,j)})_{1 \leq j \leq k} \right)_{n \in \mathbb{Z}}.$$

We better use the more standard expression of Volterra expansions, below.

Example 7.1.1 To understand why the previous definition involves arrays $((\xi_n^{(k,j)})_{1 \leq j \leq k})_{n \in \mathbb{Z}}$ of independent identically distributed random variables, it seems better to consider the simplest example of second degree polynomials

$$X_n = \sum_{i=-\infty}^{\infty} \sum_{j=-\infty}^{\infty} a_{i,j} \xi_i \xi_j.$$

The previous expansion holds if we set

$$X_n^{(2)} = \sum_{i<j} (a_{i,j} + a_{j,i}) \xi_{n-i} \xi_{n-j}$$

$$X_n^{(1)} = \sum_i a_{i,i} (\xi_{n-i}^2 - \sigma^2)$$

$$X_n^{(0)} = \left(\sum_i a_{i,i} \right) \sigma^2.$$

with $\sigma^2 = \mathbb{E}\xi_0^2$.

Consider now Volterra models with higher order Appell polynomials $A_s(\xi_n)$, see Sect. 7.1.2. Remark that $\xi_n^2 - \sigma^2$ takes into account the repetitions in the diagonal terms.

Exercise 34 For $\mathbb{L}^2$-Volterra processes suppose, without loss of generality, that $\mathbb{E} \left| \xi_n^{(k,j)} \right|^2 = 1$.

Then

$$\begin{aligned}
\mathbb{E}X_0^{(k)}X_n^{(l)} &= 0, \qquad \text{if} \quad k \neq l \\
\mathbb{E}X_0^{(k)}X_0^{(k)} &= \sum_{j_1<j_2<\cdots<j_k} \left|c_{j_1,\ldots,j_k}^{(k)}\right|^2 \\
\mathbb{E}X_0^{(k)}X_n^{(k)} &= \sum_{j_1<j_2<\cdots<j_k} c_{j_1,\ldots,j_k}^{(k)} c_{n+j_1,\ldots,n+j_k}^{(k)}.
\end{aligned}$$

Those calculations yield explicit expressions for the covariance of the process $(X_n)_{n\in\mathbb{Z}}$ from a simple summation in case

$$\sum_{j_1<j_2<\cdots<j_k} \left|c_{j_1,\ldots,j_k}^{(k)}\right|^2 < \infty.$$

Remark 7.1.2 (*Local stationarity*) The notion sketched in Remark 4.2.1 for the spectral approach and in Sect. 6.6 for linear processes still fits the present framework with now

$$c_{j_1,\ldots,j_k}^{(k,n)} = c_{j_1,\ldots,j_k}^{(k)}\left(\frac{k}{n}\right).$$

7.1.2 Appell Polynomials

Analogously to the special case of the Gaussian laws, which yields the construction of Hermite chaos, one may define orthogonal polynomials associated with a fixed distribution on the real line $\mathbb{R}$. Let ξ_0 be a real valued random variable.

Definition 7.1.2 Let $m \in \mathbb{N}^*$, we assume that $\mathbb{E}|\xi_0|^m < \infty$.

The Appell polynomials $A_0, \ldots, A_m$ associated with the distribution of ξ_0, are defined recursively by $A_0(x) = 1$ and

$$A_k'(x) = kA_{k-1}(x), \qquad \sum_{j=0}^{k} \mathbb{E}\xi_0^j \cdot A_j(0) = 0, \qquad 1 \le k \le m.$$

Hence

$$\begin{aligned}
A_0(x) &= 1 \\
A_1(x) &= x - \mathbb{E}\xi_0 \\
A_2(x) &= x^2 - 2\mathbb{E}\xi_0 x + 2(E\xi_0)^2 - \mathbb{E}\xi_0^2 \\
\cdots &\quad \cdots \\
\cdots &\quad \cdots \\
A_k(x) &= x^k + \cdots
\end{aligned}$$

If the Laplace transform of ξ_0's distribution is analytic around 0, this entails

$$\sum_{k=0}^{\infty} \frac{z^k}{k!} A_k(x) \mathbb{E} e^{z\xi_0} = e^{zx}.$$

Let P be a polynomial with degree $d = d^\circ P$:

$$P(x) = \sum_{k=0}^{d} \frac{c_k}{k!} A_k(x).$$

Reasoning on the degree allows us to derive uniqueness in the above representation.

Assume that the cumulative distribution function F of ξ_0's distribution (defined for $x \in \mathbb{R}$ by $F(x) = \mathbb{P}(\xi_0 \le x)$) is continuously differentiable, then we denote by $f = F'$ the density of this law. Then under higher order differentiability assumptions,

$$c_k = \mathbb{E} P^{(k)}(\xi_0) = (-1)^k \int_{-\infty}^{\infty} P(x) f^{(k)}(x) dx.$$

An important property of those Appell polynomials is

$$\mathbb{E} A_k(\xi_0) P(\xi_0) = 0, \quad \text{if} \quad d^\circ P < k.$$

Set $g(x) = f(x)P(x)$, then

$$\mathbb{E} A_k(\xi_0) P(\xi_0) = \int_{-\infty}^{\infty} A_k(x) g(x) dx.$$

Since the function g admits k derivatives then k integrations by parts prove this identity. Set $g_l(x) = f^{(l)}/f$ then, analogously to the proof for the Gaussian chaos:

$$\mathbb{E} A_k(\xi_0) g_l(\xi_0) = \begin{cases} 1, & \text{if } k = l, \\ 0, & \text{if } k \ne l. \end{cases}$$

Remark 7.1.3 Extensions to more general functions are much more complicated than the previous Gaussian theory! To consider non-polynomial functions, Kazmin (1969) assumes that the function

$$x \mapsto A(z) = \frac{1}{\mathbb{E} e^{z\xi}}$$

is analytic and it does not vanish on the open disk

$$D(0,\sigma)=\left\{z\in\mathbb{C}/\,|z|<\sigma\right\}.$$

Then each function g, analytic on a disk $D(0,\tau)$, admits a representation

$$g(z)=\sum_{n=0}^{\infty}\frac{c_n}{n!}A_n(z),\quad \limsup_{n\to\infty}|c_n|^{\frac{1}{n}}<\tau,$$

for series which converge uniformly over compact subsets of the disk $D(0,\tau)$. Conversely for a sequence such that

$$\limsup_{n\to\infty}|c_n|^{\frac{1}{n}}<\tau,$$

the function g defined this way is proved to be analytic on $D(0,\tau)$.

Under those assumption the series defining g is convergent and

$$c_n=\mathbb{E}g^{(n)}(\xi),$$

this proves uniqueness of the expansion of analytic functions. Those results are far from representing all the $\mathbb{L}^2$-functions as in the Gaussian case.

To justify the representation of Volterra processes in Definition 7.1.1, the notion of Appell polynomials needs multivariate extension.

Multivariate Appell Polynomials

If now $\xi=(\xi_1,\dots,\xi_k)\in\mathbb{R}^k$ is a vector valued random variable it is easy to define analogously $A_{n_1,\dots,n_k}(x_1,\dots,x_k)$ through relations

$$\frac{\partial}{\partial x_i}A_{n_1,\dots,n_k}(x_1,\dots,x_k)=n_iA_{n_1,\dots,n_k}(x_1,\dots,x_k),\qquad 1\le i\le k$$

and

$$\mathbb{E}A_{n_1,\dots,n_k}(\xi)=\begin{cases}1, & \text{if } n_1+\cdots+n_k=0,\\ 0, & \text{otherwise.}\end{cases}$$

If the random variables $\xi_1,\dots,\xi_k$ are independent and admit respective distributions $\nu_1,\dots,\nu_k$, then

$$A_{n_1,\dots,n_k}(x_1,\dots,x_k)=A_{n_1}^{(\nu_1)}(x_1)\cdots A_{n_k}^{(\nu_k)}(x_k).$$

These multivariate polynomials allow the orthogonality property in Definition 7.1.1. General polynomial chaotic expansions can be written as orthogonal Volterra series.

Unfortunately Remark 7.1.3 does not allow such chaotic representations of stationary processes.

7.2 Memory Models

This section considers some *few* models with explicit Volterra expansions.

As usual these memory models will be excited by iid innovations with values in the measurable space $(\mathbb{R}, \mathcal{B}(\mathbb{R}))$.

They are solutions of some recursion:

$$X_t = M(X_{t-1}, X_{t-2}, \ldots, \xi_t).$$

for iid inputs an some explicit function $\mathbb{R}^{\mathbb{N}} \times \mathbb{R} \to \mathbb{R}$. More general spaces may also be used, both for the innovations and for the states, but this section is restricted to real values for simplicity. Multivariate extensions are immediate and left to the reader.

Remark 7.2.1 Here again, locally stationary models (see Remark 4.2.1 and Sect. 6.6, and Example 6.6.1) may be defined in the space domain by replacing M by a parametric family $u \mapsto M_u$, now

$$X_t = M_{\frac{t}{n}}(X_{t-1}, X_{t-2}, \ldots, \xi_t), \qquad 1 \le t \le n.$$

Exercise 35 (*tvAR(1)*) Dahlhaus (2012) defines, among others, time varying AR(1)-models from a recursion for large sample sizes n. Suppose that $X_{0,n} = x$ and

$$X_{t,n} = a\left(\frac{t}{n}\right) X_{t-1} + \xi_t, \qquad 1 \le t \le n,$$

then

1. $X_{t,n} = x + \sum_{k=0}^{t-1} \xi_{t-k} \prod_{j=0}^{k-1} a\left(\frac{t-j}{n}\right)$, for $1 \le t \le n$.
2. Assume that $a : [0,1] \to \mathbb{R}$ is a C^1-function with a bounded derivative such that $\alpha = \sup_u |a(u)| < 1$.
   Fix $u \in [0,1]$ and set $X^{(u)}$ as the stationary solution of the equation:

$$X_t^{(u)} = a(u) X_{t-1}^{(u)} + \xi_t.$$

   We now suppose that $X_{0,n} = X_0^{(u)}$ then derive that for each $1 \le t \le n$:

$$|X_{t,n} - X_t^{(u)}| \le \sum_{k=0}^{\infty} k\alpha^{k-1} \|a'\|_\infty \left(|\Delta| + \frac{1}{n}\right).$$

3. Deduce that for some $o(\frac{1}{n})$ in probability and in $\mathbb{L}^p$, as $n \to \infty$ and for each fixed t

$$X_{t,n} = X_t^{(u)} + \left(\frac{t}{n} - u\right) \frac{d}{du} X_t^{(u)} + o\left(\frac{1}{n}\right).$$

Hint (see Dahlhaus 2012, *Sect. 3).*

1. As in the proof of Proposition (7.2.1) a simple recursion leads to:

$$\begin{aligned}
X_{t,n} &= \xi_t + a\left(\frac{t}{n}\right) X_{t-1,n} \\
&= \xi_t + a\left(\frac{t}{n}\right) \xi_{t-1} + a\left(\frac{t}{n}\right) a\left(\frac{t-1}{n}\right) X_{t-2,n} \\
&= \xi_t + a\left(\frac{t}{n}\right) \xi_{t-1} + a\left(\frac{t}{n}\right) a\left(\frac{t-1}{n}\right) \xi_{t-2} \\
&\quad + a\left(\frac{t}{n}\right) a\left(\frac{t-1}{n}\right) a\left(\frac{t-2}{n}\right) X_{t-3,n} \\
&= \dots
\end{aligned}$$

   Due to the condition $X_{0,n} = x$, only t iterations are possible in the above display.
2. Use the fact that

$$X_t^{(u)} = \sum_{k=0}^{\infty} \xi_{t-k} a^k(u),$$

   then with $\Delta = \dfrac{t}{n} - u$:

$$\begin{aligned}
&\left| \prod_{j=0}^{k-1} a\left(\frac{t-j}{n}\right) - a^k(u) \right| \\
&\qquad \le \alpha^{k-1} \sum_{j=0}^{k-1} \left| a\left(\frac{t-j}{n}\right) - a(u) \right| \\
&\qquad \le \alpha^{k-1} \|a'\|_\infty \left(k|\Delta| + \sum_{j=0}^{k-1} \frac{j}{n} \right) \le k\alpha^{k-1} \|a'\|_\infty \left(|\Delta| + \frac{1}{n} \right).
\end{aligned}$$

   Summing up yields the requested bound.
3. A Taylor expansion in the above point yields the result.

Many developments of those non-stationarity are processed and an example exhibiting together periodic behaviour is Bardet and Doukhan (2017).

7.2.1 Bilinear Models

We first consider a very simple bilinear model.

Proposition 7.2.1 *Consider the Markov bilinear model*

$$X_n = (a + b\xi_n)X_{n-1} + \xi_n, \tag{7.1}$$

Assume that for some $p \geq 1$,

$$\alpha^p = \mathbb{E}|a + b\xi_0|^p < 1.$$

Then there exists stationary solution of this Markov recursion, this solution is in $\mathbb{L}^p$; it can be written:

$$X_n = \sum_{k=0}^{\infty} \xi_{n-k} \prod_{j=0}^{k-1} (a + b\xi_{n-j}).$$

Proof It is simple to check that the previous series is normally convergent in $\mathbb{L}^p$ since independence entails

$$\left\| \xi_{n-k} \prod_{j=0}^{k-1} (a + b\xi_{n-j}) \right\|_p = \|\xi_0\|_p \|a + b\xi_{n-j}\|_p^k.$$

To check the result, write

$$X_n = \sum_{k=0}^{m} \xi_{n-k} \prod_{j=0}^{k-1} (a + b\xi_{n-j}) + X_{n-m} \prod_{j=0}^{m-1} (a + b\xi_{n-j}).$$

Then the previous remark implies that the main term in this equality converges as $m \uparrow \infty$, and its $\mathbb{L}^p$-norm is bounded above by some $A > 0$.

This also entails $(1 - \alpha)\|X_0\|_p \leq A$.

Bilinear models (7.1) behave quite analogously to some white noises. The associated sequence of covariances presents some bumps and then rapidly decays.

In Fig. 7.1 we present empirical covariances; the convergence of these expressions is considered later.

Exercise 59 proves the consistence of such estimates for models listed in Example 9.1.3, extending bilinear models.

For such bilinear models, the sequence of covariances also fits a recursion.

Exercise 36 Assume that $\mathbb{E}\xi_0 = 0$, $\mathbb{E}\xi_0^2 = 1$ and consider the $\mathbb{L}^2$-strictly stationary solution (X_t) of (7.1).

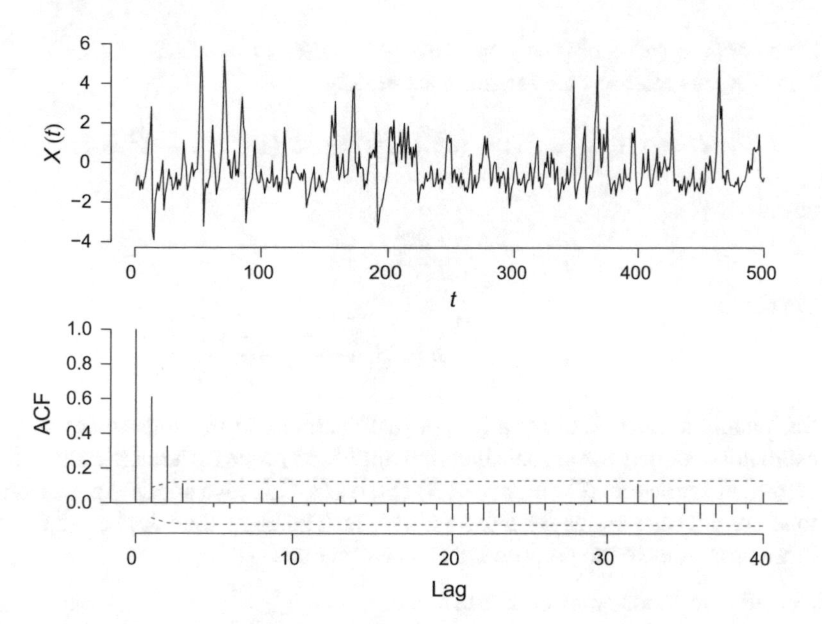

Fig. 7.1 Simulated trajectory of an bilinear process and sample autocorrelation function. Here, $X_t = 0.75X_{t-1} + \varepsilon_{t-1} + 0.6X_{t-1}\varepsilon_{t-1}$ with $\varepsilon_t \sim \mathcal{N}(0, 1)$

Set also the notations $M = \mathbb{E}X_0^2$ and $C = \mathrm{Cov}\,(X_0, X_1)$.

1. Prove that

$$\mathbb{E}X_0 = 0, \quad M = \frac{1}{1-(a^2+b^2)}, \quad C = \frac{a}{1-(a^2+b^2)}.$$

2. From empirical estimators of the previous expressions

$$\widehat{M} = \frac{1}{n}\sum_{k=1}^{n} X_k^2, \qquad \widehat{C} = \frac{1}{n-1}\sum_{k=2}^{n} X_k X_{k-1},$$

deduce that the following estimators of parameters a, b in the model are consistent:

$$\widehat{a} = \frac{\widehat{M}}{\widehat{C}}, \qquad \widehat{b} = \sqrt{\frac{\widehat{M}^2 - \widehat{C}^2 - \widehat{M}}{\widehat{M}}}.$$

Hints for Exercise 36.

1. From independence of ξ_t with X_{t-1} and (7.1):

$$\mathbb{E}X_1 = a\mathbb{E}\xi_0, \quad \mathbb{E}X_1^2 = \mathbb{E}(a + b\xi_0)^2\mathbb{E}X_0^2 + \mathbb{E}\xi_0^2$$

hence $M(1-(a^2+b^2))=1$, moreover $C=\mathbb{E}X_0X_1=aM$.

2. The previous relations are rewritten accurately:

$$C=aM, \qquad M^2(1-(a^2+b^2))=M^2(1-b^2)-C^2=M$$

hence

$$Mb^2=M^2-C^2-M.$$

We obtain:

$$a=\frac{M}{C}, \qquad b=\sqrt{\frac{M^2-C^2-M}{M}}.$$

See results in Sect. 7.3.3 for a formal justification and the consistency of these estimators, namely the ergodic theorem applies to prove a.s. consistency of these empirical estimators (Corollary 9.1.3) and a $\sqrt{n}$-CLT also applies to get asymptotic confidence bounds for these estimators. The Δ-method applies to transfer those properties to the proposed plug-in estimators.

This entails the conclusions of this exercise.

Exercise 37 (*Resampling bilinear models*) Let (X_t) be the stationary solution of (7.1). Prove the equation

$$\xi_n=\frac{X_n-aX_{n-1}}{1+bX_{n-1}}.$$

As in Sect. 4.6 use Exercise 36 to resample this model.

Hint. Consider observations over the epochs Obs $=\{1,\ldots,T\}$. These observations are divided into three disjointed parts Obs $=A\cup B\cup C$ with $A=\{1,\ldots,N\}$, $B=\{N+1,\ldots,N+q\}$, and setting $T=N+q+p$, with $C=\{N+q+1,\ldots,N+q+p\}$, and

- $1\ll q\ll p\wedge N$ is designed to make the two parts A and C almost independent (the part B is used to make them almost independent),
- Exercise 36 provides the estimation of the coefficients a,b in this model over observations $(Y_t)_{t\in A}$,
- Residuals are fitted through observations $(Y_t)_{t\in C}$ from the above relation

$$\widehat{\xi}_n=\frac{X_n-\widehat{a}X_{n-1}}{1+\widehat{b}X_{n-1}}, \qquad \forall n\in C.$$

Then consider an iid sequence with marginal distribution

$$\widehat{\nu}=\frac{1}{p}\sum_{s\in C}\delta_{\widehat{\xi}_s},$$

to complete the resampling procedure.

Remark 7.2.2 (*Centring innovations*) A more adequate resampling will be performed with the centred form

$$\widetilde{\xi}_s = \widehat{\xi}_s - \frac{1}{p}\sum_{u\in C}\widehat{\xi}_u$$

of $\widehat{\xi}_s$, and

$$\widetilde{\nu} = \frac{1}{p}\sum_{s\in C}\delta_{\widetilde{\xi}_s}.$$

Thanks to a personal communication with Patrice Bertail (Paris, Nanterre), that this conditional centring of innovations, improves the computational performances of the resampling procedure.

An extension of the above-mentioned bilinear models is sketched below:

Exercise 38 One variant for the model (7.1) is

$$X_n = h(\xi_n)X_{n-1} + \xi_n$$

and, in case $\mathbb{E}|h(\xi_0)| < 1$, a stationary solution is

$$X_n = \sum_{k=0}^{\infty} \xi_{n-k}\prod_{j<k} h(\xi_{n-j}).$$

Notice that such expressions may be provided under more complicated assumptions for models like

$$X_n = H_n X_{n-1} + \xi_n$$

with H_n some adapted and stationary sequence.

Assume that:

$$\sum_{k=0}^{\infty} \left\| \xi_{n-k}\prod_{j<k} H_{n-j}\right\|_p < \infty$$

then the following series is convergent in $\mathbb{L}^p$.

An $\mathbb{L}^p$ and strictly stationary solution of the previous recursion writes as:

$$X_n = \sum_{k=0}^{\infty} \xi_{n-k}\prod_{j<k} H_{n-j}.$$

Hints for Exercise 38. Use the recursion

$$
\begin{aligned}
X_n &= \xi_n + H_n X_{n-1} \\
&= \xi_n + H_n(\xi_{n-1} + H_{n-1} X_{n-2}) \\
&= \xi_n + \xi_{n-1} H_n + X_{n-2} H_n H_{n-1} \\
&= \xi_n + \xi_{n-1} H_n + (\xi_{n-2} + H_{n-2} X_{n-3}) H_n H_{n-1} \\
&= \xi_n + \xi_{n-1} H_n + \xi_{n-2} H_n H_{n-1} + H_{n-2} X_{n-3} H_n H_{n-1} \\
&= \cdots
\end{aligned}
$$

This allows us first to prove that this series is normally convergent in the Banach space $\mathbb{L}^p$, and then to check that the remainder term tends to 0. The expansion of the stationary solution is thus proved.

Remark 7.2.3 Note in Exercise 38 that if $H_n = h(\xi_n, \xi_{n-1}, \ldots, \xi_{n-r+1})$ is an r-dependent sequent.

Then H_n and X_{n-1} are not independent anymore which needs additional moment conditions,

$$
\begin{aligned}
\left\| \xi_{n-k} \prod_{j<k} H_{n-j} \right\|_p &= \left\| \xi_{n-k} \prod_{k-r<j<k} H_{n-j} \right\|_p \left\| \prod_{j=0}^{k-r} H_{n-j} \right\|_p \\
&\leq \left\| \xi_{n-k} \prod_{k-r<j<k} H_{n-j} \right\|_p \|H_0\|_{pr}^{(\ell-1)r},
\end{aligned}
$$

if $k = \ell r$.

Indeed it follows from the Hölder inequality (Proposition A.2.2) that if $k = \ell r$,

$$
\prod_{j=0}^{(\ell-1)r} H_{n-j},
$$

can be written as the product of r products of $(\ell - 1)$ independent terms.

Assumptions $\|H_0\|_{pr} < 1$, $\|H_0 \cdots H_{r-1} \xi_r\|_p < \infty$, together ensure the $\mathbb{L}^p$-convergence of the previous series.

The last relation holds if

$$
\|\xi_0\|_{qr} < \infty, \quad \text{and} \quad \|H_0\|_{q'pr} < \infty,
$$

for $q, q' \in [1, +\infty]$ with $\dfrac{1}{q} + \dfrac{1}{q'} = 1$.

7.2.2 LARCH(∞)-Models

General stationary non-Markov models are introduced hereafter, see for example Doukhan et al. (2007b) and Giraitis et al. (2012) for analogous LRD models (see Sect. 4.3).

Theorem 7.2.1 *Assume that $(\xi_k)_{k\in\mathbb{Z}}$ is an iid real valued sequence. Consider the recurrence equation LARCH(∞)-equation:*

$$X_n = \left(b_0 + \sum_{j=1}^{\infty} b_j X_{n-j}\right)\xi_n.$$

Under condition

$$\|\xi_0\|_p \sum_{k=1}^{\infty} |b_k| < 1,$$

an $\mathbb{L}^p$-valued strictly stationary solution of this recursion, called linear auto-regressive conditionally heteroskedastic with infinite order, LARCH(∞), can be written as

$$\begin{aligned} X_n &= b_0\,\xi_n \sum_{k=0}^{\infty}\sum_{l_1=1}^{\infty}\cdots\sum_{l_k=1}^{\infty} b_{l_1}\cdots b_{l_k}\xi_{n-l_1}\xi_{n-l_1-l_2}\cdots\xi_{n-(l_1+\cdots+l_k)} \\ &= b_0\,\xi_n \sum_{k=0}^{\infty}\sum_{0<j_1<\cdots<j_k=1}^{\infty} b_{j_1}b_{j_2-j_1}\cdots b_{j_k-j_{k-1}}\xi_{n-j_1}\xi_{n-j_2}\cdots\xi_{n-j_k} \end{aligned}$$

(here we set 1 for the empty sum obtained for $k = 0$).

Hints. Indeed, it is easy to derive from the independence of those factors that: $\|\xi_{n-j_1}\xi_{n-j_2}\cdots\xi_{n-j_k}\|_p = \|\xi_0\|_p^k$.

If now the variables ξ_n are centred and admit a finite variance the representation of Theorem 7.2.1 still holds in $\mathbb{L}^2$ if, only,

$$\mathbb{E}\xi_0^2 \sum_{k=1}^{\infty} b_k^2 < 1.$$

This assumption allows long-range dependent behaviours as proved in the volume (Giraitis et al. 2012).

A vector valued variant of this model as well as a random field variant have both been developed.

Usual ARCH-models $(Y_n)_{n\in\mathbb{Z}}$ are such that squares $X_n = Y_n^2$ satisfy the previous equation.

They are defined through a sequence of non-negative real numbers (b_j) with $b_j = 0$ if j is large enough or a centred sequence of independent identically distributed random variables (ξ_j)

$$Y_n = \sqrt{b_0 + \sum_{j=1}^{J} b_j Y_{n-j}^2} \cdot \xi_n. \tag{7.2}$$

In this case the vector valued model $Y_n = (X_n, \ldots, X_{n-J+1})$ is a Markov process with values in $\mathbb{R}^J$. Remark that the general model is not J-Markov for any $J > 0$.

Exercise 39 (*Resampling LARCH(J)-models*) To resample (see Sect. 4.6) the model (7.2), use the equation

$$\xi_n = \sqrt{b_0 + \sum_{j=1}^{J} b_j Y_{n-j}^2} \Big/ Y_n.$$

Hint. One considers an observation sample over epochs Obs $= \{1, \ldots, T\}$. These observations are divided into three disjointed parts Obs $= A \cup B \cup C$ with $A = \{1, \ldots, N\}$, $B = \{N+1, \ldots, N+q\}$ and $C = \{N+q+1, \ldots, N+q+p\}$ where $T = N + q + p$ and

- $1 \ll q \ll p \wedge N$ is designed to make both parts A, C almost independent.
- Whittle estimation allows us to fit coefficients $b_0, \ldots, b_J$ over observations $(Y_t)_{t \in A}$.
- Residuals are fitted through observations $(Y_t)_{t \in C}$ from the above relation

$$\widehat{\xi}_n = \sqrt{\widehat{b}_0 + \sum_{j=1}^{J} \widehat{b}_j Y_{n-j}^2} \Big/ Y_n.$$

Then consider an iid sequence with marginal distribution

$$\frac{1}{p} \sum_{s \in C} \delta_{\widehat{\xi}_s}$$

to complete the resampling procedure.

Again Remark 7.2.2 is valuable to improve the practical performance of resampling.

7.3 Stable Markov Chains

Proposition 7.6 of Kallenberg (1997) proves that any Markov chain (homogeneous in time) (X_t) with values in $\mathbb{R}^d$ for some $d \geq 1$ may be represented as the solution of a recursion or iterative random model or autoregressive model assuming Condition 1 below:

$$X_t = M(X_{t-1}, \xi_t). \tag{7.3}$$

Condition 1 *$(\xi_t)_{t\in\mathbb{Z}}$ an independent identically distributed sequence with values in a measurable space $(E, \mathcal{E})$ for a measurable function M is a (measurable) kernel*

$$M : (\mathbb{R}^d, \mathcal{B}(\mathbb{R}^d)) \times (E, \mathcal{E}) \to (\mathbb{R}^d, \mathcal{B}(\mathbb{R}^d)).$$

For several models the innovation space has to be specified differently. Sometimes it will be $\mathbb{R}^d$ but sometimes a product space, the one associated with thinning operators, or a point process distribution, associated with Poisson processes, see Definition A.2.5, may be needed.

This section exhibits simple sufficient conditions for such iterative models to admit a stationary solution.

Further we will see that such solutions can be written as Bernoulli shifts (7.15). A contraction argument is used.

Suppose that (ξ_t) is an independent identically distributed sequence with values in a space E. Moreover for $d \geq 1$ and for a measurable space $(E, \mathcal{E})$ we denote by $\|\cdot\|$ some norm on $\mathbb{R}^d$.

We shall assume as in Duflo (1996) that the model is contractive:

Condition 2 *The Markov kernel $M(u, z)$ fits the conditions (7.4) and (7.5). There exist $a \in [0, 1)$, and $u_0 \in \mathbb{R}^d$, such that for all $u, v \in \mathbb{R}^d$,*

$$\mathbb{E}\|M(u, \xi_0) - M(v, \xi_0)\|^p \leq a^p \|u - v\|^p, \tag{7.4}$$

$$\mathbb{E}\|M(u_0, \xi_0)\|^p < \infty. \tag{7.5}$$

Theorem 7.3.1 *Assume that conditions (7.5) and (7.4) hold for some $p \geq 1$.*

Equation (7.3) admits a stationary condition in $\mathbb{L}^p$ such that for each $t \in \mathbb{Z}$, X_t is measurable wrt to the σ-algebra $\mathcal{F}_t = \sigma(\xi_s / \, s \leq t)$.

Proof Define $(U_t^{(n)})_{t\in\mathbb{Z}}$ a Markov chain such that

$$\begin{aligned} U_t^{(n)} &= u_0, && \text{if } t \leq -n, \\ U_t^{(n)} &= M(U_{t-1}^{(n)}, \xi_t), && \text{if } t > -n. \end{aligned}$$

The Lipschitz condition implies with independence of inputs:

$$\mathbb{E}\left\| U_0^{(n)} - U_0^{(n+1)} \right\|^p \leq a^p \mathbb{E}\left\| U_0^{(n-1)} - U_0^{(n)} \right\|^p.$$

From a recursion

$$\mathbb{E}\left\| U_0^{(n)} - U_0^{(n+1)} \right\|^p \leq a^{np} \mathbb{E}\left\| M(u_0, \zeta_0) - u_0 \right\|^p.$$

Hence $U_0^{(n)} \to U_0$ $(n \to \infty)$ converges in $\mathbb{L}^p$ to a random variable $U_0 \in \mathbb{L}^p$.

Moreover $U_0^{(n)}$ is measurable wrt the σ-algebra generated by $\{\xi_t/\,t \le 0\}$ hence this is also the case for U_0. U_0 may also be represented as a function $U_0 = H(\xi_0, \xi_{-1}, \dots)$ of this sequence.

Then the sequence $X_t = H(\xi_t, \xi_{t-1}, \xi_{t-2}, \dots)$ is a stationary solution of the previous recursion.

Now the sequences $(U_t^{(0)})_t$ and $(U_t^{(1)})_t$, satisfy

$$\begin{aligned} U_0^{(0)} &= u_0, \\ U_1^{(0)} &= M(u_0, \xi_1) &&= H(\xi_1, 0, 0, \dots) \\ U_2^{(0)} &= M(M(u_0, \xi_1), \xi_2) &&= H(\xi_2, \xi_1, 0, 0, \dots) \end{aligned}$$

and from a recursion for each $t > 0$,

$$U_t^{(0)} = V(\xi_t, \xi_{t-1}, \dots, \xi_1, 0, 0, 0, \dots).$$

Analogously

$$U_t^{(1)} = H(\xi_t, \xi_{t-1}, \dots, \xi_1, \xi_0, 0, 0, \dots).$$

Hence

$$\gamma_n = \left\| U_n^{(0)} - U_n^{(1)} \right\|_p \le a\gamma_{n-1},$$

and thus:

$$\gamma_n \le a^n \gamma_0 = a^n \left\| M(u_0, \zeta_0) - u_0 \right\|_p \tag{7.6}$$

decays exponentially to 0 since $a < 1$.

In fact the assumption that the function $u \mapsto M(u, e)$ admits a fixed point may simply be replaced by assumption (7.5).

Set $U_{-n}^{(n)} = M(u_0, \xi_{-n})$ to conclude.

Set:

Condition 3 (Fixed point) *Suppose (7.4) holds and that, for some $e \in E$, the function $u \mapsto M(u, e)$ admits a fixed point u_0 (if E is a vector space a simple change allows to suppose $e = 0$).*

We also obtain:

Proposition 7.3.1 *The stochastic equation (7.3) admits a strictly stationary solution in $\mathbb{L}^p$ ($p \ge 1$), if Conditions 1 and 3 hold.*

Example 7.3.1 Diaconis and Freedman (1995) provide a nice series of examples for which the previous technique applies. One may also refer to Doukhan (1994), Doukhan and Louhichi (1999), as well as to the volume (Dedecker et al. 2007).

More general models with infinite memory may in fact be considered, see Theorem 7.4.1.

7.3.1 AR-ARCH-Models

Proposition 7.3.2 *Let $d = 1$, $E = \mathbb{R}$ and set*

$$M(u, z) = A(u) + B(u)z \tag{7.7}$$

for Lipschitz functions $A(u)$, $B(u)$, $u \in \mathbb{R}$.
If

$$Lip(A) = \sup_{u \neq v} \frac{|A(u) - A(v)|}{|u - v|}$$

then the stability conditions in Sect. 7.3 hold if moreover $\xi_0 \in \mathbb{L}^p$.

- *If $p \geq 1$ in case*

$$a = Lip(A) + \|\xi_0\|_p Lip(B) < 1.$$

- *If $p = 2$ and $\mathbb{E}\xi_t = 0$ with*

$$a^2 = (Lip(A))^2 + \mathbb{E}\xi_0^2\,(Lip(B))^2 < 1.$$

Remark 7.3.1 For $p = 2$ and in case $\mathbb{E}\xi_0 = 0$, then the second assumption improves on the first one, indeed

$$\begin{aligned}(\mathrm{Lip}(A) + \|\xi_0\|_2 \mathrm{Lip}(B))^2 &= (\mathrm{Lip}(A))^2 + \mathbb{E}\xi_0^2\,(\mathrm{Lip}(B))^2 + 2\|\xi_0\|_2 \mathrm{Lip}(A)\,\mathrm{Lip}(B)\\ &\geq (\mathrm{Lip}(A))^2 + \mathbb{E}\xi_0^2\,(\mathrm{Lip}(B))^2\,.\end{aligned}$$

Moreover, the inequality is strict except in case $)\|\xi_0\|_2\mathrm{Lip}(A\mathrm{Lip}(B) \neq 0$.

Proofs. Note that the Minkowski inequality (Corollary A.2.1) implies that for $p \geq 1$,

$$\|M(u_0, \xi_0)\|_p \leq \|A(u_0)\| + \|B(u_0)\|\,\|\xi_0\|_p,$$

and,

$$\|M(u, \xi_0) - M(v, \xi_0)\|_p \leq \|A(u) - A(v)\| + \|B(u) - B(v)\|\,\|\xi_0\|_p,$$

which allows us to derive the second point of the proposition.
If $p = 2$ then

$$\begin{aligned}\mathbb{E}(M(u, \xi_0) - M(v, \xi_0))^2 &= (A(u) - A(v))^2 + (B(u) - B(v))^2\mathbb{E}\xi_0^2\\ &\quad + 2(A(u) - A(v))(B(u) - B(v))\mathbb{E}\xi_0\end{aligned}$$

and the last rectangle term simply vanishes from $\mathbb{E}\xi_0 = 0$; improving the previous bound, as noted in Remark 7.3.1.

These relations yield a simple way to conclude.

Example 7.3.2 (*Some special cases*) Specializing the above model yields very classical models (see e.g. Doukhan 1994)

- Non-linear AR(1)-models (with $B = 1$) satisfy the equation

$$X_n = A(X_{n-1}) + \xi_n.$$

- Stochastic volatility models (with $A = 0$) are solutions of the equation

$$X_n = B(X_{n-1})\xi_n.$$

- The AR-ARCH(1)-classical model is solution of the equation

$$X_n = \alpha X_{n-1} + \sqrt{\beta + \gamma^2 X_{n-1}^2} \cdot \xi_n.$$

Here $A(u) = \alpha u$ and $B(u) = \sqrt{\beta + \gamma^2 u^2}$ for $\alpha, \beta, \gamma \geq 0$.
The Lipschitz constant can be written $a = \alpha^2 + \mathbb{E}\zeta_0^2 \gamma$ from a direct calculation of the derivatives $A'(u) = \alpha$ and

$$|B'(u)| = \frac{\gamma^2 |u|}{\sqrt{\beta + \gamma^2 u^2}} = \gamma \cdot \frac{\sqrt{\gamma^2 u^2}}{\sqrt{\beta + \gamma^2 u^2}} \leq \gamma.$$

This model is defined conditionally wrt to its past history:

$$X_t \mid \mathcal{F}_{t-1} \sim \mathcal{N}(\alpha X_{t-1}, \beta + \gamma^2 X_{t-1}^2);$$

remark that the above recursion is just the simplest way to get such conditional distributions for Gaussian innovations.
- ARCH(2)-models are solutions of the equations

$$X_t = \sigma_t \xi_t, \qquad \sigma_t^2 = \alpha^2 + \beta^2 X_{t-1}^2 + \gamma^2 X_{t-2}^2.$$

Their trajectories may be seen in Fig. 7.2.

Exercise 40 Consider the ARCH(1)-model $X_t = \sigma_t \xi_t$, with $\sigma_t^2 = \alpha^2 + \beta^2 X_{t-1}^2$ for $\mathbb{E}\xi_t = 0$ and $\mathbb{E}\xi_t^2 = 1$. Then

$$X_t^2 = \alpha^2 + \beta^2 X_{t-1}^2 + \eta_t, \qquad \text{with} \qquad \eta_t = \sigma_t^2(\xi_t^2 - 1).$$

Determine μ such that $Z_t = X_t^2 - \mu$ is the solution of the AR(1) equation

$$Z_t = \beta^2 Z_{t-1} + \eta_t.$$

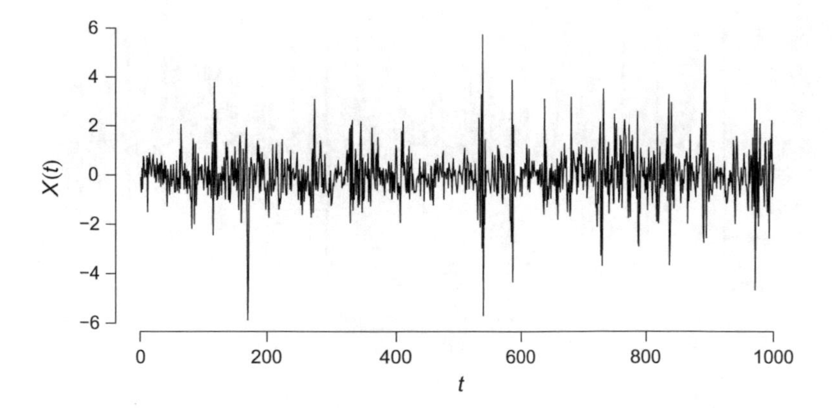

Fig. 7.2 Simulated trajectory of an ARCH(2) process.
Here $X_t = \sqrt{\sigma_t^2}\xi_t$ with $\sigma_t^2 = \alpha^2 + \beta^2 X_{t-1}^2 + \gamma^2 X_{t-2}^2$ and $\xi_t \sim \mathcal{N}(0, 1)$. We used $\alpha = 0.5, \beta = 0.6$ and $\gamma = 0.7$

This is a special case of the point that an ARCH model may be rewritten as an AR process excited by a weak white noise, here η_t is not iid but it is a martingale increment. It is simple to extend this representation to ARCH(p) models.

Note that $\mathbb{E}\eta_t^2 < \infty \Leftrightarrow \beta^2\text{Var}\,(\xi_0^2) < 1$ from Proposition 7.3.2.

- GARCH(1,1)-models are solutions of the equations

$$X_t = \sigma_t \xi_t, \qquad \sigma_t^2 = \alpha^2 + \beta^2 X_{t-1}^2 + \gamma^2 \sigma_{t-1}^2.$$

It is clear through iterations that one may rewrite such models as

$$\sigma_t^2 = \alpha^2 + \sum_{k=1}^{\infty} \beta_k^2 X_{t-1}^2,$$

and such models have also been designed for financial purposes for the associated clustering properties; trajectories may be seen in Fig. 7.3.

Shumway and Stoffer (2011), Example 5.5, p. 288 propose a GARCH(1,1) model for the NYSE returns, as this seems reasonable from the Figs. 7.3 and 7.4.

7.3.2 *Moments of ARCH(1)-Models*

We are interested here in checking that recursive models without low order moments may be generated from inputs with all finite moments. Consider the simplest ARCH-model

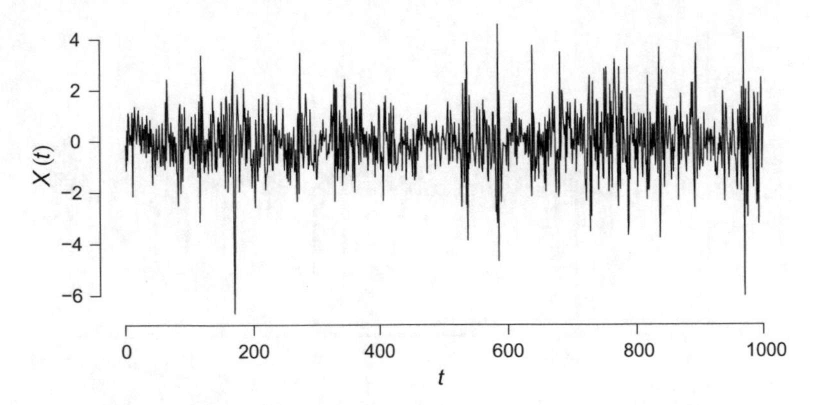

Fig. 7.3 Simulated trajectory of an GARCH(1,1).
Here, $X_t = \sqrt{\sigma_t^2}\xi_t$ with $\sigma_t^2 = \alpha^2 + \beta^2 X_{t-1}^2 + \gamma^2\sigma_{t-1}^2$ and $\xi_t \sim \mathcal{N}(0, 1)$. We used $\alpha = 0.5, \beta = 0.6$ and $\gamma = 0.7$

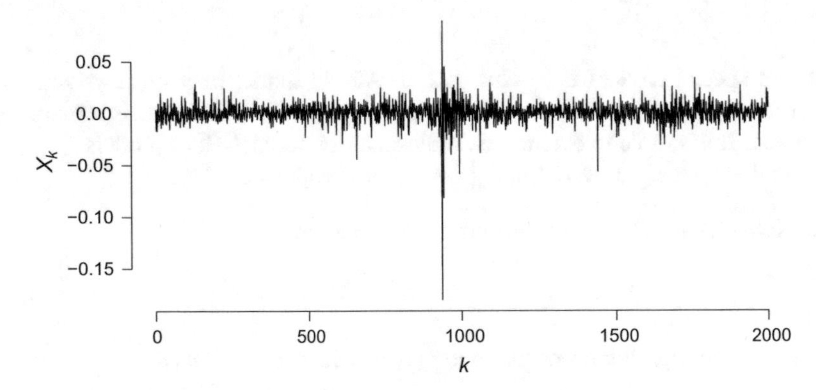

Fig. 7.4 NYSE returns.
Source: Shumway and Stoffer (2011), p. 7.
The data are daily value weighted market returns from February 2, 1984 to December 31, 1991 (2000 trading days). The crash of October 19, 1987 occurs at $t = 938$

$$X_t = \sqrt{\beta + \gamma^2 X_{t-1}^2} \cdot \xi_t. \tag{7.8}$$

We check that the function $p \mapsto \|\xi_0\|_p$ is monotonically non-decreasing from Jensen's inequality (Proposition A.2.1) applied with $t \mapsto t^r$ for $r \geq 1$. If $|\xi_0|$ is not constant a.s. this function is strictly increasing. E.g. if $\mathbb{P}(|\xi_0| \notin \{0, a\}) = 0$ then

$$\|\xi_0\|_p = \left(1 + a^p \mathbb{P}(|\xi_0| = a)\right)^{\frac{1}{p}}.$$

Remark 7.3.2 This result is used in Szewcsak (2012) to derive a central limit theorem with the unusual $\sqrt{n \log n}$ rate in this case.

For the AR-LARCH models with centred inputs the limit condition $\alpha^2+\gamma^2\mathbb{E}\xi_0^2=1$, analogously implies that any solution of this equation does not have second order moment.

Lemma 7.3.1 *Let $Z \ge 0$ be a non-negative, and non a.s. constant, random variable such that $\mathbb{E}Z^m < \infty$ for some $m > 0$, then the function $p \mapsto \|Z\|_p$ defined on $(0, m] \to \mathbb{R}^+$ is strictly monotonic.*

Proof With $Z = |\xi_0|^p$ we need to prove that if $p' > p$ and $r = p'/p$ then $\mathbb{E}Z \le \|Z\|_r$.

As in the proof of (A.2) Jensen's inequality for $g(u) = u^r$ with $r = p'/p > 1$, we consider an affine minorant $f(u) = au + b$ for the function g with $f(z) = g(z)$, for some z to be defined.

Indeed $a = rz^{r-1}$ makes $f'(z) = g'(z)$ and $b = (1-r)z^r$, then implies $f(z) = g(z)$.

Now if $u \ne z$ then $f(u) < g(u)$ hence $\mathbb{E}f(Z) < \mathbb{E}g(Z)$ because Z is not a.s. a constant.

Let now $z = \mathbb{E}Z$, then

$$\mathbb{E}f(Z) = (\mathbb{E}Z)^r < \mathbb{E}g(Z) = \mathbb{E}Z^r.$$

This is enough to conclude.

Proposition 7.3.3 *$\gamma\|\xi_0\|_p < 1$ is a necessary condition for (7.8) to admit an $\mathbb{L}^p$ and strictly stationary solution (X_t).*

Moreover if $\gamma\|\xi_0\|_2 = 1$ and $|\xi_0|$ is not constant a.s., then the stationary solution of (7.8) satisfies $\mathbb{E}X_t^2 = \infty$ and $\mathbb{E}|X_t|^p < \infty$.

Proof The first statement follows from Proposition 7.3.2.

If $\gamma\|\xi_0\|_2 = 1$, from Lemma 7.3.1 the previous equation admits a strictly stationary solution in $\mathbb{L}^p$ for each $p < 2$. Moreover this solution is not $\mathbb{L}^2$-integrable.

Otherwise indeed:

$$\begin{aligned}
\mathbb{E}X_t^2 &= (\beta + \gamma^2\mathbb{E}X_{t-1}^2)\|\xi_t\|_2^2 \\
&= \beta\|\xi_t\|_2^2 + \mathbb{E}X_{t-1}^2 \\
&> \mathbb{E}X_{t-1}^2 = \mathbb{E}X_t^2.
\end{aligned}$$

Also there exists a $\mathbb{L}^p$-solution of this equation in case p is small enough and if $|\xi_0|$ is not constant.

7.3.3 *Estimation of LARCH(1)-Models*

This section describes some important features of LARCH(1)-models in order to provide some simple estimators of their parameters as sketched in Exercise 36.

The ideas are essentially from the Yule–Walker equations, Sect. 6.3, and the main point is an MA-representation with $\mathbb{L}^2$-weak-white noise inputs.

Also (ξ_t) is an iid real valued sequence with $\mathbb{E}|\xi_0|^p < \infty$ for some $p > 0$ and

$$Z_t = (\beta + \delta Z_{t-1})\xi_t. \tag{7.9}$$

Remark that even though covariances of the model appear to decay quite rapidly, the behaviour of the trajectory looks erratic.

Lemma 7.3.2 *Let $p > 0$ be a fixed positive number. Then the assumption $|\delta| \cdot \|\xi_0\|_p < 1$ implies that a unique stationary solution exists and it is in $\mathbb{L}^p$.*

Proof Remark that $|\delta| \cdot \|\xi_0\|_p < 1$ is the contraction constant in this case. Now the solution of the equation is the limit of a polynomial in the innovations and can be written as a Bernoulli shift in $\mathbb{L}^p$.

A first estimator of the parameter $\theta = (\beta, \delta)$ is described in Sect. 4.4.2; this is the Whittle estimator based on a minimization of the periodogram.

The latter estimate needs explicit expressions of Z's spectral density, or equivalently of all the covariances of Z. This may be quite heavy (Fig. 7.5).

In Sect. 4.5 the QMLE of the Markov chains

$$Z_t = \xi_t \sigma_\theta(X_{t-1}),$$

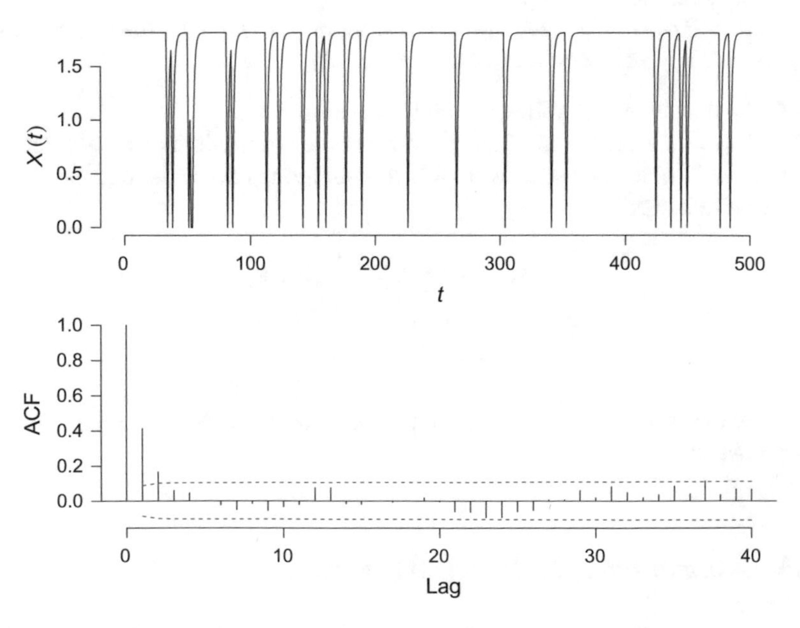

Fig. 7.5 Simulated trajectory and simple correlogram of an LARCH(1,1) process. Here $X_t = \varepsilon_t(1 + \beta_1 x_{t-1})$ with $\varepsilon_t \sim \mathcal{B}(0.95)$. We used $\beta_1 = 0.45$

is considered in case $\xi_t \sim \mathcal{N}(0, 1)$; here the transition probability density is

$$\pi_\theta(x, y) = \frac{1}{\sqrt{2\pi(\beta + \delta x)^2}} \exp\left(-\frac{1}{2}\frac{y^2}{(\beta + \delta x)^2}\right).$$

The QMLE of (Z_t) is now the couple $\theta = (\beta, \delta)$ minimizing the expression

$$L_\theta(Z_1, \ldots, Z_n) = \sum_{t=2}^{n} \frac{Z_t^2}{(\beta + \delta Z_{t-1})^2} + \log(\beta + \delta Z_{t-1})^2$$

This estimator is considered under general situations in the monograph (Straumann 2005).

In our simple situation we choose a more direct way to estimate the parameters. It will result in simple empirical estimators.

Lemma 7.3.3 (Close expressions of moments) *Let (Z_t) be the stationary solution of (7.9), and assume that $Z_t \in \mathbb{L}^p$.*

1. *Assume that $p \geq 1$ then $m = \mathbb{E}Z_0 = \dfrac{\beta\mathbb{E}\xi_0}{1 - \delta\mathbb{E}\xi_0}$.*
2. *Assume that $p \geq 2$ and $\mathbb{E}\xi_0 = 1$ and set $\nu = \mathbb{E}\xi_0^2$; then*

$$M = \mathbb{E}Z_0^2 = \frac{\nu\beta^2(1 + \delta)}{(1 - \delta)(1 - \nu\delta^2)}.$$

3. *Assume now that $p \geq 3$ and that $\mathbb{E}\xi_0 = E\xi_0^3 = 0$, then:*

$$M = \mathbb{E}Z_0^2 = \frac{\nu\beta}{1 - \nu\delta^2}.$$

Set $\ell = Cov\,(Z_0, Z_1^2) = \mathbb{E}Z_0Z_1^2$ the leverage of Z then:

$$\ell = 2\nu\beta\delta M = \frac{2\nu^2\beta^2\delta}{1 - \nu\delta^2}.$$

4. *If now $p \geq 3$, $\mathbb{E}\xi_0 = 0$ and $\mathbb{E}\xi_0^3 \neq 0$, then $M = \mathbb{E}Z_0^2 = \dfrac{\nu\beta}{1 - \nu\delta^2}$.*
Set $\eta = \mathbb{E}\xi_0^3$ then:

$$P = \mathbb{E}Z_0^3 = \frac{\eta\beta(\beta^2 + 3\delta^2 M)}{1 - \eta\delta^3}.$$

Proofs.

1. From $\mathbb{L}^1$-stationarity and independence: $\mathbb{E}Z_0 = \mathbb{E}\xi_0(\beta + \delta\mathbb{E}Z_0)$.

2. From independence and $\mathbb{L}^2$-stationarity:

$$\begin{aligned}\mathbb{E}Z_0^2 &= \nu\mathbb{E}\{(\beta+\delta Z_0)^2\}\\ &= \nu(\beta^2+2\beta\delta\mathbb{E}Z_0+\delta^2\mathbb{E}Z_0^2)\\ &= \nu\left(\beta^2+\frac{2\beta^2\delta}{1-\delta}+\delta^2\mathbb{E}Z_0^2\right)\\ &= \nu\left(\frac{\beta^2(1+\delta)}{1-\delta}+\delta^2\mathbb{E}Z_0^2\right).\end{aligned}$$

3. Proceeding as before, we derive $\mathbb{E}Z_0^2=\nu(\beta^2+\delta^2\mathbb{E}Z_0^2)$, since $\mathbb{E}\xi_0=0$. Thus

$$M=\mathbb{E}Z_0^2=\frac{\nu\beta}{1-\nu\delta^2}.$$

We have here $\mathbb{E}\xi_0^3=0$, hence:

$$\begin{aligned}\ell &= \nu\mathbb{E}Z_0(\beta+\delta Z_0)^2 \;=\; \nu(2\beta\delta\mathbb{E}Z_0^2+\delta^2\mathbb{E}Z_0^3)\\ &= 2\nu\beta\delta\mathbb{E}Z_0^2\\ &= \frac{2\nu^2\beta^2\delta}{1-\nu\delta^2}.\end{aligned} \tag{7.10}$$

4. From independence,

$$P=\mathbb{E}Z_1^3=\mathbb{E}\xi_1^3\mathbb{E}(\beta+\delta Z_0)^3.$$

The binomial formula yields the result.

All possible cases when moments exist may be considered.

Lemma 7.3.4 *A.s. consistent estimators of* $m=\mathbb{E}Z_0$, $M=\mathbb{E}Z_0^2$, $P=\mathbb{E}Z_0^3$ *and* ℓ *are provided if respectively* $p\ge 1,2,3$ *by*

$$\widehat{m}=\frac{1}{n}\sum_{k=1}^n Z_k,\qquad \widehat{M}=\frac{1}{n}\sum_{k=1}^n Z_k^2,$$

$$\widehat{P}=\frac{1}{n}\sum_{k=1}^n Z_k^3,\qquad \widehat{\ell}=\frac{1}{n-1}\sum_{k=1}^n Z_kZ_{k+1}.$$

Proof From Proposition 7.2.1, the process Z_t admits an explicit chaotic expansion with respect to the iid sequence (ξ_t). Hence it is ergodic from the examples following Corollary 9.1.3. The ergodic theorem (Corollary 9.1.3) entails the a.s. convergence of those empirical expressions, as $n\to\infty$.

From the above results we derive the consistency of the estimators from the inversion of a function; in Lemma 7.3.4:

Corollary 7.3.1 *Assume that $|\delta| \cdot \|\xi_0\|_1 < 1$ and $\beta = 1$ then an a.s. consistent estimator of δ is*

$$\widehat{\delta} = 1 - \frac{1}{\widehat{m}}.$$

Corollary 7.3.2 *Assume that $|\delta| \cdot \|\xi_0\|_2 < 1$ and that $\mathbb{E}\xi_0 = 1$, $\mathbb{E}\xi_0^2 = \nu$, then a.s. consistent estimators of β, δ are:*

$$\widehat{\delta} = \sqrt{\frac{\nu\widehat{M} - \widehat{m}^2}{\nu(\widehat{M} - \widehat{m}^2)}}, \qquad \widehat{\beta} = \left(1 - \sqrt{\frac{\nu\widehat{M} - \widehat{m}^2}{\nu(\widehat{M} - \widehat{m}^2)}}\right)\widehat{m}.$$

Remark 7.3.3 (*Δ-method*) The Δ-method drives the construction of empirical confidence intervals from a central limit theorem for the empirical moments.

Namely if a constant vector μ and a sequence of random vectors μ_n are such that $\sqrt{n}(\mu_n - \mu) \xrightarrow{\mathcal{L}}_{n\to\infty} \mathcal{N}_d(0, \Sigma)$.

If $G : \mathbb{R}^d \to \mathbb{R}^p$ is a $\mathcal{C}^1$-differentiable function, then the following asymptotic holds outside of a set with small probability:

$$G(\mu_n) - G(\mu) \sim DG(\mu)(\mu_n - \mu),$$

thus

$$\sqrt{n}\Big(G(\mu_n) - G(\mu)\Big) \xrightarrow{\mathcal{L}}_{n\to\infty} \mathcal{N}_p(0, DG(\mu)'\Sigma DG(\mu)).$$

Remark 7.3.4 Applying the previous results to the ARCH(1)-model

$$X_t = \sqrt{\beta + \delta X_{t-1}^2} \cdot \zeta_t$$

is simple since $Z_t = X_t^2$ is a LARCH(1)-model with innovations $\xi_t = \zeta_t^2$ hence $\mathbb{E}\xi_0 \neq 0$ and may be chosen equal to 1 and $\nu = \mathbb{E}\zeta_0^4$.

Proof It is easy to derive that $\beta = (1 - \delta)m$ and:

$$\frac{m^2(1 - \delta^2)}{1 - \nu\delta^2} = M,$$

and $m^2(1 - \delta^2) = (1 - \nu\delta^2)M$ implies $\nu\delta^2(\nu M - m^2) = (M - m^2)$, and thus $\mathrm{Var}\, Z_0 = M - m^2 \geq 0$.

Also, even though the Cauchy–Schwarz inequality entails $\nu \geq 1$ the above relation implies $M - \nu m^2 = \nu m^2 \mathrm{Var}\, \xi_0 \geq 0$ and the following expression is well defined:

$$\delta = \sqrt{\frac{M - \nu m^2}{\nu(M - m^2)}}, \quad \beta = \left(1 - \sqrt{\frac{M - \nu m^2}{\nu(M - m^2)}}\right) m.$$

The corresponding estimators $\widehat{\beta}$, and $\widehat{\delta}$ are consistent.

Corollary 7.3.3 *Assume that $|\delta|\|\xi_0\|_3 < 1$, $\mathbb{E}\xi_0 = 0$, $\mathbb{E}\xi_0^2 = \nu$, and $\mathbb{E}\xi_0^3 = 0$, then a.s. consistent estimators of β, and δ are:*

$$\widehat{\delta} = -1 + \sqrt{1 + \nu\widehat{\ell}}\,, \qquad \widehat{\beta} = \frac{\widehat{M}}{\nu}\left(2\sqrt{1+\nu\widehat{\ell}} - (1+\nu\widehat{\ell})\right).$$

Remark 7.3.5 As a special case of situation (3) in Lemma 7.3.3, note that for *the symmetric innovations* with third order finite moments, we have indeed $\mathbb{E}\xi_0 = \mathbb{E}\xi_0^3 = 0$.

In the special case $\mathbb{P}(\xi_0 = \pm 1) = \frac{1}{2}$ of Rademacher-distributed inputs and $\beta = 1$, Doukhan et al. (2009) prove that the model is not strong mixing if $\delta \in \left]\frac{3-\sqrt{5}}{2}, \frac{1}{2}\right]$.

Moreover the polynomial equation $\ell\delta^2 + 2\delta - 1 = 0$ only admits the solution $\delta = -1 + \sqrt{1+\ell}$ in $]-1, 1[$. Indeed the other solution of the previous second degree equation is not in this set $]-1, 1[$.

Proof Relations $\beta, \delta > 0$ imply with its existence that $\ell > 0$. Now (7.10) together with $\nu\beta = M(1-\delta^2)$ entails $\ell(1-\nu\delta^2) = 2\delta M$. Then δ is the positive solution of the second order equation

$$\nu\ell\delta^2 + 2\delta M - \ell = 0.$$

Hence $\delta = -1 + \sqrt{1+\nu\ell}$ and

$$\beta = \frac{M}{\nu}\left(1 - \left(-1 + \sqrt{1+\nu\ell}\right)^2\right) = \frac{M}{\nu}\left(2\sqrt{1+\nu\ell} - (1+\nu\ell)\right).$$

The plug-in empirical estimator takes the same form as above.

The sign of ℓ, relies on the sign of the product $\beta\cdot\delta$; thus leverage $\ell \le 0$ is deduced from the equation: $\delta = -1 - \sqrt{1+\nu\ell} < 0$.

Remark 7.3.6 Assume that $|\delta|\|\xi_0\|_3 < 1$, $\mathbb{E}\xi_0 = 0$, $\mathbb{E}\xi_0^2 = \nu$, $\mathbb{E}\xi_0^3 = \eta \neq 0$ and $\beta, \delta > 0$ then a.s. consistent estimators of β, δ can be written analogously by solving equations (4) in Lemma 7.3.3 and replacing M, P by their empirical counterparts $\widehat{M}, \widehat{P}$.

The above equation in Lemma 7.3.3(3) is easy to solve:

$$\beta = \frac{M}{\nu(1-\nu\delta^2)}.$$

In the definition of P and solving the remaining equation wrt δ

$$P = \frac{\eta\beta(\beta^2 + 3\delta^2 M)}{1-\eta\delta^3}.$$

Unfortunately the resulting equation appears as a polynomial of degree 3 wrt to the variable δ^2. Hence the solution results of the Cardan formula, which provides the roots of 3rd degree polynomials.

Empirical use of financial data, see e.g. Giraitis et al. (2012), leads to:

Definition 7.3.1 The stationary process (X_t) is said to have leverage of order $k \geq 1$ in case $\ell_j = \mathrm{Cov}\,(X_0, X_j^2) < 0$ for $1 \leq j \leq k$.

Exercise 41 (*Asymmetric ARCH-model*) Assuming that the iid real valued sequence $(\xi_t)_t$ admits pth order moment, consider the following equation:

$$X_t = \sqrt{(aX_{t-1} + b)^2 + c^2} \cdot \xi_t, \qquad a > 0.$$

1. The previous equation admits a stationary solution in $\mathbb{L}^p$ if $|a| \cdot \|\xi_0\|_p < 1$.
2. Prove that for each $\alpha \in]0, 1[$, there exists $A > 0$ such that

$$|(ax + b)^2 + c^2|^{\frac{p}{2}} \geq (1 - \alpha)|ax|^p - A.$$

3. If $a\|\xi_0\|_p > 1$ and $a\|\xi_0\|_q < 1$ for some $q < p$, deduce that the $\mathbb{L}^q$-stationary solution is not in $\mathbb{L}^p$.
4. If $p = 2$ and $a^2\mathbb{E}|\xi_0|^2 = 1$, and $a\|\xi_0\|_q < 1$ for some $q < 2$, prove that $\mathbb{E}X_0^2 = \infty$, if moreover $b^2 + c^2 \neq 0$. Hence the result in the first question is essentially tight.
5. Assume that $p \geq 1$ then the previous stationary solution satisfies $\mathbb{E}X_0 = 0$.
6. If now $p \geq 2$, $\mathbb{E}\xi_0 = 0$, and $\mathbb{E}\xi_0^2 = 1$, then

$$\mathbb{E}X_0^2 = \frac{b^2 + c^2}{1 - a^2}.$$

7. Assume that $p = 3$ and ξ_0 admits a symmetric distribution, then $\mathbb{E}X_0^3 = 0$.
8. Leverage $\ell_t = \mathrm{Cov}\,(X_0, X_t^2)$ measures the asymmetry properties of a random process. Prove that if ξ_0 is symmetric and if again $p = 3$, then for $t \geq 1$:

$$\ell_t = \frac{2a^{2t-1}b(b^2 + c^2)}{1 - a^2}.$$

Deduce that if $ab < 0$ this asymmetric ARCH(1) model admits leverage at order k for each integer k.
9. Fit such asymmetric ARCH(1) models.
10. If $p = 4$ and ξ_0 admits a symmetric distribution, set $\nu = \mathbb{E}\xi_0^4$. Prove that

$$\mathbb{E}X_0^4 = \frac{6a^2b^2(b^2 + c^2) + (1 - a^2)(b^4 + c^4)}{(1 - a^2)(1 - \nu a^4)}.$$

Hints. See Doukhan et al. (2016) for many further developments of such models with infinite memory.

For the case of the above model with order 1, we also fit the marginal density of innovations and we resample the model in Doukhan and Mtibaa (2016).

1. Use Proposition 7.3.2.
2. As $p \geq 1$ the function $x \mapsto |x|^p$ is convex which entails that for each $\epsilon \in]0, 1[$, writting

$$ax = (1-\epsilon)\frac{ax+b}{1-\epsilon} + \epsilon\frac{-b}{\epsilon},$$

then,

$$|ax|^p \leq (1-\epsilon)\left(\frac{|ax+b|}{1-\epsilon}\right)^p + \epsilon\left(\frac{|b|}{\epsilon}\right)^p.$$

Hence

$$\begin{aligned}(1-\epsilon)^{p-1}|ax|^p &\leq |ax+b|^p + \left(\frac{1}{\epsilon}-1\right)^{p-1}|b|^p \\ &\leq |(ax+b)^2+c^2|^{\frac{p}{2}} + \left(1-\frac{1}{\epsilon}\right)^{p-1}|b|^p.\end{aligned}$$

The inequality holds with:

$$\alpha = 1-(1-\epsilon)^{p-1}, \qquad A = \left(\frac{1}{\epsilon}-1\right)^{p-1}|b|^p.$$

3. Choosing α small enough so that $\beta = (1-\alpha)a^p\mathbb{E}|\xi_0|^p > 1$, we derive:

$$\begin{aligned}\mathbb{E}|X_1|^p &= \mathbb{E}|\xi_1|^p(\mathbb{E}(aX_0+b)^2+c^2)^{\frac{p}{2}} \\ &\geq \beta\mathbb{E}|X_0|^p - A\mathbb{E}|\xi_1|^p.\end{aligned}$$

Then $\mathbb{E}|X_1|^p - B \geq \beta(\mathbb{E}|X_0|^p - B)$, for $B = A\mathbb{E}|\xi_1|^p/(\beta-1)$.
So if we suppose that the $\mathbb{L}^q$-stationary solution is also in $\mathbb{L}^p$ we derive $\mathbb{E}|X_0|^p = 0$.
Alternatively, with $\gamma_t = \mathbb{E}|X_t|^p - B$, the proof yields $\gamma_t \geq \beta\gamma_{t-1}$ and $\gamma_t \geq \beta^t\gamma_0$ hence $\gamma_t = +\infty$ if $\mathbb{E}|X_0|^p \neq 0$.
4. In case $p = 2$ and $a^2\mathbb{E}|\xi_0|^2 = 1$, we have

$$\mathbb{E}|X_1|^2 = \mathbb{E}|\xi_1|^2(\mathbb{E}(aX_0+b)^2+c^2) = \mathbb{E}|X_1|^2 + b^2 + c^2.$$

Hence $b^2 + c^2 > 0$ entails $\mathbb{E}X_0^2 = \infty$.
5. This follows from the independence of ξ_t and X_{t-1}.
6. Squaring the ARCH equation, the moment $M = \mathbb{E}X_0^2$ satisfies:

$$M = a^2M + b^2 + c^2.$$

7. If $p = 3$ and ξ_0 admits a symmetric distribution, analogously to point (4) we derive $\mathbb{E}X_1^3 = 0$ from $\mathbb{E}\xi_0^3 = 0$.
   If $p = 3$ and $\mathbb{E}\xi_0 = 0$, we derive $\mathbb{E}X_1^3 = \mathbb{E}\xi_1^3\mathbb{E}((aX_0 + b)^2 = c^2)^{\frac{3}{2}}$.
8. Take into account that $\mathbb{E}\xi_0^2 = 1$ and $\mathbb{E}X_0 = 0$. Then first

$$\begin{aligned}\ell_1 = \mathbb{E}X_0X_1^2 &= \mathbb{E}X_0((aX_0 + b)^2 + c^2) \cdot \mathbb{E}\xi_0^2\\ &= \mathbb{E}X_0((aX_0 + b)^2 + c^2)\\ &= a^2\mathbb{E}X_0^3 + 2ab\mathbb{E}X_0^2\\ &= \frac{2ab(b^2 + c^2)}{1 - a^2}.\end{aligned}$$

   A recursion entails:

$$\begin{aligned}\ell_t = \mathbb{E}X_0X_t^2 &= \mathbb{E}X_0((aX_{t-1} + b)^2 + c^2)\\ &= a^2\mathbb{E}X_0X_{t-1}^2 + 2ab\mathbb{E}X_0X_{t-1} + c^2\mathbb{E}X_0\\ &= a^2\mathbb{E}X_0X_{t-1}^2\\ &= \frac{2a^{2t-1}b(b^2 + c^2)}{1 - a^2}.\end{aligned}$$

9. The triplet is $(M, \ell_1, \ell_2) = F(a, b, c)$. From these relations $\ell_2/\ell_1 = a^2$ and $\ell_1 = 2abM$ imply $a = \sqrt{\ell_2/\ell_1}$, $b = \ell_1^{3/2}/2M\ell_2^{1/2}$.
   Finally

$$c^2 = M(1 - a^2) - b^2 = M\left(1 - \frac{\ell_2}{\ell_1}\right) - \frac{\ell_1^3}{4M^2\ell_2}.$$

10. Expanding the square of this expression we take advantage of the fact that all the finite odd order moments of ξ_1 vanish.
    Then setting $N = \mathbb{E}X_1^4$ we derive $N = \nu a^4 N + 6a^2b^2M + b^4 + c^4$.
    We obtain

$$N = \frac{6a^2b^2 \cdot \dfrac{b^2 + c^2}{1 - a^2} + b^4 + c^4}{1 - \nu a^4}.$$

    This yields an alternative estimation of this model.
    However it needs $a\|\xi_0\|_4 < 1$ while, even if the fourth order moment is finite, the first estimator needs the weaker condition $a\|\xi_0\|_3 < 1$ (indeed $\|\xi_0\|_3 \le \|\xi_0\|_3$ and Lemma 7.3.1 proves that the inequality is often a strict inequality.

Remark 7.3.7 In Exercise 41 we may inverse the relation relating the coefficients with the second order moment the leverages:

$$(a, b, c) = \left(\sqrt{\frac{\ell_2}{\ell_1}}, \frac{\sqrt{\ell_1^3}}{2M\sqrt{\ell_2}}, \sqrt{M\left(1 - \frac{\ell_2}{\ell_1}\right) - \frac{\ell_1^3}{4M^2\ell_2}}\right).$$

Since the above relation defines the inverse the function F, an accurate estimator of the model is:

$$(\widehat{a}, \widehat{b}, \widehat{c}) = F^{-1}\left(\widehat{M}, \widehat{\ell}_1, \widehat{\ell}_2\right).$$

Define the standard empirical estimators:

$$\widehat{M} = \frac{1}{n}\sum_{i=1}^{n} X_i^2,$$

$$\widehat{\ell}_1 = \frac{1}{n-1}\sum_{i=1}^{n-1} X_i X_{i+1}^2,$$

$$\widehat{\ell}_2 = \frac{1}{n-2}\sum_{i=1}^{n-2} X_i X_{i+2}^2.$$

Some Asymptotic Considerations.

By using the previous transforms the only asymptotic to be considered is for empirical estimators of M, ℓ_1, ℓ_2:

- The consistency of those estimators follows from the ergodic Theorem 9.1.1.
- A central limit theorem provides asymptotic confidence intervals by using the Δ-method in Remark 7.3.3. This result holds for empirical (vector-) moments since

  - θ-weak-dependence (see Definition 11.1.1) holds with a geometric decay and,
  - the moment condition $a\|\xi_0\|_{6+\epsilon} < 1$ holds for some $\epsilon > 0$.

  See Dedecker et al. (2007) for more details.

7.3.4 *Branching Models*

We introduce models branching or switching models, analogue to (7.3). We assume that different "regimes" are randomly obtained, moreover some of those "regimes" may even be explosive.

Here the Eq. (7.3) concerns a process with values in $\mathbb{R}$ ($d = 1$) and innovations $\xi_j \in E = \mathbb{R}^{D+1}$ for some $D \geq 2$ in $\mathbb{L}^2$ (hence $m = 2$).

Let $\xi_t = \left(\xi_t^{(0)}, \xi_t^{(1)}, \ldots, \xi_t^{(D)}\right)$ be such that

- $\xi_t^{(0)}$ is independent of $\left(\xi_t^{(1)}, \ldots, \xi_t^{(D)}\right)$,
- $\mathbb{E}\xi_t^{(i)}\xi_t^{(j)} = 0$, if $i \neq j$ and $i, j \geq 1$,
- $\mathbb{P}(\xi_t^{(0)} \notin \{1, 2, \ldots, D\}) = 0$.

If the functions $M_1, \ldots, M_D$ are Lipschitz on $\mathbb{R}$ and satisfy assumptions (7.4) and (7.5) with constants $a_j > 0$ for each $j = 1, \ldots, D$:

$$\begin{array}{ll} \forall u, v \in \mathbb{R}^d, & \|M_j(u, \xi_0^{(j)}) - M(v, \xi_0^{(j)})\|_p \le a_j \|u - v\|, \\ \exists u_0 \in \mathbb{R}^d, & \|M_j(u_0, \xi_0)\|_p \qquad\qquad\quad < \infty. \end{array}$$

We set

$$M\left(u, (z^{(1)}, \ldots, z^{(D)})\right) = \sum_{j=1}^{D} M_j(u, z^{(j)}) \, \mathbb{1}_{\{z^{(0)}=j\}},$$

for $(z^{(0)}, \ldots, z^{(D+1)}) \in \mathbb{R}^D$.

The previous contraction assumption holds with the Euclidean norm $\| \cdot \|$ if

$$a = \sum_{j=1}^{D} a_j \mathbb{P}(\xi_0^{(0)} = j) < 1.$$

Now in case $p = 2$ we also improve the result if $\mathbb{E}\xi_0^{(j)} = 0$ and we denote

$$a^2 = \sum_{j=1}^{D} a_j^2 \, \mathbb{P}(\xi_0^{(0)} = j) < 1.$$

For example in case $M_j(u, z) = A_j(u) + z$ we have $a_j = \text{Lip}\, A_j$:

- Let $\xi_t^{(1)} \sim b(p)$ be an iid sequence Bernoulli-distributed and independent of the centred iid sequence $\xi_t^{(2)} \in \mathbb{L}^2$. Prove that $p < 1$ implies the stationarity of an $\mathbb{L}^2$-solution of:

$$X_n = \begin{cases} X_{n-1} + \xi_n^{(2)}, & \text{if } \xi_n^{(1)} = 1, \\ \xi_n^{(2)}, & \text{if } \xi_n^{(1)} = 0. \end{cases} \tag{7.11}$$

Its trajectories are simulated in Fig. 7.6.

Exercise 42 Estimate the parameters (p, μ), $p = \mathbb{P}(\xi_t^{(1)} = 1)$ and $\mu = \mathbb{E}\xi_t^{(2)}$, in the model (7.11).

Hints. Use a moment method for $m = \mathbb{E}X_0$ and $M = \mathbb{E}X_0^2$.

Set $q = 1 - p$ and $\nu = \mathbb{E}(\xi_t^{(2)})^2$, which we assumed to be known, then

$$m = q\mu + p(\mu + m) = \mu + pm \Longrightarrow m = \frac{\mu}{q},$$

and $M = q\nu + p(M + 2\mu m + \nu) = pM + \nu + 2\mu^2 \cdot \dfrac{p}{q}$, then

$$M = \frac{\nu}{q} + 2\mu^2 \cdot \frac{\nu}{q^2}.$$

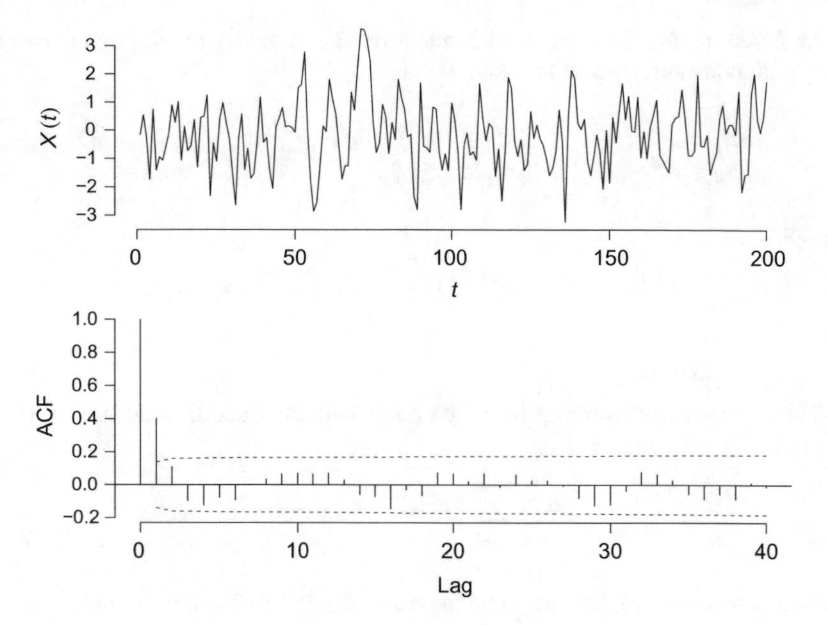

Fig. 7.6 Simulated trajectory and simple correlogram of a switching process.
Here, $X_t = \xi_t^{(1)} X_{t-1} + \xi_t^{(2)}$ with $\xi_t^{(1)} \sim \mathcal{B}(0.5)$ and $\xi_t^{(2)} \sim \mathcal{N}(0, 1)$. This model switches between a random walk and an iid behaviour

The relation $q = \dfrac{\mu}{m}$ entails $M = \dfrac{\nu}{m}\mu + 2\nu m^2$, so that:

$$\mu = \frac{\nu m}{M - 2\nu m^2}, \qquad p = 1 - \frac{\nu}{M - 2\nu m^2}.$$

Those expressions are respectively fitted by

$$\widehat{\mu} = \frac{\nu \widehat{m}}{\widehat{M} - 2\nu \widehat{m}^2}, \qquad \widehat{p} = 1 - \frac{\widehat{\nu}}{\widehat{M} - 2\nu \widehat{m}^2}.$$

The ergodic Theorem 9.1.1 entails the consistency of such estimators.

- If $D = 3$ and $\xi_0^{(1)} = 1 - \xi_0^{(2)} \sim b(p)$ is again independent of the centred random variable $\xi_0^{(3)} \in \mathbb{L}^2$, we get random regime models if $A_3 = 1$ and the contraction conditions can be written as $\mathbb{E}|\xi_0^{(3)}|^2 < \infty$ and

$$a = p\,(\mathrm{Lip}(A_1))^2 + (1-p)\,(\mathrm{Lip}(A_2))^2 < 1.$$

This model is defined through the recursion

$$X_n = \begin{cases} A_1(X_{n-1}) + \xi_n^{(3)}, & \text{if } \xi_n^{(1)} = 1, \\ A_2(X_{n-1}) + \xi_n^{(3)}, & \text{if } \xi_n^{(1)} = 0. \end{cases}$$

7.3.5 Integer Valued Autoregressions

Definition 7.3.2 Let $(\mathbf{P}(a))_{a\in\mathbb{R}}$ denote a family of integer valued distributions where $\mathbf{P}(a)$ admits the mean a.

The Steutel–van Harn (or thinning) operator is defined if $x \in \mathbb{N}$ as

$$a \circ x = \begin{cases} \sum_{i=1}^{x} Y_i, & \text{for } x \geq 1, \\ 0, & \text{otherwise.} \end{cases}$$

for a sequence of independent identically distributed random variables with marginal distribution $Y_i \sim \mathbf{P}(a)$. The random variables Y_i are also assumed to be context free, i.e. independent of any past history.

Remark 7.3.8 In any case $(a, x) \mapsto Z(a, x) = a \circ x$ is a stochastic process (see Exercise 49).

The previous "context free" assumption means in fact that this process is independent from the past history.

Example 7.3.3 (*Integral distributions*) The following distributions admit integer supports.

- Bernoulli distributions are generated from a an iid sequence of uniform random variables $U_1, U_2, \ldots \sim U[0, 1]$, and simultaneously through the relation

$$a \circ x = \mathbb{I}_{\{U_1 \leq a\}} + \cdots + \mathbb{I}_{\{U_x \leq a\}}.$$

- For the Poisson case, assume that P is a Poisson process:

$$a \circ x = P(ax).$$

- For any random variable $Z \in [0, 1]$, $a \circ x = P(axZ)$ yields a very general class of random variables.
  A simple example is if $Z \sim b(p)$ then

$$a \circ x \sim \mathcal{P}(ax)$$

  with the probability p and it is 0, otherwise.

For example the Galton–Watson process with immigration (naturally called INAR(1)-model) fits the simple recursion

$$X_t = a \circ X_{t-1} + \zeta_t.$$

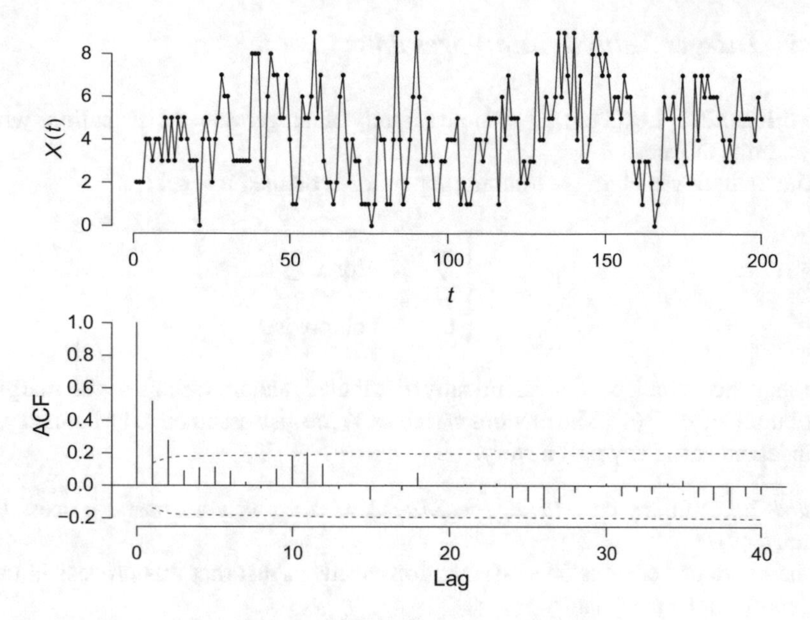

Fig. 7.7 Simulated trajectory and simple correlogram of INAR(1).
Here, process satisfying $X_t = \alpha \circ X_{t-1} + \zeta_t$ with $\zeta_t \sim \mathcal{P}(2)$ and $\mathcal{B}(0.5)$ thinning operator

Here (ζ_t) denotes another independent identically distributed and integer valued sequence, which is also independent of the thinning operators. This model is simulated in Fig. 7.7.

Now, for an independent identically distributed triangular array $(Y_{t,i})_{t\in\mathbb{Z},i\in\mathbb{N}}$, we have

$$X_t = Z_t(a, X_{t-1}) + \zeta_t, = \sum_{i=1}^{X_{t-1}} Y_{t,i} + \zeta_t.$$

Hence we again write this as a model with independent and identically distributed innovations $(\xi_t)_{t\in\mathbb{Z}}$

$$X_t = M(X_{t-1}, \xi_t), \qquad \text{with} \qquad \xi_t = ((Y_{t,i})_{i\geq 1}, \zeta_t).$$

Here $M(0, \xi_0) = \zeta_0$ hence $\|M(0, \xi_0)\|_p = \|\zeta_0\|_p$.

For $y > x$ and $p \geq 1$ we derive

$$M(y, \xi_0) - M(x, \xi_0) = \sum_{i=x+1}^{y} Y_i,$$

and

$$\|M(y, \xi_0) - M(x, \xi_0)\|_p \leq a|y - x|.$$

Many other integer models contain the same idea. A simple example is the following bilinear model:

$$X_t = a \circ X_{t-1} + b \circ (X_{t-1}\zeta_t) + \zeta_t.$$

The following exercise is immediate:

Exercise 43 Prove that the contraction assumption (on a, b and on ζ_0's distribution) in the previous theorem, if $p = 1$, can be written as:

$$a\mathbb{E}\zeta_0 + b < 1.$$

Integer valued extensions of AR(p) models are also easy to define as well as vector valued models.

Definition 7.3.3 (*INMA-models*) Define a sequence of iid thinning operators (Z_t) as above. Let $a_0, \ldots, a_m \geq 0$. Define integer moving averages with order m as

$$X_t = Z_t(a_0, 1) + \cdots + Z_{t-m}(a_m, 1).$$

Exercise 44 The above model is again strictly stationary; determine both its mean and its variance in the Poisson case, $a \circ x = P(ax)$. Define this distribution in terms of the distributions determined in Exercise 91.

Exercise 45 Let $(\zeta_t)_{t\in\mathbb{Z}}$ be an iid integer valued sequence, define

$$X_t = b_0 Z_t(a_0, \zeta_t) + \cdots + b_m Z_{t-m}(a_m, \zeta_{t-m}).$$

- Prove that this model is another stationary time series model with integer values. It is also an m-dependent sequence. This means that the sigma-fields

$$\sigma(X_i/i \leq k), \text{ and } \sigma(X_i/i \geq k + m)$$

  are independent for each k.
- Assume now that $\zeta_i = 1$ is constant and that $Z_i(a, 1)$ are independent unit Poisson processes. Describe the marginal distribution of X_0 (refer to Exercise 91).

7.3.6 Generalized Linear Models

Another way to produce attractive classes of integer valued models follows the same lines as for AR-ARCH models.

Generalized Linear Models (GLM) are easily produced from the rich monograph (Kedem and Fokianos 2002).

Definition 7.3.4 Assume that $(V(u))_{u\in\mathbb{U}}$ is a process defined on a Banach space $\mathbb{U}$ equipped with a norm $\|\cdot\|$ and $f : E \times \mathbb{U} \to E$ is a measurable function. Then

$$X_t|\mathcal{F}_{t-1} \sim V(\lambda_t), \qquad \lambda_t = f(X_{t-1}, \lambda_{t-1})$$

and $\mathcal{F}_{t-1} = \sigma(Z_s / s < t)$ denotes the historical filtration associated with the process $Z_t = (X_t, U_t)$.

Remark 7.3.9 GARCH(p, q) models are analogously solution of the above equation with $\lambda_t = f(X_{t-1}, \ldots, X_{t-p}, \lambda_{t-1}, \ldots, , \lambda_{t-q})$.

Example 7.3.4 Some examples of this situation follow.

- Let $U \sim U[0, 1]$ follow a uniform distribution, a simple example of such a process is
$$V(u) = \mathbb{I}_{\{U \le u\}}.$$
This provides us with Bernoulli distributed GLMs. A first example of this situation is developed in Example 1.1.3. Such models are nice for modelling categorical data.
- The usual way to define ARCH-models follows with $\mathbb{U} = \mathbb{R}$, $V = W$ (the Brownian motion) and
$$f(x, u) = \sqrt{\beta + \gamma^2 x^2} \cdot u.$$
- Set $\mathcal{P}(\lambda)$ the Poisson distribution with parameter λ. Consider a unit Poisson processes (see Definition A.2.5). Poisson GLM models (integer valued) are defined as:
$$X_t \,|\, \mathcal{F}_{t-1} \sim \mathcal{P}(\lambda_t), \qquad \lambda_t = f(X_{t-1}, \lambda_{t-1}).$$

The simple equations fitting the above constraints can be written as the recursive system

$$X_t = P_t(\lambda_t), \qquad \lambda_t = f(X_{t-1}, \lambda_{t-1}) \tag{7.12}$$

for some independent identically distributed sequence P_t of unit Poisson processes.

Note that X_t is not Markov and that either λ_t nor $Z_t = (X_t, \lambda_t)$ are Markov processes, *equivalently random iterative systems*, $X_t = M(X_{t-1}, \xi_t)$. As an exercise on may check the existence of $\mathbb{L}^1$-solutions of those processes determined with the affine function $f(x, \ell) = a + bx + c\ell$ in Fig. 7.8.

A main point relies on the fact that, for a homogeneous unit Poisson process:

$$|P(u) - P(v)| \sim P(|u - v|).$$

Consider the bivariate model $Z_t = (X_t, \lambda_t)$ on $\mathbb{R}^+ \times \mathbb{N} \subset \mathbb{R}^2$ equipped with the norm $\|(u, \ell)\| = |u| + \epsilon|\ell|$ for a given parameter $\epsilon > 0$.

Then this GLM model can be written as

$$M((x, \ell), P) = (P(f(x, \ell)), f(x, \ell)).$$

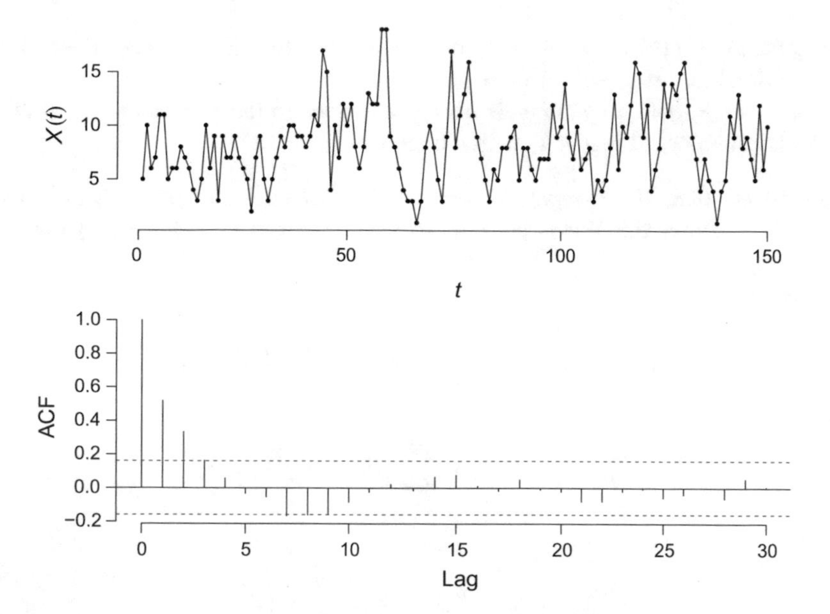

Fig. 7.8 Simulated trajectory and simple correlogram of INGARCH. Here, $X_t \sim \mathcal{P}(\lambda_t)$ with $\lambda_t = 0.5 + 0.25X_{t-1} + 0.5\lambda_{t-13}$

Here for $Z \in \mathbb{R}^2$ a random vector, we set $\|Z\|_1 = \mathbb{E}\|Z\|$. Then it is possible to check the contraction assumptions in Sect. 7.3:

- $M((0,0), P)$ is a vector with a first random coordinate $\mathcal{P}(f(0,0))$ and a deterministic second coordinate $f(0,0)$; it admits moments e.g. with order 1.
- Increments are considered as follows:

$$M((x,\ell), P) - M((x',\ell'), P) \\ = (P(f(x,\ell)) - P(f(x',\ell')), f(x,\ell) - f(x',\ell')).$$

We derive:

$$\|M((x,\ell), P) - M((x',\ell'), P)\|_1 = (1+\epsilon)|f(x,\ell) - f(x',\ell')|.$$

If the function f is Lipschitz with

$$|f(x,\ell) - f(x',\ell')| \le a|x - x'| + b|\ell - \ell'|,$$

then relations $(1+\epsilon)a < 1$ and $(1+\epsilon)b < \epsilon$ together imply the relation (7.4) with $k = a + b < 1$ by choosing $\epsilon = b/a$.

Then, some cases may be considered:

- The stability holds if Lip $f < \frac{1}{2}$; for this set $\epsilon = 1$.

- If $f(x, \ell) = g(x)$ only depends on x (analogous to ARCH-cases), the stability condition holds if $\mathrm{Lip}\, g < 1$ ($\epsilon = 0$).
- If $f(x, \ell) = g(\ell)$ only depends on ℓ (analogous to the MA-case), the stability condition holds if $\mathrm{Lip}\, g < 1$ (with a large $\epsilon > 0$).

Exercise 46 (*ARCH(∞)-representation of GARCH-models*) Define X_t as in Definition 7.3.4. Prove that this is possible to write $\lambda_t = g(X_{t-1}, X_{t-2}, \ldots)$ for some function g.

Assume that

$$|f(x', \ell') - f(x, \ell)| \le a|x' - x| + b|\ell' - \ell|.$$

Prove that

$$|g(y) - g(x)| \le \sum_{p=1}^{\infty} a^{p-1} b\, |y_p - x_p|.$$

Note that

$$\sum_{p=1}^{\infty} a^{p-1} b = \frac{b}{1-a} < 1 \Longleftrightarrow a + b < 1.$$

This last point also proves that assumptions in Theorem 7.4.1 thus hold in case $a + b < 1$.

Hints. First $g(x_1, x_2, \ldots) = f(g(x_2, x_3, \ldots), x_1)$. Thus iteratively $g(x_1, x_2, \ldots)$ is an explicit function of $(x_1, x_2, \ldots, x_p)$ in case $x_k = 0$ for $k > p$. The idea is that iterating p times provides an explicite form of g as a function of $(x_1, x_2, \ldots, x_p)$.

Now for $x = (x_1, x_2, \ldots)$ and $y = (y_1, y_2, \ldots)$ we derive recursively:

$$\begin{aligned}
|g(y) - g(x)| &= |f(g(y_2, y_3, \ldots), y_1) - f(g(x_2, x_3, \ldots), x_1)| \\
&\le a|g(y_2, y_3, \ldots) - g(x_2, x_3, \ldots)| + b|y_1 - x_1| \\
&\le a^2|g(y_3, y_4, \ldots) - g(x_3, x_4, \ldots)| + ab|y_2 - x_2| + b|y_1 - x_1| \\
&\quad \ldots\ldots \\
&\le a^p|g(y_{p+1}, y_{p+2}, \ldots) - g(x_{p+1}, x_{p+2}, \ldots)| \\
&\quad + a^{p-1}b|y_p - x_p| + \cdots + ab|y_2 - x_2| + b|y_1 - x_1|.
\end{aligned}$$

The proof is complete.

Exercise 47 (*GARCH-Poisson model*) In Eq. (7.12) consider the function $f(x, \ell) = a + bx + c\ell$. Assume that coefficients are such that a stationary solution of the equation $(X_t, \lambda_t)_t$ exists in $\mathbb{L}^2$. Then

$$\mathbb{E}X_0 = \mathbb{E}\lambda_0 = \frac{a}{b+c}.$$

Set $\mu = \mathbb{E}X_0$ then $\mu = a + (b+c)\mu$. As before such considerations are useful to fit the model.

Exercise 48 (*ARCH-Bernoulli model*) Let (U_k) be an iid uniform random variables on $[0, 1]$ and U be uniform too. Set $\beta_k(x) = \mathbb{1}_{\{U_k \le x\}}$, $\beta(x) = \mathbb{1}_{\{U \le x\}}$ and $a, b \ge 0$, consider the equation:

$$X_k = \beta_k\Big(aX_{k-1} + b\Big). \tag{7.13}$$

1. Write (7.13) as $X_k = F(X_{k-1}, \beta_k)$. Give a condition on $a, b > 0$ such that (7.13) admits a stationary solution; also prove that $\mathbb{P}(X_k \notin \{0, 1\}) = 0$.
2. If $b > 0$ prove that 0 is not solution of (7.13).
3. If $a + b \le 1$, then derive $\mathbb{E}F(x, \gamma) = ax + b$ for $x \in \{0, 1\}$.
4. Suppose from now on that $a \ge 0, b > 0, a + b \le 1$. Compute $m = \mathbb{E}X_k$, $c = \mathbb{E}X_k X_{k+1}$ as a function of a, b.
5. How may parameters a, b be fitted?

The Example 1.1.3 exhibits some advantages of analogue models.

Hints.

1. $\mathbb{E}F(x, \gamma) = (ax + b) \wedge 1$. Now $\mathbb{E}|F(y, \beta) - F(x, \beta)| \le (b - a)x$ if $y \ge x$, thus $a < 1$.
2. $\mathbb{P}(F(x, \beta) = 1) \ge b > 0$.
3. Immediate.
4. $m = am + b$ implies $m = b/(1 - a)$,
   $c = \mathbb{E}X_k X_{k+1} = \mathbb{E}X_k(aX_k + b) = (a + b)m = b(a + b)/(1 - a)$.
5. $b = m(1 - a)$ hence $c(1 - a) = m(1 - a)(m(1 - a) + a)$.
   Then $(1 - a)(m(m(1 - a) + a) - c) = 0$ which implies

   $$a = 1, \qquad \text{or} \qquad m(1 - m)a = c - m^2;$$

   condition $a + b < 1$ excludes $a = 1$ and then $a = (c - m^2)/(m(1 - m))$. a, b are fitted by plug-in through empirical estimates of m, c.

Exercise 49 Consider the simplex:

$$\Pi = \left\{ p = (p_k)_{k \ge 0} \Big/ \; p_k \ge 0, \; \sum_{k=0}^{\infty} p_k = 1 \right\}.$$

Use the simulation trick in (A.4) to define an iid sequence of processes $Z_n(p)$ where $p = (p_k)_{k \ge 0} \in \Pi$ with a fixed support $\{x_k / \; k \ge 0\}$ in case one generates an iid sequence $(U_n)_{n \in \mathbb{Z}}$ with uniform marginals.

Remark 7.3.10 For simplicity let the support be $\mathbb{N}$ ($x_k = k$ for $k \ge 0$) then $Z_n(p) \in \mathbb{N}$ is integer valued.

For a function $g : \mathbb{R}^+ \to \Pi$ one goes back to a sequence of random processes $Y_n(\lambda)$ analogously to the Poisson case by setting $p = g(\lambda)$.

The simplest case of this situation is a Bernoulli distributed processes

$$Y_n(\lambda) = \mathbb{1}_{\{U_n \le \lambda\}}.$$

Such models allow to define GLM-models, also introduced in Remark 7.3.8.

7.3.7 Non-linear AR(d)-Models

The (real valued) non linear auto-regressive model with order d is:

$$X_t = r(X_{t-1}, \ldots, X_{t-d}) + \xi_t, \tag{7.14}$$

The vector valued sequence $U_n = (X_n, X_{n-1}, \ldots, X_{n-d+1})$ can be written as a Markov models with values in $\mathbb{R}^d$.

Here $E = \mathbb{R}$ and

$$M(u_1, \ldots, u_d, z) = A(u_1, \ldots, u_d) + (1, 0, \ldots, 0)z,$$

where

$$A(u_1, \ldots, u_d) = \big(r(u_1, \ldots, u_d), u_1, \ldots, u_{d-1}\big).$$

Theorem 7.3.2 *Assume that* $\mathbb{E}|\xi_0|^m < \infty$ *and*

$$|r(u_1, \ldots, u_d) - r(v_1, \ldots, v_d)| \le \sum_{i=1}^d a_i |u_i - v_i|,$$

for $a_1, \ldots, a_d \ge 0$ *such that*

$$\alpha^d = \sum_{i=1}^d a_i < 1.$$

Then the Eq. (7.14) *admits a stationary solution and this solution is in* $\mathbb{L}^m$.

Proof Define a norm on $\mathbb{R}^d$ by

$$\|(u_1, \ldots, u_d)\| = \max\{|u_1|, \alpha|u_2|, \ldots, \alpha^{d-1}|u_d|\}.$$

For $u = (u_1, \ldots, u_d)$, $v = (v_1, \ldots, v_d) \in \mathbb{R}^d$ set $w_j = |u_j - v_j|$ for $j = 1, \ldots, d$:

$$\begin{aligned}\|A(u) - A(v)\| &\le \max\left\{\alpha^d \max\{w_1, \ldots, w_d\}, \alpha w_1, \ldots, \alpha^{d-1} w_{d-1}\right\}\\ &\le \alpha \max\left\{\alpha^{d-1} \max\{w_1, \ldots, w_d\}, w_1, \ldots, \alpha^{d-2} w_{d-1}\right\}\\ &\le \max\{w_1, \alpha w_2, \ldots, \alpha^{d-1} w_d\} = \alpha\|u - v\|.\end{aligned}$$

The Duflo condition (7.4) holds with $a = \alpha^m < 1$.

Remark 7.3.11 As an alternative proof, Theorem 7.4.1 directly implies the result.

7.4 Bernoulli Schemes

The following approach to time series modelling is definitely simpler and sharper but it is also less intuitive so that it appears only at the end of the chapter.

7.4.1 *Structure and Tools*

Definition 7.4.1 (*Informal definition*) The model

$$X_n = H(\xi^{(n)}), \qquad \text{with} \qquad \xi^{(n)} = (\xi_{n-t})_{t\in\mathbb{Z}}. \tag{7.15}$$

is defined through the function H defined on $E^{\mathbb{Z}} \to \mathbb{R}^1$ and $\xi^{(n)} = (\xi_{n-k})_{k\in\mathbb{Z}}$ is again an iid sequence with a shifted time index.

Suppose that $\{\xi_k / \; k \in \mathbb{Z}\}$ takes values in a measurable space $(E, \mathcal{E})$.

We consider some examples of such situations.

An important special case is that of causal processes $H : E^{\mathbb{N}} \to \mathbb{R}$ and we write in a simpler formulation

$$X_n = H(\xi_n, \xi_{n-1}, \xi_{n-2}, \ldots).$$

Such a stationary process is said causal since the history of X before the epoch n is included in that of ξ. This means

$$\sigma(X_s/s \le n) \subset \sigma(\xi_s/s \le n).$$

Fix some $e \in E$ we denote $\widehat{\xi}(n) = (\widehat{\xi}_j(n))_j$ the sequence with current element $\widehat{\xi}_j(n) = \xi_j$, if $|j| \le n$ and $\widehat{\xi}_j(n) = e$ if $|j| > n$. Let $m \ge 1$, a simple condition to define such models is

$$\sum_{n=1}^{\infty} \zeta_n < \infty, \tag{7.16}$$

with

$$\zeta_n^p = \mathbb{E}\left|H\left(\widehat{\xi}(n)\right) - H\left(\widehat{\xi}(n-1)\right)\right|^p. \tag{7.17}$$

[1]Few such functions H may be rigorously defined; this question is solved through the "formal definition" 7.4.2.

Due to the completeness of the space $\mathbb{L}^p$, a normally convergent series is convergent and:

Proposition 7.4.1 *Let $p \geq 1$ be such that relation* (7.16) *holds then the sequence $(X_n)_{n\in\mathbb{Z}}$ defined this way is stationary and $\mathbb{L}^p$-valued.*

Proof The relation (7.16) indeed implies the convergence $\mathbb{L}^p$ of the well-defined sequence $H\left(\left(\xi_j\right)_{|j|\leq n}\right)$.

A bit more is needed to conclude; one extends the previous remark to the random variable $Z_n = (X_{n+1}, \ldots, X_{n+s}) \in \mathbb{R}^s$.

This is the limit of a sequence of $\mathbb{R}^s$-valued random variables with a distribution independent of n.

Example 7.4.1 (*Bernoulli shifts*)

- Let $H : \mathbb{R}^m \to \mathbb{R}$ then the process $X_n = H(\xi_n, \ldots, \xi_{n-m+1})$ is an m-dependent sequence, i.e. $\sigma\{X_j / \; j < a\}$ and $\sigma\{X_j / \; j > a + m\}$ are independent σ-algebras.
- Infinite moving averages

$$H(u_0, u_1, \ldots) = a_0 u_0 + a_1 u_1 + \cdots$$

yield $\zeta_n = |a_n| \cdot \|\xi_0\|_p$, which again confirms Lemma 6.1.1. This simple example is a case for which the function H is not always properly defined, in case infinitely many coefficients a_k do not vanish. E.g. let $N = \{k / \; a_k \neq 0\}$ then set $u_k = 1/a_k$ if $k \in N$ and 0 otherwise, then $H(u_0, u_1, \ldots) = \text{Card } N = \infty$.
- *Stochastic volatility model.* Let $Y_n = H(\xi_n, \xi_{n-1}, \ldots)$ be a causal Bernoulli scheme such that the independent identically distributed innovations $\xi_n \in \mathbb{L}^2$ are centred.
Set

$$X_n = \xi_n Y_{n-1} = \xi_n H(\xi_{n-1}, \xi_{n-2}, \ldots).$$

The sequence X_n is orthogonal and

$$\text{Var}\,(X_n|\mathcal{F}_{n-1}) = Y_{n-1}^2.$$

This property indicates possible rapid changes adapted to model the stocks exchange.
- All the previous sections of the present chapter provide us with a series of examples of this situation.

The previous Definition 7.4.1 is really adapted to deal with the previous chaotic examples for which tails may be bounded above.

A more general setting is adapted to prove the existence of a stationary process.

Definition 7.4.2 (*Formal definition*) Let μ be a probability distribution on a measurable space $(E, \mathcal{E})$. Consider an iid sequence $(\xi_n)_{n\in\mathbb{Z}}$ with marginal law μ. Set $\nu = \mu^{\otimes\mathbb{Z}}$ the law of $(\xi_n)_{n\in\mathbb{Z}}$ on the space $(E^{\mathbb{Z}}, \mathcal{E}^{\otimes\mathbb{Z}})$.

Then $\mathbb{L}^p(\nu)$ is the space of measurable functions ν-a.s defined on $E^{\mathbb{Z}}$ and such that

$$\mathbb{E}\Big|H\Big((\xi_n)_{n\in\mathbb{Z}}\Big)\Big|^p < \infty.$$

Analogously, we set $\nu^+ = \mu^{\otimes\mathbb{N}}$ for the law of $(\xi_n)_{n\in\mathbb{N}}$ on the space $(E^{\mathbb{N}}, \mathcal{E}^{\otimes\mathbb{N}})$.

A Bernoulli shift is an $\mathbb{L}^p$-stationary process defined as

$$X_n = H\big((\xi_{n-j})_{j\in\mathbb{Z}}\big), \quad \text{for some } H \in \mathbb{L}^p(\nu).$$

A causal Bernoulli shift is associated with $H \in \mathbb{L}^p(\nu^+)$.

Remark 7.4.1 The spaces $\mathbb{L}^p(\nu)$ and $\mathbb{L}^p(\nu^+)$ are Banach spaces (complete normed vector spaces) equipped respectively with the norms:

$$\begin{aligned} \|H\|_p &= \left(\mathbb{E}|H\left((\xi_n)_{n\in\mathbb{Z}}\right)|^p\right)^{\frac{1}{p}}, \qquad \text{for the general case,} \\ &= \left(\mathbb{E}|H\left((\xi_n)_{n\in\mathbb{N}}\right)|^p\right)^{\frac{1}{p}}, \qquad \text{for the causal case.} \end{aligned}$$

The definition of Bernoulli shifts is as in the informal Definition 7.4.1 and applies to non-causal or causal schemes for elements $H \in \mathbb{L}^p(\nu)$ or $\mathbb{L}^p(\nu^+)$ respectively. Moreover condition (7.16) implies with Proposition 7.4.1, a simple sufficient condition for functions of infinitely many variables to exist in these huge spaces.

Warning about notations. We denote

$\|Z\|_p = (\mathbb{E}|Z|^p)^{\frac{1}{p}}$ for a random variable $Z \in \mathbb{L}^p(\Omega, \mathcal{A}, \mathbb{P})$,

$\|H\|_p = \Big(\mathbb{E}|H\left((\xi_n)_{n\in\mathbb{Z}}\right)|^p\Big)^{\frac{1}{p}}$ for $H \in \mathbb{L}^p(\nu, \mathcal{B}_{\mathbb{R}}^{\otimes\mathbb{Z}}, \mathbb{R}^{\mathbb{Z}})$,

$\|H\|_p = \Big(\mathbb{E}|H\left((\xi_n)_{n\in\mathbb{N}}\right)|^p\Big)^{\frac{1}{p}}$ for $H \in \mathbb{L}^p(\nu^+, \mathcal{B}_{\mathbb{R}}^{\otimes\mathbb{N}}, \mathbb{R}^{\mathbb{N}})$.

These are a bit confusing but they have the advantage of simplicity. Recall that confusion is avoided once one understands H as a function and Z as a random variable!

The next subsection also proves that these assumptions are relevant to check short-range conditions.

A quite simple and elegant proof relies on the previous notions and proves moreover that there exists a unique element $H \in \mathbb{L}^p(\nu^+)$ such that a stationary solution of (7.3) is

$$X_t = H(\xi_t, \xi_{t-1}, \xi_{t-2}, \ldots).$$

Exercise 50 (*Fixed-point criterion*)

1. Prove again Theorem 7.3.1.

2. Show that there exists some $C > 0$ and a sequence of Bernoulli shifts $X_{n,t} = H_n(\xi_t, , \dots, \xi_{t-n})$ such that

$$\|X_t - X_{n,t}\|_p \le Ca^n.$$

Hints for Exercise 50.

1. To this end consider the application

$$\Phi : \mathbb{L}^p(\nu^+) \to \mathbb{L}^p(\nu^+), \qquad H \mapsto \Phi(H) = K,$$

with $K(v_0, v_1, \dots) = M(H(v_1, v_2, \dots), v_0)$.
Conditions (7.5) and (7.4) allow to prove that prove that $K \in \mathbb{L}^p(\nu^+)$ if $H \in \mathbb{L}^p(\nu^+)$; to prove this use conditionning wrt ξ_0 and the triangular inequality.
Below, the fixed point e is considered as an element of $\mathbb{L}^p$:

$$\|K\|_p = \mathbb{E}^{\frac{1}{p}}|M(H(\xi_1, \dots), \xi_0)|^p \le \mathbb{E}^{\frac{1}{p}}|M(e, \xi_0)|^p + a\|H - e\|_p.$$

Now if $H, H' \in \mathbb{L}^p(\nu^+)$ then again conditioning with respect to $\xi_1, \xi_2, \dots$ implies

$$\|K - K'\|_p \le a\|H - H'\|_p.$$

The Banach–Picard fixed point theorem (see Choquet 1973) classically implies that Φ admits a unique fixed point $H^\star$.
2. This theorem also implies that the iterates $H_n = \Phi^n \circ H_0$ converge to this fixed point $H^\star$ of Φ. In other words $X_{t,n} = H_n(\xi_t, \xi_{t-1}, \xi_{t-2}, \dots)$ converge in $\mathbb{L}^p$ to the stationary solution of the previous recursion as $n \uparrow \infty$ for each value of t.
Moreover the convergent rate is geometric:

$$\|X_t - X_{t,n}\|_p = \|H^\star - H_n\|_p \le Ca^n,$$

for suitable constants $C > 0$ and with $a \in [0, 1[$ from the assumptions.

More generally, the fixed point theorem in the Banach space $\mathbb{L}^p$ implies:

Theorem 7.4.1 (Doukhan and Wintenberger 2008) *Let B be a Banach space, $(E, \mathcal{E})$ be a measurable space and let be $(\xi_n)_{n\in\mathbb{Z}}$ be an iid E-valued sequence.*
Suppose $M : B^{\mathbb{N}^*} \times E \to B$, satisfies for some $p \ge 1$, and some $x_0 \in B^{\mathbb{N}^*}$,

$$\|M(x_0, \xi_0)\|_p < \infty. \tag{7.18}$$

Suppose also that there exists some sequence $a_j \ge 0$ $(j \ge 1)$ such that, for all $x = (x_j)_{j\ge1}, y = (y_j)_{j\ge1} \in B^{\mathbb{N}^*}$:

$$\|M(y, \xi_0) - M(x, \xi_0)\|_p \le \sum_{j\ge1} a_j \|y_j - x_j\|. \tag{7.19}$$

$$a = \sum_{j\ge1} a_j < 1,$$

Then there exists a strictly stationary solution in $\mathbb{L}^p$ *for the infinite recursion*

$$X_t = M(X_{t-1}, X_{t-2}, X_{t-3}, \ldots, \xi_t), \qquad \forall t \in \mathbb{Z}.$$

Uniqueness holds if $X_t = H(\xi_t, \xi_{t-1}, \ldots)$ *for* $H \in \mathbb{L}^p(E^{\mathbb{N}}, \mathcal{E}^{\otimes \mathbb{N}}, \nu^+)$ *with* $\nu^+ = \mu^{\otimes \mathbb{N}}$ *for* μ *the distribution of* ξ_0.

Hint for Theorem 7.4.1. Proceed as in Exercise 50 with adequate changes in notations; since now the model is B-valued and no longer real valued, thus $|\cdot|$ needs to be replaced by $\|\cdot\|$.

Now:

$$K(v_0, v_1, \ldots) = M\big(H(v_1, v_2, \ldots), H(v_2, v_3, \ldots), H(v_3, v_4, \ldots), \ldots; v_0\big).$$

Set $K = \Phi(H)$ and $K' = \Phi(H')$.

Conditioning wrt $\xi_1, \xi_2, \ldots$, we may write the above assumption (7.19) with

$$x = \Big(H(\xi_1, \xi_2 \ldots), H(\xi_2, \xi_3, \ldots), H(\xi_3, \xi_4, \ldots), \ldots\Big) \quad = (x_0, x_1, x_2 \ldots),$$
$$y = \Big(H'(\xi_1, \xi_2, \ldots), H'(\xi_2, \xi_3, \ldots), H'(\xi_3, \xi_4, \ldots), \ldots\Big) = (y_0, y_1, y_2 \ldots).$$

Here $x_i = H(\xi_{i+1}, \xi_{i+2}, \ldots)$, $y_i = H'(\xi_{i+1}, \xi_{i+2}, \ldots)$ are random variables and $\|x_i - y_i\|_p = \|H - H'\|_p$ (here again the notation $\|\cdot\|_p$ may be troublesome since it admits two different meanings).

Then after de-conditioning:

$$\|K - K'\|_p \le \sum_{i=0}^{\infty} a_i \|x_i - y_i\|_p = a\|H - H'\|_p.$$

The special case $H' = 0$ yields with relation (7.18) and the triangle inequality that indeed $\Phi(H) \in \mathbb{L}^p(\nu^+)$.

The assumptions are associated with this proof. The operator $\Phi : \mathbb{L}^p(\nu^+) \to \mathbb{L}^p(\nu^+)$ is contractive.

Example 7.4.2 (*Poisson GLM-models*) As in Sect. 7.3.6 one may consider GLM-models $Y_t = P_t(\lambda_t)$ with infinite memory.

Setting $X_t = (Y_t, \lambda_t)$, we assume that those models admit an infinite memory:

$$\lambda_t = g(X_{t-1}, X_{t-2}, X_{t-3}, \ldots).$$

Assume that for all $x = (x_j)_{j\ge1}$ with $x_j = (y_j, \ell_j)$, and $x' = (x'_j)_{j\ge1}$ with $x'_j = (y'_j, \ell'_j)$.

We obtain:

$$|g(x') - g(x)| \le \sum_{j=1}^{\infty}(a_j|y'_j - y_j| + b_j|\ell'_j - \ell_j|)$$

then $M(x; P_0) = (P_0(g(x)), g(x))$, and we set again a norm on $B = \mathbb{R}^2$ as $\|(u, \lambda)\| = |u| + \epsilon|\lambda|$ for some $\epsilon \in \mathbb{R}$.

Now

$$\|M(x'; P_0) - M(x; P_0)\|_p \le (1+\epsilon)\sum_{j=1}^{\infty}(a_j|y'_j - y_j| + b_j|\ell'_j - \ell_j|)$$

the assumption in Theorem 7.4.1 holds in case there exists $0 < k < 1$ such that for each x, x':

$$(1+\epsilon)\sum_{j=1}^{\infty}(a_j|y'_j - y_j| + b_j|\ell'_j - \ell_j|) \le k\sum_{j=1}^{\infty}\|x'_j - x_j\|.$$

This holds if for each $j \ge 1$, $(1+\epsilon)a_j \le k$ and $(1+\epsilon)b_j \le k\epsilon$.

- If $b_j = 0$ for each j set $\epsilon > 0$ as small as wanted, then the condition

$$\sum_{j=1}^{\infty} a_j < 1$$

  implies contraction.
- If $a_j = 0$ for each j set $\epsilon > 0$ as large as wanted, the condition

$$\sum_{j=1}^{\infty} b_j < 1$$

  implies contraction.
- In the general case $\epsilon = 1$ implies contraction if

$$\sum_{j=1}^{\infty}(a_j + b_j) < \frac{1}{2}.$$

Remark 7.4.2 Contrary to the GARCH(1,1)-Poisson case in Equation (7.12), the proof above does not directly imply that the condition $a + b < 1$ yields contraction. In the present memory model of the GARCH(p, q)-type, indeed, ϵ would need to depend on $j \in [1, p]$, which does not make sense.

Exercise 46 also proves that both the GARCH(1,1) and the ARCH(∞) representations admit exactly the same contractive properties.

The extension[2] of Exercise 46 to GARCH(p, q) allows to skip the above additional factor $\frac{1}{2}$ in the general case $p \geq 2$.

Remark 7.4.3 Set $M_n((x_0, \ldots, x_n), v) = M((x_0, \ldots, x_n, 0, \ldots), v)$. Prove that there exists a $(n+1)$-Markov stationary process with

$$X_{n,t} = M_n(X_{t-1}, , \ldots, X_{t-n-1}; \xi_t),$$

such that

$$\|X_t - X_{n,t}\|_p \leq C \sum_{i=n+1}^{\infty} a_i.$$

The existence of the $(n+1)$-Markov stationary process follows from Theorem 7.4.1. The approximation through Markov models is a special case of Lemma 5.5 in Doukhan and Wintenberger (2008) for the special Orlicz function $\Phi(u) = u^p$. It relies on the respective $\mathbb{L}^p$-approximations of the functionals on $\mathbb{L}^p(\mu)$ denoted Φ in the previous point and where we denote Φ_n the functional associated with M_n.

Exercise 51 Prove that LARCH(∞) models in Sect. 7.2.2 satisfy the assumptions (7.18) and (7.19) in case

$$\|\xi_0\|_p \sum_{i=1}^{\infty} |a_i| < 1.$$

Exercise 52 Prove that NLARCH(∞)-models (NL for non-linear)

$$X_t = \xi_t \left(a_0 + \sum_{k=1}^{\infty} a_k(X_{t-k}) \right),$$

satisfy the assumptions (7.18) and (7.19) in case

$$\|\xi_0\|_p \sum_{i=1}^{\infty} \operatorname{Lip} a_i < 1.$$

Exercise 53 Use the Steutel–van Harn operator (Definition 7.3.2) in order to prove that INLARCH(∞)-models (IN for INteger)

$$X_t = \xi_t \left(a_0 + \sum_{k=1}^{\infty} a_k \circ X_{t-k} \right),$$

[2]Work on progress with Konstantinos Fokianos and Joseph Rynkiewicz.

satisfy the assumptions (7.18) and (7.19) if the following condition holds:

$$\|\xi_0\|_p \sum_{k=1}^{\infty} a_k < 1.$$

7.4.2 Couplings

This section discusses ways to couple such Bernoulli shifts. Decorrelation rates are also deduced. This will allow to derive quantitative laws of large numbers for expressions of statistical interest. These ideas are widely developed in Chap. 9 in order to understand how to derive limit theorems in distribution. Let $(\xi'_k)_{k\in\mathbb{Z}}$ be another independent identically distributed sequence, independent of $(\xi_k)_{k\in\mathbb{Z}}$, and with the same distribution. For $n \geq 0$ set $\widetilde{\xi}(n) = (\widetilde{\xi}(n)_k)_{k\in\mathbb{Z}}$ with

$$\widetilde{\xi}(n)_k = \begin{cases} \xi_k, & \text{if } |k| \leq n, \\ \xi'_k, & \text{if } |k| > n. \end{cases}$$

Then we set

$$\delta_n^p = \mathbb{E}\left|H\left(\widetilde{\xi}(n)\right) - H\left(\xi\right)\right|^p. \tag{7.20}$$

Definition 7.4.3 Assume that a Bernoulli-shift satisfies $\lim_n \delta_n^{(p)} = 0$ with the above definition (7.20) then it will be called $\mathbb{L}^p$-dependent.

Remark 7.4.4 Replace $\widetilde{\xi}(n)$ by

$$\widehat{\xi}(n)_k = \begin{cases} \xi_k, & \text{if } |k| \neq n, \\ \xi'_k, & \text{if } |k| = n. \end{cases}$$

This leads to the fruitful physical measure of dependence by Wei Biao Wu, see Wu and Rosenblatt (2005).

Here

$$\widehat{\delta}_n^p = \mathbb{E}\left|H\left(\widehat{\xi}(n)\right) - H\left(\xi\right)\right|^p. \tag{7.21}$$

The two previous proposals are **couplings** in the sense that they leave unchanged the marginal distribution of the Bernoulli-shift.

An alternative is to set

$$\xi'(n)_k = \begin{cases} \xi_k, & \text{if } |k| \leq n, \\ 0, & \text{if } |k| > n. \end{cases}$$

which is essentially the same as $\widetilde{\xi}(n)$.

In this case set:

$$\delta_n'^p = \mathbb{E}\left|H\left(\xi'(n)\right) - H\left(\xi\right)\right|^p. \tag{7.22}$$

This makes it easy to define functions of infinitely many variables as limits of functions of finitely many variables in the Banach space $\mathbb{L}^p(\nu^+)$ from the fact that $H(\xi'(n))$ is a Cauchy sequence in case

$$\sum_n \|H(\xi'(n)) - H(\xi'(n-1))\|_p < \infty.$$

Remark 7.4.5 We introduced three different coupling conditions (7.20)–(7.22). Simple relations link them but a simple exercise is useful to understand the situation.

Exercise 54 Set

$$H(x) = \sum_{i=0}^{\infty} a_i x_i$$

then bounds of the previous coupling coefficients are:

$$\delta_n = 2\|\xi_0\|_p \sum_{i>n} |a_i|,$$
$$\delta'_n = 2\|\xi_0\|_p \sum_{i>n} |a_i|,$$
$$\widehat{\delta_n} = 2|a_n| \cdot \|\xi_0\|_p.$$

Convergence of the series is a sufficient condition for the existence of linear processes while $\widehat{\delta_n}$ allows to deal with long-range dependent (LRD) linear series, see Sect. 4.3.

Now as an introduction to the weak-dependence conditions in Sect. 11.4 (see Doukhan and Louhichi 1999), we remark that:

Proposition 7.4.2 (Decorrelation) *If the stationary process $(X_n)_{n\in\mathbb{Z}}$ satisfies $\mathbb{E}|X_0|^p < \infty$ for $p \geq 2$ and is as before dependent as before if H is unbounded or bounded we have:*

$$|Cov\,(X_0, X_k)| \leq 4\|X_0\|_p \delta_{[k/2]},$$
$$\leq 4\|H\|_\infty \delta^2_{[k/2]}.$$

If the Bernoulli scheme is causal the previous inequalities can be written as:

$$|Cov\,(X_0, X_k)| \leq 2\|X_0|\|_p \delta_k,$$
$$\leq 2\|H\|_\infty \delta^2_k.$$

Remark 7.4.6 Such results imply the short-range dependence of the process X in the sense of Definition 4.3.1, in case the above covariances are summable.

Proof For this, use Hölder's inequality (Proposition A.2.2) after the relation:

$$\text{Cov}\,(X_0, X_k) = \text{Cov}\,(X_0 - X_{0,l}, X_k) + \text{Cov}\,(X_{0,l}, X_k - X_{k,l}),$$

which holds if $2l \le k$ when setting $X_{k,l} = H\left(\widetilde{\xi}(l)^{(k)}\right)$. Recall that $\widetilde{\xi}(l)^{(k)}$ denotes the sequence whose jth element is ξ_{k-j} if $|j| \le l$ and ξ'_{k-j} if $|j| > l$. If the Bernoulli scheme is causal the relation simplifies since

$$\mathrm{Cov}\,(X_0, X_k) = \mathrm{Cov}\,(X_0, X_k - X_{k,k}).$$

Now factors 4 and 2 arise from the fact that covariances are expectations of a product minus the product of expectations; the same bounds apply to both terms.

An important question is the heredity of such properties through instantaneous images $Y_k = g(X_k)$.

Denote the corresponding expressions by $\delta_{k,Y}$ and $\delta_{k,X}$, then:

Lemma 7.4.1 *Assume that $m \ge 1$. Consider a Lipschitz function $g : \mathbb{R} \to \mathbb{R}$ such that $\mathrm{Lip}\, g = L < \infty$.*

Set $Y_k = g(X_k)$ then:

$$\delta_{k,Y} \le L \cdot \delta_{k,X}.$$

The above study is justified since many statistics write as empirical mean

$$\frac{1}{n}\sum_{i=1}^{n} g(X_i, X_{i+1}, \ldots, X_{i+d}).$$

Some examples follow.

Exercise 55 Kernel density estimators are introduced in Definition 3.3.1. Assume that $(X_t)_{t\in\mathbb{Z}}$ is strictly stationary and admits bounded marginal densities for couples (X_0, X_k) uniformly wrt k.

Then the kernel density estimator satisfies for suitable constant $c > 0$:

$$\mathrm{Var}\,\widehat{f}(x) \le \frac{c}{nh}\left(1 + \sum_{k=1}^{n-1} \delta_k^{\frac{1}{3}}\right).$$

In case the previous series are convergent this is the same bound as for the iid case (see Theorem 3.3.2).

Hints. One may write

$$\widehat{f}(x) - \mathbb{E}\widehat{f}(x) = \frac{1}{nh}\sum_{k=1}^{n} U_h(X_k),$$

with $\mathrm{Lip}\, U_h \le c/h$, hence Proposition 7.4.2 with $p = 2$ entails with the above bound:

$$|\mathrm{Cov}\,(U_h(X_0), U_h(X_k))| \le \delta_k \frac{C}{h} \mathrm{Lip}\, K.$$

The case $k = 0$ corresponds to independence and is already taken into account in Theorem 3.3.2,

$$|\text{Cov}\,(U_h(X_0), U_h(X_k))| \le ch.$$

On the other hand, it is simple to check from a change of variable that

$$|\text{Cov}\,(U_h(X_0), U_h(X_k))| \le ch^2,$$

for some constant $c > 0$.

With the boundedness of the kernel we obtain:

$$|\text{Cov}\,(U_h(X_0), U_h(X_0))| \le ch\left(\frac{\delta_k}{h^2} \wedge h\right) \le ch\delta_k^{\frac{1}{3}}.$$

The last bound comes from Exercise 12.18 for $\alpha = \frac{1}{3}$. Resuming the bounds yields the result.

Simple indicators $g_x(u) = 1\!\!1_{\{u \le x\}}$ are the functions designed to derive bounds for the empirical process.

Indicator functions are the simplest discontinuous functions; they are classes of functions with only one singularity.

Under an additional concentration condition we are able to fix this problem:

Lemma 7.4.2 *If $p = 2$ and if there exist constants $c, C > 0$ such that on each interval $\mathbb{P}(X \in [a, b]) \le C|b - a|^c$, then the process defined by $Y_{x,n} = 1\!\!1_{\{X_n \le x\}}$ satisfies:*

$$\delta_{k,Y_x} \le 2(2C)^{\frac{2}{c+2}} \delta_{k,X}^{\frac{c}{c+2}}.$$

Proof Consider the continuous function such that:

$$g_{x,\epsilon} = \begin{cases} 1, & \text{if } \ u \le x - \epsilon, \\ 0, & \text{for } u \ge x, \\ \text{affine}, & \text{otherwise.} \end{cases}$$

Consider $Y_{x,\epsilon,n} = g_{x,\epsilon}(X_n)$.

Then

$$|g_{x,\epsilon}(u) - g_{x,\epsilon}(v)| \le \frac{|u - v|}{\epsilon}, \text{ and } \delta_{k,Y_{x,\epsilon}} \le \frac{\delta_{k,X}}{\epsilon}.$$

Moreover

$$|\delta^2_{k,Y_{x,\epsilon}} - \delta^2_{k,Y_x}| \le 2\mathbb{P}(X_0 \in [x - \epsilon, x]) \le 2C\epsilon^c.$$

So

$$\delta^2_{k,Y_{x,\epsilon}} \le \frac{\delta^2_{k,X}}{\epsilon^2} + 2C\epsilon^c.$$

To conclude the proof, set $\epsilon^{c+2} = \dfrac{\delta_{k,X}^2}{2C}$.

Remark 7.4.7 Up to a constant the result remains valid for any function g Lipschitz-continuous on finitely many intervals.

A control for the cumulative empirical distribution follows:

Exercise 56 Prove that:

$$\text{Var } F_n(x) = \mathcal{O}\left(\frac{1}{n}\right), \qquad \text{if} \quad \sum_{k=0}^{\infty} \delta_{k,X}^{\frac{c}{c+2}} < \infty.$$

In case $c = 1$, which hold for X_0's distribution with a bounded density, the condition is

$$\sum_{k=0}^{\infty} \delta_{k,X}^{\frac{1}{3}} < \infty.$$

This holds for example in case the marginal law of X_0 admits a bounded density.

Remark 7.4.8 Higher order moment inequalities for such partial sums can also be derived as in Chap. 12, see Doukhan and Louhichi (1999) and Doukhan et al. (2011).

Chapter 8
Associated Processes

The notion of association, or positive correlation, was naturally introduced in two different fields: **reliability** (Esary et al. 1967) and **statistical physics** (Fortuin et al. 1971) to model a tendency that the coordinates of a vector valued random variable admit such behaviours. We refer the reader to Newman (1984) for more details. This notion deserves much attention since it provides a class of random variables for which independence and orthogonality coincide. Another case for which this feature holds is the Gaussian case, see Chap. 5. The notion of independence is more related to σ-algebras but in those two cases it is related to the geometric notion of orthogonality. Those remarks are of interest for modelling dependence as this is the aim of Chap. 9.

8.1 Association

Definition 8.1.1 A random vector $X \in \mathbb{R}^p$ is associated if, for all measurable functions $f, g : \mathbb{R}^p \to \mathbb{R}$ with $\mathbb{E}|f(X)|^2 < \infty$ and $\mathbb{E}|g(X)|^2 < \infty$ such that f, g are coordinatewise non-decreasing, we have:

$$\mathrm{Cov}(f(X), g(X)) \geq 0.$$

Definition 8.1.2 A random process $(X_t)_{t\in\mathbb{T}}$ is associated if the vector $(X_t)_{t\in F}$ is associated for each finite $F \subset \mathbb{T}$.

Remark 8.1.1 Covariances of an associated process are non-negative if this process is square integrable.

We will present the main inequality (8.1) to prove that weak-dependence conditions in Chap. 11 are related with association.

P. Doukhan, *Stochastic Models for Time Series*, Mathématiques et Applications 80,
https://doi.org/10.1007/978-3-319-76938-7_8

Exercise 57 A real random variable is always associated.

Hint for Exercise 57. Indeed if X' is an independent copy of X then calculus proves that

$$\mathrm{Cov}(f(X), g(X)) = \frac{1}{2}\mathbb{E}(f(X) - f(X')(g(X) - g(X')).$$

Hence for f, g monotonic this expression is non-negative.

More generally:

Theorem 8.1.1 (Newman 1984) *Independent vectors on* $\mathbb{R}^p$ *are associated.*

Proof A recursion is needed. The simple Exercise 57 considers the case of dimension 1.

A careful conditioning is needed. For this first derive from Cauchy-Schwartz inequality that:

Lemma 8.1.1 *Let* $Z = (X, Y) \in \mathbb{R}^{p+q}$ *and* $f, g : \mathbb{R}^{p+q} \to \mathbb{R}$ *be such that* $f(Z)$ *and* $g(Z) \in \mathbb{L}^2$.

If X, Y *are independent vectors then*

$$F(x) = \mathbb{E}f(x, Y) \text{ and } G(x) = \mathbb{E}g(x, Y) \in \mathbb{L}^2 \text{ for, a.s., each } x \in \mathbb{R}^p.$$

Exercise 58 Let $Z = (X, Y) \in \mathbb{R}^{p+q}$ be a random vector. Assume that the vectors X and Y are independent. Setting $U(x) = \mathrm{Cov}(f(x, Y), g(x, Y))$, prove that:

$$\mathrm{Cov}(f(Z), g(Z)) = \mathbb{E}U(x) + \mathrm{Cov}(F(X), G(X)).$$

Hint. From the Cauchy–Schwarz inequality one easily derives that both random variables $F(X), G(X) \in \mathbb{L}^2$.

End of the proof of Theorem 8.1.1. Use Lemma 8.1.1 with $p = 1$ for the recursion. The decomposition in Exercise 58 ends the proof.

Theorem 8.1.2 (Newman 1984) *A limit in distribution of a sequence of associated vectors is associated.*

Proof In the definition of association first restrict to bounded coordinatewise non decreasing functions f, g, then if $X_n \to X$ from boundedness of $f, g, f \times g$ we get $\mathbb{E}f(X_n) \to \mathbb{E}f(X), \mathbb{E}g(X_n) \to \mathbb{E}g(X)$ and $\mathbb{E}f(X_n)g(X_n) \to \mathbb{E}f(X)g(X)$.

For $M > 0$ and for any function f, coordinatewise monotonic,

$$f_M(x) = f(x) \wedge M \vee (-M)$$

is again monotonic.

Moreover from the tightness of previous distributions for each $\epsilon > 0$ there is some M, n_0 such that

$$\mathrm{Cov}(f_M(X_n), f_M(X_n)) \geq -\epsilon, \qquad \forall n \geq n_0.$$

Combining with the previous relations yields the result.

8.2 Associated Processes

Definition 8.2.1 A random process $(X_t)_{t\in\mathbb{T}}$ is associated if for each $S \subset \mathbb{T}$ finite, the vector $(X_t)_{t\in S}$ is associated.

Remark 8.2.1 Heredity properties of association are very important to handle applications involving associated processes.

Example 8.2.1 The following examples inherit the association properties:

- Non-decreasing images of associated sequences are associated.

  This heredity property admits many consequences:

- For example monotonic images of independent sequences are associated.

- LARCH(∞)-models with non-negative coefficients $a_j \ge 0$ and inputs $\xi_j \ge 0$ are associated:

$$X_t = \Big(a_0 + \sum_{j=1}^{\infty} a_j X_{t-j}\Big)\xi_t.$$

  characterize To check this, use a recursion, the point that a linear function

$$(x_1, \ldots, x_p) \mapsto \sum_{j=1}^{p} b_j x_j,$$

  with non-negative coefficients $b_j (=a_j \xi_t)$ is non-decreasing and the fact that association is stable under limits in distribution.

- Autoregressive process. Solutions of an equation

$$X_t = r(X_{t-1}, \ldots, X_{t-p}) + \xi_t,$$

  are associated if the function $r : \mathbb{R}^p \to \mathbb{R}$ is a coordinatewise non-decreasing function.

- INAR-models

$$X_t = a \circ X_{t-1} + \epsilon_t,$$

  are associated.

- More general integer bilinear models

$$X_t = a \circ X_{t-1} + b \circ (\epsilon_{t-1} X_{t-1}) + \epsilon_t,$$

are associated if $\epsilon_t \geq 0$ is iid and integer valued, and if $a\circ$ and $b\circ$ are both thinning operators with non-negative random variables.
Indeed one may write $(X_1, \ldots, X_n)$ as a monotonic function of independent sequences (thus it is associated).

- GLM-Poisson models in Example 7.4.2 can be written as

$$Y_t = P_t(\lambda_t), \qquad \lambda_t = g(Y_{t-1}, \lambda_{t-1}, Y_{t-2}, \lambda_{t-2}, \ldots).$$

If the function g is coordinatewise non-decreasing then the solution of the above equation is again an associated process.

8.3 Main Inequality

A new concept is needed

Definition 8.3.1 Let $f, f_1 : \mathbb{R}^p \to \mathbb{R}$ then we set $f \ll f_1$ if both functions $f \pm f_1$ are coordinatewise non-decreasing.

Example 8.3.1 Assume that the function f satisfies

$$|f(y) - f(x)| \leq a_1|y_1 - x_1| + \cdots + a_p|y_p - x_p|,$$

for all vectors $x = (x_1, \ldots, x_p)$, $y = (y_1, \ldots, y_p) \in \mathbb{R}^p$.
Then $f \ll f_1$ if one sets

$$f_1(x) = a_1x_1 + \cdots + a_px_p.$$

Proof In order to prove this only work out inequalities by grouping terms invoking x's or y's only:

$$\begin{aligned} -a_1(y_1 - x_1) - \cdots - a_p(y_p - x_p) & \\ \leq f(y) - f(x) & \\ \leq a_1(y_1 - x_1) + \cdots + a_p(y_p - x_p). & \end{aligned}$$

The previous inequalities apply to vectors x, y such that $x_i = y_i$ except for only one index $1 \leq i \leq p$.
The corresponding inequalities exactly write $f \ll f_1$.

An essential inequality follows:

Lemma 8.3.1 (Newman 1984) *Let* $X \in \mathbb{R}^p$ *be an associated random vector and* f, g, f_1, g_1 *be measurable functions* $\mathbb{R}^p \to \mathbb{R}$ *then:*

$$|Cov(f(X), g(X))| \leq Cov(f_1(X), g_1(X)),$$

if those function are such that $f(X), g(X), f_1(X), g_1(X) \in \mathbb{L}^2$ *and* $f \ll f_1, g \ll g_1$.

Proof The four covariances

$$\mathrm{Cov}\Big(f(X) + af_1(X), g(X) + bg_1(X)\Big),$$

are non negative if $a, b = -1$ or 1, then adding them two by two yields the result.

For this, we consider separately cases

- $ab = -1$, and
- $ab = 1$,

which correspond to the couples $(a, b) = (-1, 1), (1, -1)$ and $(1, 1), (-1, -1)$, respectively.

A simple byproduct of the above lemma is with Example 8.3.1 the result:

Theorem 8.3.1 *Let* $(Y, Z) \in \mathbb{R}^u \times \mathbb{R}^v$ *be an associated vector in* $\mathbb{L}^2$. *If for some constants* $a_1, \ldots, a_u, b_1, \ldots, b_v \geq 0$, *the functions* f *and* g *satisfy respectively:*

$$|f(y) - f(y')| \leq \sum_{i=1}^{u} a_i |y_i - y_i'|, \quad \forall y, y' \in \mathbb{R}^u,$$

$$|g(z) - g(z')| \leq \sum_{j=1}^{v} b_j |z_j - z_j'|, \quad \forall z, z' \in \mathbb{R}^v,$$

then:

$$|Cov(f(Y), g(Z))| \leq \sum_{i=1}^{u} \sum_{j=1}^{v} a_i b_j Cov(Y_i, Z_j). \tag{8.1}$$

Remark 8.3.1 We derive that for each associated random vector $(Y, Z) \in \mathbb{R}^{u+v}$ in $\mathbb{L}^2$:

- **Independence**
  If the vectors Y, Z admit pairwise-orthogonal components then they are stochastically independent as for the Gaussian case.
  The fact that the class Λ of such Lipschitz functions is rich enough to characterize distributions (see Exercise 85) allows to derive the following equality of distributions
  $$\mathcal{L}(Y, Z) = \mathcal{L}(Y) \otimes \mathcal{L}(Z).$$

• **Quasi-independence**

$$
\begin{aligned}
|\mathrm{Cov}(f(Y), g(Z))| &\le \mathrm{Lip} f \cdot \mathrm{Lip} g \sum_{i=1}^{u} \sum_{j=1}^{v} \mathrm{Cov}(Y_i, Z_j) \\
&\le uv \mathrm{Lip} f \cdot \mathrm{Lip} g \max_{1\le i\le u} \max_{1\le j\le v} \mathrm{Cov}(Y_i, Z_j). \qquad (8.2)
\end{aligned}
$$

This inequality means that the asymptotic dependence structure of an associated random vector relies on its second order structure.

Remark 8.3.2 (Bibliographical comments) This inequality in fact led us to the definition of weak-dependence in Doukhan and Louhichi (1999). It incidentally proves that κ-weak-dependence holds for associated models (see Chap. 11).

I am especially grateful to Alexander Bulinski for discussing those association concepts in the early 1990s.

The idea of weak dependence was introduced in Sana Louhichi's PhD thesis in 1996. Pr. Bulinski was a referee for this defence and he successfully developed this concept in the area of random fields, see Bulinski and Sashkin (2007). The relation (8.2) yields a convenient definition for quasi-independent random fields.

8.4 Limit Theory

Newman (1984) proved the following elegant and powerful weak invariance principle.

Theorem 8.4.1 (Newman 1984) *Assume that the real valued random process $(X_n)_{n\in\mathbb{Z}}$ is stationary, centred and has a finite variance. If the condition*

$$
\sigma^2 = \sum_{n=-\infty}^{\infty} Cov(X_0, X_n) < \infty,
$$

holds for the stationary and associated process $(X_n)_{n\in\mathbb{Z}}$ then

$$
\frac{1}{\sqrt{n}} \sum_{k=1}^{[nt]} X_k \to_{n\to\infty} \sigma W_t, \quad \text{in the Skohorod space } \mathcal{D}[0, 1],
$$

where (W_t) denotes a standard Brownian motion.

Remark 8.4.1 Note that the condition precisely extends that obtained for the independent identically distributed case since it reduces to $\mathbb{E}X_0^2 < \infty$ in this case. The assumption can thus not be improved. Under dependence the useful Lemma 11.5.1 makes use of higher order assumptions. In order to use the assumptions of Theorem 11.5.1 it is hard to avoid moment conditions of higher order.

Example 8.4.1 Consider a sequence such that $a_j \geq 0$. As an application, the above result implies the invariance principle for the associated MA(∞) processes

$$X_k = \sum_{j=-\infty}^{\infty} a_j \xi_{k-j},$$

under the conditions $\mathbb{E}\xi_0^2 < \infty$ the iid inputs, and $\sum_{j=-\infty}^{\infty} |a_j| < \infty$.

The general case $a_j \in \mathbb{R}$ follows by considering the two associated MA(∞) processes $(X_k^{\pm})_k$ built with $a_j^+ = a_j \vee 0 \geq 0$ and $a_j^- = -(a_j \wedge 0) \geq 0$. Note that $X_k = X_k^+ - X_k^-$. The convergence in $\mathcal{D}[0, 1]$ of the two processes

$$Z_n^+(t) = \frac{1}{\sqrt{n}} \sum_{k=1}^{[nt]} X_k^+, \qquad Z_n^-(t) = \frac{1}{\sqrt{n}} \sum_{k=1}^{[nt]} X_k^-.$$

to Gaussian distributions follows. It is also easy to prove that the finite distributions of this process converge to those of a Brownian motion, under the same $\mathbb{L}^2$-condition. To proceed consider linear combinations

$$S_n = \sum_{i=1}^{k} Z_n(t_i),$$

and, use the decomposition of partial sums S_n as sums of in dependent rvs from Exercise 62; Remark 2.1.2, using Bardet and Doukhan (2017), entails the existence of an Orlicz function with $\|\xi_0\|_\psi < \infty$ and from linearity this extends to the terms of the decomposition, hence Lindeberg Lemma 2.1.1 allows to conclude to fdd convergence. The Remark B.2.3 proves the convergence of the sequence of processes $Z_n^+(t) - Z_n^-(t)$ in $\mathcal{D}[0, 1]$, under J_1-topology.

Merlevède et al. (2006); Dedecker et al. (2007) give further results. It is even proved in Peligrad and Utev (2006) that the Donsker invariance principle holds for moving averages with summable coefficient sequences and such that the innovations satisfy a Donsker invariance principle, which is an alternative proof oF the previous result.[1]

[1]Thanks to Florence Merlevède, François Roueff, and Wei Biao Wu for fruitful discussions.

Part III
Dependence

The first chapter in this part begins with the ergodic theorem which asserts that the strong law of large numbers (SLLN) works for the partial sum process of most of the previously introduced models.

Assume that $\theta = \mathbb{E}X_0$ is an unknown parameter for a stationary sequence $(X_n)_{n\in\mathbb{Z}}$, then the ergodic theorem can be written as:

$$\bar{X}_n = \frac{1}{n}(X_1 + \cdots + X_n) \to_{n\to\infty} \theta, \qquad a.s.$$

The question of convergence rates in this results is solved in the following dependence types for stationary sequences.

Two additional chapters detail as much as possible more precise asymptotic results useful for statistical applications.

According to whether they are LRD or SRD, very different asymptotic behaviours will be seen to occur, including corresponding rates.

$$n^{\alpha}(\bar{X}_n - \theta) \underset{n\to\infty}{\overset{\mathcal{L}}{\to}} Z,$$

For some $\alpha = 1/2$ or $>1/2$ according to whether SRD or LRD holds, asymptotic confidence bounds may now be derived.

Namely, set a confidence level $\tau > 0$, and then in case there exists $z_\tau > 0$ with $\mathbb{P}(Z \leqslant z_\tau) = 1 - \tau$:

$$\mathbb{P}\left(\theta \in \left[\bar{X}_n - \frac{z_\tau}{n^\alpha},\ \bar{X}_n + \frac{z_\tau}{n^\alpha}\right]\right) \to_{n\to\infty} \tau.$$

This also yields goodness-of-fit tests for the mean parameter θ.

Chapter 9
Dependence

We propose an overview of the notions of dependence in this chapter, good references are Doukhan et al. (2002b) for long-range dependence, and Doukhan (1994) and Dedecker et al. (2007) for weak-dependence.

9.1 Ergodic Theorem

The present presentation comes[1] from Dedecker et al. (2007).

Definition 9.1.1 A transformation $T : (\Omega, \mathcal{A}) \to (\Omega, \mathcal{A})$ defined on a probability space $(\Omega, \mathcal{A}, \mathbb{P})$ is bijective bi-measurable and $\mathbb{P}$-invariant if it is bijective, measurable, and if it admits a measurable inverse and moreover $\mathbb{P}(T(A)) = \mathbb{P}(A)$ for all $A \in \mathcal{A}$.

Note

$$\mathcal{I} = \left\{A \in \mathcal{A} \;/\; T(A) = A\right\}$$

the sub-sigma algebra of $\mathcal{A}$ containing all the T-invariant events.

This transformation is *ergodic* if $A \in \mathcal{I}$ implies $\mathbb{P}(A) = 0$ or 1.

Remark 9.1.1 (Link to stationary processes) Let $X = (X_n)_{n\in\mathbb{Z}}$ be a real valued stationary process defined on the probability space $(\Omega, \mathcal{A}, \mathbb{P})$.

Then the image $\mathbb{P}_X$ is a probability on the space $\left(\mathbb{R}^{\mathbb{Z}}, \mathcal{B}(\mathbb{R}^{\mathbb{Z}})\right)$. The sigma-algebra $\mathcal{B}(\mathbb{R}^{\mathbb{Z}})$ is generated by elementary events:

$$A = \prod_{k\in\mathbb{Z}} A_k, \quad \text{with } A_k = \mathbb{R}, \quad \text{except for finitely many indices } k.$$

[1] Special thanks are due to Jérôme Dedecker for the present proof.

P. Doukhan, *Stochastic Models for Time Series*, Mathématiques et Applications 80,
https://doi.org/10.1007/978-3-319-76938-7_9

The transformation T defined by

$$T(x)_i = x_{i+1}, \qquad \text{for} \qquad x = (x_i)_{i\in\mathbb{Z}},$$

in short $Tx = (x_{i+1})_{i\in\mathbb{Z}}$, satisfies

$$T\left(\prod_{k\in\mathbb{Z}} A_k\right) = \prod_{k\in\mathbb{Z}} A_{k+1}.$$

It is bijective bi-measurable and $\mathbb{P}$-invariant; it is called the **shift operator**.

Note $\mathcal{J} = X^{-1}(\mathcal{I})$ the sigma-algebra image of $\mathcal{I}$ through X.

T is ergodic $\iff \left\{\mathbb{P}(A) = 0 \text{ or } 1, \forall A \in \mathcal{J}\right\}$.

In this case, the process $X = (X_n)_{n\in\mathbb{Z}}$ is ergodic.

This means that shift-invariant events are either almost sure, or almost impossible.

Example 9.1.1 (A non-ergodic process) A very simple example of a non-ergodic process is $X_t = \zeta$ for each t and for a non-constant rv ζ. Indeed there exists $b \in \mathbb{R}$ such that $0 < \mathbb{P}(A) < 1$ if $A = (\zeta \le b)$. Then this is clear that $A \in \mathcal{J}$.

Refining it to

$$X_t = \zeta \cdot \xi_t, \qquad \text{for each} \qquad t \in \mathbb{Z},$$

provides a non-trivial example if (ξ_t) is independent identically distributed and independent of ζ, and if ζ is not a.s. constant.

- In order to prove this, assume that $\xi_t > 0$ a.s. and $\mathbb{P}(\zeta = \pm 1) = \frac{1}{2}$, then $A = \bigcap_t (X_t > 0) = (\zeta = 1) \in \mathcal{J}$, so that $0 < \mathbb{P}(A) = \frac{1}{2} < 1$, in contradiction with the ergodicity.
- Moreover the ergodic theorem (Corollary 9.1.3) proves that it is not ergodic in case only ξ_n, ζ are both integrable; indeed the empirical mean then converges to the non-constant rv $\zeta \cdot \mathbb{E}\xi_0$.

Many other examples may be found in Kallenberg (1997).

Proposition 9.1.1 *Let T be a bijective and bi-measurable $\mathbb{P}$-invariant transformation.*

Let moreover $f : (\Omega, \mathcal{A}) \to (\mathbb{R}, \mathcal{B}_{\mathbb{R}})$ be measurable with $\mathbb{E}f^2 < \infty$ then:

$$R_n(f) = \frac{1}{n}\sum_{k=1}^{n} f \circ T^k \xrightarrow{\mathbb{L}^2}_{n\to\infty} \mathbb{E}^{\mathcal{I}} f.$$

Proof of Proposition 9.1.1. Consider

$$\mathcal{H} = \left\{\sum_{i=1}^{I} a_i x_i \Big/ \ a_i \ge 0, \ x_i \in E, \ \sum_{i=1}^{I} a_i = 1, \ I \ge 1\right\},$$

the convex hull $\mathcal{H}$ of $E = \{f \circ T^k \ / \ k \in \mathbb{Z}\}$. Let $C = \overline{\mathcal{H}}$ denote the closure in $\mathbb{L}^2(\Omega, \mathcal{A}, \mathbb{P})$ of $\mathcal{H}$.

From the orthogonal projection theorem (see e.g. théorème 3.81, page 124 in Doukhan and Sifre 2001), there exists a unique $\overline{f} \in C$ with

$$\|\overline{f}\|_2 = \inf\{\|g\|_2 \ / \ g \in C\}.$$

If one proves

$$\|R_n(f)\|_2 \to_{n\to\infty} \|\overline{f}\|_2$$

then the proof of the projection theorem implies also

$$\|R_n(f) - \overline{f}\|_2 \to_{n\to\infty} 0.$$

Moreover

$$R_n(f) = f + R_{n-1}(f) \circ T.$$

Hence

$$\|\overline{f} \circ T - \overline{f}\|_2 \le \|\overline{f} \circ T - R_{n-1}(f) \circ T\|_2 + \frac{1}{n}\|f\|_2 + \|R_n(f) - \overline{f}\|_2.$$

$\mathbb{P}$-invariance of T implies that the first term in the right-hand member of this inequality becomes $\|\overline{f} - R_{n-1}(f)\|_2 \to 0$.

Hence

$$\overline{f} \circ T = \overline{f}.$$

$\overline{f}$ is $\mathcal{I}$-measurable.

Since

$$R_n(f) \to \overline{f}, \quad \text{in} \quad \mathbb{L}^2$$

we also deduce

$$\mathbb{E}^{\mathcal{I}} R_n(f) \to \mathbb{E}^{\mathcal{I}} \overline{f} = \overline{f}.$$

The fact that $\mathbb{E}^{\mathcal{I}} R_n(f) = \mathbb{E}^{\mathcal{I}} f$ allows to conclude.

In order to prove

$$\|R_n(f)\|_2 \to_{n\to\infty} \|\overline{f}\|_2,$$

consider any convex combination

$$g = \sum_{|j|\le k} a_j f \circ T^j \in C, \quad \text{with} \quad \|g\|_2 \le \|\overline{f}\|_2 + \epsilon.$$

With the invariance of T we derive

$$\|R_n(g)\|_2 \le \|g\|_2 \le \|\overline{f}\|_2 + \epsilon.$$

On the other hand

$$\begin{aligned}\|R_n(f-g)\|_2 &= \Big\| \sum_{j=-k}^{k} a_j(R_n(f) - R_n(f \circ T^j)) \Big\|_2 \\ &\le \sum_{j=-k}^{k} a_j \|R_n(f) - R_n(f \circ T^j)\|_2\end{aligned}$$

and using again T's invariance,

$$\begin{aligned}\|R_n(f) - R_n(f \circ T^j)\|_2 &\le \frac{1}{n} \sum_{i=k+1}^{k+j} (\|f \circ T^j\|_2 + \|f \circ T^{-j}\|_2) \\ &\le \frac{2j}{n} \|f\|_2. \end{aligned} \tag{9.1}$$

Using the two above inequalities implies:

$$\|R_n(f-g)\|_2 \le \sum_{|j|\le k} \frac{2ja_j}{n} \|f\|_2 \le \frac{2k}{n} \|f\|_2 \to_{n\to\infty} 0.$$

Hence

$$\|\overline{f}\|_2 \le \limsup_n \|R_n(f)\|_2 \le \|\overline{f}\|_2 + \epsilon$$

yielding the result since $\epsilon > 0$ is arbitrary.

Corollary 9.1.1 *If we only assume* $\mathbb{E}|f| < \infty$, *then*

$$R_n(f) \xrightarrow{\mathbb{L}^1}_{n\to\infty} \mathbb{E}^{\mathcal{I}} f.$$

Proof There exists a sequence $g_m \in \mathbb{L}^2$ such that $\|g_m - f\|_1 \to_{m\to\infty} 0$. It is even possible to assume that $g_m \in \mathbb{L}^\infty$.

Then

$$\begin{aligned}\|R_n(f) - \mathbb{E}^{\mathcal{I}} f\|_1 &\le \|R_n(f - g_m)\|_1 + \|R_n(g_m) - \mathbb{E}^{\mathcal{I}}(g_m)\|_1 \\ &\quad + \|\mathbb{E}^{\mathcal{I}}(g_m - f)\|_1 \\ &\le 2\|f - g_m\|_1 + \|R_n(g_m) - \mathbb{E}^{\mathcal{I}}(g_m)\|_1.\end{aligned}$$

The previous proposition implies

$$\limsup_n \|R_n(f) - \mathbb{E}^{\mathcal{I}} f\|_1 \le 2\|f - g_m\|_1.$$

The conclusion follows from a limit argument $m \to \infty$.

The ergodic theorem is also based upon the next inequality:

Lemma 9.1.1 (Hopf maximal inequality) *Let T be a bijective bi-measurable and $\mathbb{P}$-invariant transformation.*

For $f \in \mathbb{L}^1$ set $S_0(f) = 0$ and, for $k \ge 1$ set:

$$S_k(f) = \sum_{j=1}^{k} f \circ T^j, \qquad S_n^+(f) = \max_{0 \le k \le n} S_k(f).$$

Then:

$$\mathbb{E}\left(f \circ T \cdot \mathbb{1}_{S_n^+(f)>0}\right) \ge 0.$$

Proof of Lemma 9.1.1. If $1 \le k \le n+1$ then

$$S_k(f) \le f \circ T + S_n^+(f) \circ T.$$

Moreover if $S_n^+(f) > 0$ then

$$S_n^+(f) = \max_{1 \le k \le n} S_k(f).$$

Thus

$$S_n^+(f)\mathbb{1}_{\{S_n^+(f)>0\}} \le f \circ T \mathbb{1}_{\{S_n^+(f)>0\}} + S_n^+(f) \circ T \, \mathbb{1}_{\{S_n^+(f)>0\}}.$$

This entails

$$f \circ T \, \mathbb{1}_{\{S_n^+(f)>0\}} \ge (S_n^+(f) - S_n^+(f) \circ T) \, \mathbb{1}_{S_n^+(f)>0}.$$

Now

$$\mathbb{E} f \circ T \, \mathbb{1}_{\{S_n^+(f)>0\}} \ge \mathbb{E} S_n^+(f) - \mathbb{E} S_n^+(f) \circ T = 0.$$

Corollary 9.1.2 *If the assumptions in Lemma 9.1.1 hold then*

$$\mathbb{P}\left(\sup_{n\ge 1} |R_n(f)| > c\right) \le \frac{\mathbb{E}|f|}{c}, \qquad \forall c > 0.$$

Proof Apply Lemma 9.1.1 to $f - c$:

$$\mathbb{E}(f - c) \circ T \mathbb{1}_{\{S_n^+(f-c)>0\}} \ge 0,$$

and
$$\frac{\mathbb{E}(f \vee 0)}{c} \geq \frac{\mathbb{E} f \circ T \mathbb{1}_{S_n^+(f-c)>0}}{c} \geq \mathbb{P}(S_n^+(f-c) > 0).$$

We obtain:

$$S_n^+(f-c) = 0 \vee \max_{1 \le k \le n} \Big(k\,(R_k(f) - c)\Big) \geq \max_{1 \le k \le n} (R_k(f) - c)\,.$$

Hence
$$\frac{\mathbb{E}(f \vee 0)}{c} \geq \mathbb{P}\left(\max_{1 \le k \le n} (R_k(f) - c) > 0\right).$$

Replacing f by $-f$ one proves analogously:

$$\frac{-(\mathbb{E}(f \wedge 0))}{c} \geq \mathbb{P}(S_n^+(f+c) < 0) \geq \mathbb{P}\left(\max_{1 \le k \le n} (R_k(f) + c) < 0\right).$$

The result follows from summing the previous inequalities and for $n \to \infty$. Indeed $|f| = f \vee 0 - f \wedge 0$ and $\mathbb{P}(R - c > 0) + \mathbb{P}(R + c < 0) = \mathbb{P}(|R| > c)$ for each random variable R.

Theorem 9.1.1 (Ergodic theorem) *Let T bijective bi-measurable and $\mathbb{P}$-invariant. Let $f \in \mathbb{L}^1$ then*
$$R_n(f) \to_{n\to\infty} \mathbb{E}^{\mathcal{I}} f, \quad a.s.$$

In case the process is ergodic, then the limit is constant almost-everywhere for any integrable f.

Proof of Theorem 9.1.1.

- Assume first that g is bounded.
  If $n, m \geq 1$ then

$$\left|R_n(g) - \mathbb{E}^{\mathcal{I}} g\right| \leq |R_n(g - R_m(g))| + \left|R_n(R_m(g) - \mathbb{E}^{\mathcal{I}} g)\right|.$$

  Using the same idea as to derive inequality (9.1) we obtain

$$\|R_n(g) - R_n(g \circ T^j)\|_\infty \leq \frac{2j}{n}\|g\|_\infty.$$

  Hence
$$|R_n(g - R_m(g))| \leq \frac{\|g\|_\infty}{nm}\sum_{j=1}^m 2j = \frac{(m+1)\|g\|_\infty}{n}.$$

Also

$$\limsup_n \left|R_n(g) - \mathbb{E}^{\mathcal{I}} g\right| \le \sup_{n\ge 1} \left|R_n(R_m(g) - \mathbb{E}^{\mathcal{I}} g)\right|$$
$$\le \left|R_n(R_m(g) - \mathbb{E}^{\mathcal{I}} g)\right|, \text{ a.s.}$$

With Corollary 9.1.2 we derive

$$\mathbb{P}\left(\limsup_n \left|R_n(g) - \mathbb{E}^{\mathcal{I}} g\right| > c\right) \le \frac{1}{c}\mathbb{E}\left|R_m(g) - \mathbb{E}^{\mathcal{I}} g\right| \to_{m\to\infty} 0.$$

So

$$\mathbb{P}\left(\limsup_n \left|R_n(g) - \mathbb{E}^{\mathcal{I}} g\right| = 0\right) = 1.$$

- For the general case, $g \in \mathbb{L}^1$, there exists a sequence of bounded functions g_m which satisfies $\|f - g_m\|_1 \to_{m\to\infty} 0$. Then

$$\left|R_n(f) - \mathbb{E}^{\mathcal{I}} f\right| \le |R_n(f - g_m)| + \left|R_n(g_m) - \mathbb{E}^{\mathcal{I}} g_m\right| + \left|\mathbb{E}^{\mathcal{I}}(g_m - f)\right|.$$

Hence

$$\limsup_n \left|R_n(f) - \mathbb{E}^{\mathcal{I}} f\right| \le \sup_{n\ge 1} |R_n(f - g_m)| + \left|\mathbb{E}^{\mathcal{I}}(g_m - f)\right|, \ a.s.$$

We now derive two relations.

1. Markov's inequality implies $\mathbb{E}^{\mathcal{I}}(g_m - f) \xrightarrow{\mathbb{P}}_{m\to\infty} 0$.
Indeed
$$\mathbb{E}\left|\mathbb{E}^{\mathcal{I}}(g_m - f)\right| \le \frac{1}{c}\|g_m - f\|_1.$$

2. Let $A_m = \sup_{n\ge 1} |R_n(f - g_m)|$ then from Lemma 9.1.1:
$$\mathbb{P}(A_m > c) \le \frac{1}{c}\|f - g_m\|_1.$$

The relations 1. and 2. imply

$$\mathbb{P}\left(\limsup_n \left|R_n(f) - \mathbb{E}^{\mathcal{I}} f\right| > c\right) = 0.$$

This holds for each $c > 0$ which implies the result.

In the case of stationary processes this theorem is reformulated with the shift operator T.

Corollary 9.1.3 *Let $(X_n)_{n\in\mathbb{Z}}$ be a stationary process. If $f:\mathbb{R}^{\mathbb{Z}}\to\mathbb{R}$ is measurable and $\mathbb{E}|f(X)|<\infty$ then:*

$$\frac{1}{n}\sum_{k=1}^{n} f\circ T^k(X)\to_{n\to\infty}\mathbb{E}^{\mathcal{J}}f(X),\quad \text{a.s. and in } \mathbb{L}^1.$$

If now $\mathbb{E}f^2(X)<\infty$, then the convergence also holds in $\mathbb{L}^2$.

Proof The only point to notice is that $\mathbb{E}^{\mathcal{J}}f(X)=\mathbb{E}^{\mathcal{I}}_{\mathbb{P}_X}f$.

Example 9.1.2 Exercise 68 provides us with a non-ergodic sequence satisfying anyway a law of large numbers.

Remark 9.1.2 If the process X is ergodic, then:

$$\frac{1}{n}\sum_{k=1}^{n} f\circ T^k(X)\longrightarrow_{n\to\infty}\mathbb{E}f(X),\quad \text{if } \mathbb{E}|f(X)|<\infty.$$

Ergodicity may also be omitted if $\mathbb{E}f^2(X)<\infty$ and

$$\frac{1}{n}\sum_{k=1}^{n} f\circ T^k(X)\to_{n\to\infty}\mathbb{E}f(X)\text{ a.s.}$$

$$\iff\quad \frac{1}{n}\sum_{k=1}^{n}\mathrm{Cov}(f(X),f\circ T^k(X))\to 0.$$

Moreover, as a partial converse of Theorem 9.1.1, if the above limit is constant everywhere for any integrable function f, then the system is ergodic, see Kallenberg (1997).

We now derive some examples of ergodic processes.

Example 9.1.3 (Ergodic processes)

- An independent identically distributed sequence is also a stationary and ergodic sequence. For this, use Kolmogorov $0-1$'s law.
- Hence Bernoulli schemes are also ergodic. Indeed if $X=(X_i)_{i\in\mathbb{Z}}$ is defined from an independent identically distributed sequence $\xi=(\xi_i)_{i\in\mathbb{Z}}$ and a function H through Eq. (7.15) then $f\circ T^i(X)=f\circ H\circ T^i(\xi)$. Hence, as soon as $\mathbb{E}|f(X)|<\infty$, we derive:

$$\frac{1}{n}\sum_{i=1}^{n} f\circ T^i(X)\to\mathbb{E}f(X).$$

  This is true for bounded measurable functions $\mathbb{R}^{\mathbb{Z}}$ in $\mathbb{R}$. This also entails ergodicity of X.

- If the relation $\mathrm{Cov}\,(f(X_0), f(X_n)) \to 0$ as $n \to \infty$ holds for $f \in \mathcal{F}$. This class of functions indeed generates a dense linear vector subspace of $\mathbb{L}^1$. This relation implies with the Cesaro lemma that:

$$\frac{1}{n}\sum_{k=1}^{n} f \circ T^k(X) \to_{n\to\infty} \mathbb{E} f(X), \quad \text{in} \quad \mathbb{L}^1.$$

  The result still holds for each bounded function from a density argument. Now Corollary 9.1.3 entails $\mathbb{E}^{\mathcal{J}} f(X) = \mathbb{E} f(X)$ and the ergodicity follows.

The following examples follow this scheme:

- A Gaussian stationary sequence is ergodic if its covariance satisfies $r_n \to 0$ as $n \to \infty$. This condition seems to be necessary since e.g. a constant sequence $X_n = \xi_0 \sim \mathcal{N}(0, 1)$ is not ergodic.
  Assume $X_0 \sim \mathcal{N}(0, 1)$. If the Hermite expansion of f is

$$f = \sum_{k=0}^{\infty} c_k H_k,$$

  then:

$$\mathrm{Cov}\,(f(X_0), f(X_n)) = \sum_{k=1}^{\infty} \frac{c_k^2}{k!} r_n^k \; \big(= G(r_n) \big).$$

  The function $G(r)$ defined this way is continuous on $[-1, 1]$ if one sets $G(1) = \mathbb{E} f^2(X_0)$ and $G(0) = 0$. The ergodicity follows.
- Strongly mixing sequences, and all the previous examples of weakly dependent sequences (see the Definition 11.1.1) are ergodic.
- A last example is a stationary associated sequences with $\lim_{n\to\infty} r_n = 0$. To prove this, use inequality (8.1).

Exercise 59 Let $(X_n)_{n\in\mathbb{Z}}$ be a stationary and ergodic centred sequence in $\mathbb{L}^2$.
Then

$$\widehat{r}_{n,p} = \frac{1}{n - |p|} \sum_{k=|p|+1}^{n} X_k X_{k-|p|} \tag{9.2}$$

fits $r_p = \mathbb{E} X_0 X_p$ without bias, i.e. $\mathbb{E}\widehat{r}_{n,p} = r_p$, and $\widehat{r}_{n,p} \to r_p$ *a.s.* and in $\mathbb{L}^1$, i.e. it is consistent.

Solution. For this use the previous result with $f(\omega) = \omega_0 \omega_p$.
Let $(\xi_n)_{n\in\mathbb{Z}}$ be stationary and ergodic with $\mathbb{E}\xi_0^2 < \infty$.

If $|a| < 1$ then $X_n = \sum_{k=0}^{\infty} a^k \xi_{n-k}$ is stationary and ergodic and $\mathbb{E} X_0^2 < \infty$.

Moreover

$$X_n = aX_{n-1} + \xi_n, \quad \forall n \in \mathbb{Z}.$$

The previous solution is the unique sequence such that this relation holds. It is the first order auto-regressive process.

Previous arguments imply

$$\widehat{a} = \frac{\sum_{k=2}^n X_k X_{k-1}}{\sum_{k=2}^n X_k^2} \longrightarrow_{n\to\infty} a, \qquad a.s.$$

if $\mathbb{E}\xi_0 = 0$ and $\lim_{p\to\infty} \mathbb{E}\xi_0\xi_p = 0$ for the ergodic sequence (ξ_t).

Example 9.1.4 Chap. 7 includes a wide range of results for which Theorem 9.1.1 applies. Essentially the consistency of the empirical process follows from this main result.

9.2 Range

We provide some ideas yielding definitions for the range of a process. Namely we advocate to define it according to a possible limit theorem. As this is claimed at the beginning of Part III, a definition through a limit theorem in distribution allows to define an asymptotic confidence interval for testing a mean through the simplest frequentist empirical mean. After Theorem 9.1.1 it is indeed known that such empirical estimators do converge under mild assumptions.

The classical definition of the long/short-range dependence for second order stationary sequences is based on the convergence rates to zero covariances $r_k = \text{Cov}(X_0, X_k)$, more precisely the convergence of the following series is of importance:

$$\sum_{k=-\infty}^{\infty} r_k.$$

Definition 9.2.1 ($\mathbb{L}^2$*-range*) In case the series (r_k) is absolutely convergent the process is short-range dependent (SRD) and if the series diverges the process is long-range dependent (LRD).

The proof of Proposition 4.3.2 provides an expression of the square of a convergence rate in $\mathbb{L}^2$ in the ergodic theorem under $\mathbb{L}^2$-stationarity:

$$\mathbb{E}(S_n - n\mathbb{E}X_0)^2 = \text{Var}\left(\sum_{k=1}^n g(X_k)\right) = \sum_{|k|<n} (n - |k|) r_k.$$

Based on the previous definition the partial sums

$$S_n = \sum_{k=1}^{n} X_k,$$

admit variances with order n or $\gg n$ according to either an SRD or an LRD behaviour.

A phenomenon of very short-range corresponds to $g_X(0) = 0$; in this case $\mathrm{Var}\, S_n \ll n$. It is discussed in Giraitis et al. (2012).

More generally consider L_1, L_2, L_3 slowly varying functions (typically powers of logarithms) and constants $\alpha, \beta, \gamma > 0$.

Introduce the properties

$$\sum_{k=-n}^{n} r_k \sim_{n\to\infty} n^{\alpha} L_1(n), \tag{9.3}$$

$$r_n \sim_{n\to\infty} n^{-\beta} L_2(n), \tag{9.4}$$

$$g_X(\lambda) \sim_{\lambda\to 0} |\lambda|^{-\gamma} L_3\left(\frac{1}{|\lambda|}\right). \tag{9.5}$$

One may prove (Taqqu in Doukhan et al. 2002b):

Theorem 9.2.1 (Tauber) *If the sequence (r_k) is monotonous for $k \ge k_o$ then relations (9.3), (9.4) and (9.5) are equivalent with $\alpha = 1-\beta$, $L_1 = \dfrac{2}{1-\beta} L_2$, $\gamma = 1-\beta$ and $L_3 = \dfrac{\Gamma(\alpha+1)}{2\pi} \sin \dfrac{\pi(1-\alpha)}{2} L_1$.*

This yields a convergence rate in the precise law of large numbers (Ergodic Theorem 9.1.1), but if one needs a more accurate approximation for goodness-of-fit tests, then more information is needed. This definition is quite unsatisfactory because a user is concerned with the asymptotic behaviour of functionals of a process rather than its $\mathbb{L}^2$-behaviour.

Orthogonal sequences satisfy $\mathrm{Var}\, S_n = n \mathrm{Var}\, X_0$ but they do not necessarily admit an asymptotically Gaussian behaviour:

Exercise 60 Let (ξ_n) be an independent identically distributed sequence with marginals $\mathcal{N}(0, 1)$ and let η be a real valued random variable independent of the sequence (ξ_n) then $X_n = \eta\xi_n$ is orthogonal so that it is also weakly stationary but it is not ergodic.

Hint. Since $S_n/\sqrt{n}$ admits the same distribution as $\eta\xi_0$ this expression is usually not Gaussian.

The case $\mathbb{P}(\eta = \pm 1) = \frac{1}{2}$ is special and yields a Gaussian behaviour because of the symmetry of the Normal distribution.

Another naive definition of the range is based on limit theorems relative to the partial sums:

$$S_n = X_1 + \cdots + X_n.$$

Distributional range. Let $(X_n)_{n\in\mathbb{Z}}$ be a strictly stationary and centred sequence in $\mathbb{L}^2$:

- If $\frac{1}{\sqrt{n}}S_n$ is asymptotically Gaussian then we say it is short-range dependent. Precisely we may suppose that $\operatorname{Var} S_n \sim cn$ (as $n \to \infty$), for some constant $c > 0$.
  Assume that the sequence of processes
  $$t \mapsto Z_n(t) = \frac{1}{\sqrt{\operatorname{Var} S_n}} S_{[nt]}, \qquad \text{for } t \in [0, 1]$$
  converges toward a Brownian motion in the Skorohod space $\mathcal{D}[0, 1]$ (see Definition B.2.2).
- If the sequence of processes
  $$t \mapsto Z_n(t) = \frac{1}{\sqrt{\operatorname{Var} S_n}} S_{[nt]}, \qquad \text{for } t \in [0, 1]$$
  does not converge toward a Brownian motion it would be long-range dependent.

An alternative definition omits the fact that $X_n \in \mathbb{L}^2$:

SRD holds if the previous partial sums process admits a limit with independent increments, otherwise, if the previous partial sums process admits a limit with dependent increments, then LRD holds.

This nice proposal by Herold Dehling allows to aggregate cases of heavy tail processes and Lévy processes. Recent works on progress (Doukhan et al. 2017) prove that this does not always hold, see Sect. 10.6.1. This works tends to distinguish totally $\mathbb{L}^2$ and distributional LRD.

Chapter 10
Long-Range Dependence

Long-range dependent (LRD) phenomena were first exhibited by Hurst for hydrology purposes. This phenomenon occurs from the superposition of independent sources, e.g. confluent rivers provide this behaviour (see Fig. 4.2). Such aggregation procedures provide this new phenomenon. Hurst (1951) originally determined the optimum dam sizing for the Nile river's volatile rain and drought conditions observed over a long period of time. LRD is characterized by slow decorrelation properties and the behaviour of partial sums's variances. This phenomenon is discussed above, see Sects. 4.3 and 4.4. Asymptotic properties of instantaneous functions of Gaussian processes are provided in Remark 5.2.4. Infinite moving averages models with LRD properties are provided in Sects. 6.4 and 6.6. We refer the reader to Doukhan et al. (2002b) for much more.

The present chapter is dedicated to distributional LRD properties. We address the Gaussian and linear cases as well as the case of functions of such processes where LRD phenomena occur. Due to the technical difficulties we restrict to the initial example of Rosenblatt for functions of Gaussian processes. We also describe additional extensions in a more bibliographical spirit.

The most elementary example is that of Gaussian processes. We follow the presentation in Rosenblatt (1985) who discovered such long-range dependent behaviours in distribution. He considered models of instantaneous functions of a Gaussian process.

10.1 Gaussian Processes

Let $(X_n)_{n\in\mathbb{Z}}$ be a stationary centred Gaussian sequence with $r_0 = \mathbb{E}X_0^2 = 1$ and with covariance

$$r_k \sim ck^{-\beta}, \qquad \text{as} \qquad k \to \infty,$$

for $c > 0$, $\beta > 0$.

P. Doukhan, *Stochastic Models for Time Series*, Mathématiques et Applications 80,
https://doi.org/10.1007/978-3-319-76938-7_10

Theorem 4.2.1 allows to prove that the sequence $r_k = (1+k^2)^{-\frac{\beta}{2}}$ is indeed the sequence of covariances of a stationary Gaussian process. Hence such sequences exist.

Tauber's Theorem 9.2.1 implies $g(\lambda) \sim |\lambda|^{a-1}$. Also $S_n \sim \mathcal{N}(0, \operatorname{Var} S_n)$ with

$$\operatorname{Var} S_n = n \sum_{|k|<n} \left(1 - \frac{|k|}{n}\right) r_k,$$

and

$$Z_n(t) \sim \mathcal{N}\left(0, \frac{\operatorname{Var} S_{[nt]}}{\operatorname{Var} S_n}\right).$$

- Hence if $\beta > 1$, $\operatorname{Var} S_n \sim n\sigma^2$ then the sequence is SRD and

$$\frac{1}{n} \operatorname{Var} S_{[nt]} \to t\sigma^2.$$

Now Z_n converges to a Brownian motion with variance

$$\sigma^2 = \sum_{k=-\infty}^{\infty} r_k.$$

First check that

$$\mathbb{E} Z_n(t) Z_n(s) \to (s \wedge t)\sigma^2.$$

Tightness is a consequence of the inequality

$$\mathbb{E}(Z_n(t) - Z_n(s))^2 \le C|t-s|, \quad \text{for} \quad C = \sum_{k=-\infty}^{\infty} |r_k|,$$

and from the Chentsov Lemma B.2.1.
Indeed for Gaussian processes it is immediate to prove that:

$$\mathbb{E}|Z_n(t) - Z_n(s)|^p = \mathbb{E}|N|^p \, [\mathbb{E}(Z_n(t) - Z_n(s))^2]^{\frac{p}{2}},$$

for each $p > 2$ if $N \sim \mathcal{N}(0,1)$.
- If now $\beta < 1$ the series of covariances diverges

$$\operatorname{Var} S_n \sim n^{2-\beta} \qquad \text{if} \qquad r_k \sim ck^{-\beta}.$$

Hence

$$Z_n(t) \to \mathcal{N}\left(0, ct^{2-\beta}\right),$$

does not converge to the Brownian motion; indeed contrary to the case of the Brownian motion the previous variance does not increase linearly with respect to t.

Now writing that

$$\mathbb{E}(Z_n(t) - Z_n(s))^2 \to_{\to\infty} c|t-s|^{2H}, \qquad \text{with } H = 1 - \beta/2,\ s,t \geq 0$$

implies that functional convergence holds by using the Chentsov Lemma B.2.1. Moreover the relation:

$$\mathbb{E}(Z_n(s)Z_n(t)) = \frac{1}{2}(\mathbb{E}Z_n^2(t) + \mathbb{E}Z_n^2(s) - \mathbb{E}(Z_n(t) - Z_n(s))^2),$$

demonstrates that the covariance function of Z_n converges to $c\Gamma_H$, where Γ_H denotes the covariance of the fractional Brownian motion B_H, see (5.1).

Remark 10.1.1

- For SRD sequences the assumptions of Lemma B.2.1 need $p > 2$ because $a = p/2$. The above-mentioned relations imply that this holds for a Gaussian process if it holds for $p = 2$ and for some $a > 0$.
- The long-range dependent case is more simple since for $p = 2$ one derives directly $a = 2 - \alpha > 1$.

10.2 Gaussian Polynomials

Generally assume that the process (X_n) is Gaussian, and stationary. Let this process be standard Gaussian, in the sense that $\mathbb{E}X_0 = 0$ and $\operatorname{Var} X_0 = 1$. Also suppose that $r_k \sim ck^{-\beta}$ as $k \to \infty$, and that the function g is such that $\mathbb{E}|g(X_0)|^2 < \infty$.

Then one of the following two cases may occur:

- SRD case. Note that:

$$\operatorname{Var}\left(\sum_{k=1}^{n} g(X_k)\right) = \mathcal{O}(n),$$

if $\beta \cdot m(g) > 1$ and $m(g)$ denote the Hermite rank of g.
In this first case the diagram formula (Sect. 5.2.3) allows to prove the convergence in distribution, (Breuer and Major 1983):

$$\frac{1}{\sqrt{n}} \sum_{k=1}^{[nt]} g(X_k) \to_{n\to\infty} \sigma W_t, \quad \text{in} \quad \mathcal{D}[0,1].$$

The result is also proved in a shorter way in Nourdin et al. (2011) by using the method of fourth order moment from Sect. 5.2.5.

- LRD case. Otherwise, say if $\beta \cdot m(g) < 1$, then

$$\mathrm{Var}\left(\sum_{k=1}^{n} g(X_k)\right) = \mathcal{O}\left(n^{2-m(g)\beta}\right).$$

Here $1 - \frac{m(g)\beta}{2} > \frac{1}{2}$ and convergence in law still holds

$$\frac{1}{n^{1-\frac{m(g)\beta}{2}}} \sum_{k=1}^{n} g(X_k) \xrightarrow{\mathcal{L}}_{n\to\infty} Z_r,$$

to some non-Gaussian distribution in case the rank is > 1 (Taqqu, 1975, see Dobrushin and Major 1979).
The technique is involved since, for $k > 2$, the Laplace transform for the law of X_0^k is not analytic around 0, thus characteristic functions do not determine convergence to such laws.
The case $k = 1$ is considered in the previous section and the case $k = 2$ is discussed in the next one.

10.3 Rosenblatt Process

The previous non-Gaussian asymptotic may be proved elementary “à la main” for the case $m = 2$ described in Rosenblatt (1961), see also the nice monograph (Rosenblatt 1985). Set $Y_n = X_n^2 - 1$ then the Mehler formula (Lemma 5.2.2) implies that the covariance $\mathrm{Cov}\,(Y_0, Y_k)$ equals $2r_k^2 \sim 2c^2k^{-2\beta}$. The series of these covariances is divergent in case $\beta < \frac{1}{2}$. In this case we aim to prove that

$$U_n = n^{\beta-1} \sum_{k=1}^{n} Y_k,$$

converges toward a non-Gaussian limit.

More explicitly the normalization should be written $\sqrt{n^{2\beta}}/n$.

Set R_n for the covariance matrix of the vector $(X_1, \ldots, X_n)$, then for t small enough:

$$\begin{aligned}
\mathbb{E}e^{tU_n} &= \mathbb{E}e^{tn^{\beta-1}\sum_{k=1}^{n}(X_k^2-1)} \\
&= e^{-tn^{\beta}} \int_{\mathbb{R}^n} e^{-\frac{1}{2}x^t(R_n^{-1}-2tn^{\beta-1}I_n)x} \frac{dx}{(2\pi)^{n/2}\sqrt{\det R_n}} \\
&= e^{-tn^{\beta}} \int_{\mathbb{R}^n} e^{-\frac{1}{2}y^t(I_n-2tn^{\beta-1}R_n)y} \frac{dy}{(2\pi)^{n/2}} \\
&= e^{-tn^{\beta}} \det{}^{-\frac{1}{2}}\left(I_n - 2tn^{\beta-1}R_n\right).
\end{aligned}$$

Indeed through a linear change in variable for each symmetric definite positive matrix A with order n:

$$\int_{\mathbb{R}^n} e^{-\frac{1}{2}y^t Ay} \frac{dy}{(2\pi)^{n/2}} = \frac{1}{\sqrt{\det(A)}}.$$

Now denote by $(\lambda_{i,n})_{1\le i\le n}$ the eigenvalues (≥ 0) of the symmetric and non-negative matrix R_n (thus, diagonalizable) then

$$\begin{aligned}\frac{1}{\sqrt{\det\left(I_n - 2tn^{\beta-1}R_n\right)}} &= \prod_{i=1}^n \left(1 - 2tn^{\beta-1}\lambda_{i,n}\right)^{-1/2} \\ &= \exp\left(-\frac{1}{2}\sum_{i=1}^n \log\left(1 - 2tn^{\beta-1}\lambda_{i,n}\right)\right).\end{aligned}$$

Use the following analytic expansion (valid for $|z| < 1$)

$$\log(1-z) + z = -\sum_{k=2}^{\infty} \frac{z^k}{k}.$$

The simple observation that trace$(R_n) = n$ follows from the fact that R_n's diagonal elements equal 1. We deduce that

$$e^{-tn^\beta} = \exp\left(-2tn^{\beta-1}\text{ trace } R_n\right) = \exp\left(-\sum_{i=1}^n (2tn^{\beta-1})\lambda_{i,n}\right).$$

Thus:

$$\begin{aligned}\mathbb{E}e^{tU_n} &= \exp\left(-\frac{1}{2}\sum_{i=1}^n \left\{\log\left(1 - 2tn^{\beta-1}\lambda_{i,n}\right) + 2tn^{\beta-1}\right\}\right) \\ &= \exp\left(\frac{1}{2}\sum_{k=2}^{\infty} \frac{1}{k}(2tn^{\beta-1})^k \text{ trace } R_n^k\right).\end{aligned}$$

Now:

$$\begin{aligned}\frac{n^{k\beta}}{n^k}\text{trace } R_n^k &= n^{k(a-1)} \sum_{i_1=1}^n \cdots \sum_{i_k=1}^n r_{i_1-i_2} r_{i_2-i_3} \cdots r_{i_{k-1}-i_k} r_{i_k-i_1} \\ &\sim \frac{c^k}{n^k} \sum_{i_1=1}^n \cdots \sum_{i_k=1}^n \frac{1}{\left|\frac{i_1}{n} - \frac{i_2}{n}\right|^\beta} \frac{1}{\left|\frac{i_2}{n} - \frac{i_3}{n}\right|^\beta} \cdots \frac{1}{\left|\frac{i_k}{n} - \frac{i_1}{n}\right|^\beta}.\end{aligned}$$

Hence through the discretization of a multiple integral by Riemann sums we derive:

$$\frac{n^{k\beta}}{n^k}\text{ trace } R_n^k \to_{n\to\infty} c_k > 0,$$

with

$$c_k = c^k \int_0^1 \cdots \int_0^1 \frac{1}{|x_1 - x_2|^\beta} \times \cdots \times \frac{1}{|x_{k-1} - x_k|^\beta} \times \frac{1}{|x_k - x_1|^\beta} dx_1 \cdots dx_k.$$

For this, simple upper and lower bounds for integrals over cubes with volume n^{-k} allow to derive this convergence; indeed the function to be integrated is locally monotonic with respect to each coordinate.

More generally (Polya and Szegö 1970) prove the validity of such approximations for generalized integrals, in the case of functions monotonic around their singularity as here. Hence for $k = 2$, one obtains that for each $1 \le i \le n$,

$$\begin{aligned} 0 \le \lambda_{i,n}^2 \le \sum_{j=2}^n \lambda_{j,n}^2 &= \text{trace } R_n^2 \\ &\le n \sum_{j=0}^n r_j^2 \le n \sum_{j=0}^n j^{-2\beta} = \mathcal{O}(n) = o(n^{2(1-\beta)}). \end{aligned}$$

Indeed $2(1-\beta) < 1$ since $\beta > \frac{1}{2}$ and we derive $2|t|\,n^{\beta-1}\lambda_{i,n} < 1$ if $|t| < c$ for some constant $c > 0$.

The previous bound proves that the **necessary convergences hold** make the above-mentioned calculations rigorous if n is large enough.

Hence for each t:

$$\mathbb{E}e^{tU_n} \to_{n\to\infty} \exp\left(\frac{1}{2}\sum_{k=2}^{\infty} (2t)^k \cdot \frac{c_k}{k}\right).$$

Hence this sequence converges in distribution to a non-Gaussian law (this distribution is therefore called Rosenblatt's distribution).

Indeed the logarithm of its Laplace transform is not a polynomial with order 2.

Remark 10.3.1 This technique does not extend to polynomials with degree more than 2 since their Laplace transform is not analytic. Indeed it is easy to prove that if $N \sim \mathcal{N}(0,1)$ then

$$\mathbb{E}\exp(t|N|^3) = 2\int_0^\infty \exp\left(tx^3 - \frac{1}{2}x^2\right)\frac{dx}{\sqrt{2\pi}} = \infty, \quad \text{if} \quad t > 0,$$

and the method of moments does not apply to prove convergence in law (see Theorem 12.1.1). Dobrushin and Major (1979) introduced weaker convergences for sequences of multiple Ito integrals in order to derive "non-central limit theorems".

10.4 Linear Processes

Consider linear processes with iid inputs such that $\mathbb{E}\xi_0 = 0$ and $\mathbb{E}\xi_0^2 = 1$,

$$X_n = \sum_{k=0}^{\infty} c_k \xi_{n-k},$$

for which $c_k \sim_{k\to\infty} ck^{-\beta}$ with $\frac{1}{2} < \beta < 1$ are easily proved to satisfy

$$r_k = \sum_l c_l c_{l+k} \sim_{k\to\infty} ck^{1-2\beta} \int_0^\infty \frac{ds}{(s(s+1))^\beta}$$

(use approximations of an integral by Riemann sums), hence

$$\text{Var}\,(X_1 + \cdots + X_n) \sim c' n^{2-2\beta}$$

and it is possible to prove

$$n^{\beta-1} \sum_{k=1}^{[nt]} X_k \xrightarrow{\mathcal{L}}_{n\to\infty} B_H(t),$$

with convergence in law in the Skorohod space $\mathcal{D}[0,1]$ of right-continuous functions with limit on the left (càdlàg functions, see Definition B.2.2):

Theorem 10.4.1 (Davydov, 1970) *Let (X_n) be a linear process. Set*

$$S_n = X_1 + \cdots + X_n.$$

If Var $S_n = n^{2H} L(n)$ for a slowly varying function L and $0 < H < 1$ then

$$\frac{1}{n^H L(n)} \sum_{k=1}^{[nt]} X_k \xrightarrow{\mathcal{L}}_{n\to\infty} B_H(t).$$

Hint. This result also relies on the Lindeberg Theorem 2.1.1 and use the following decomposition in formula (10.3) with $\gamma_k(a) = a\xi_k$.

The end of the chapter is more a sequence of bibliographic comments than a sequence of formal rigorous results, due to their highly technical proofs.

10.5 Functions of Linear Processes

A martingale-based technique was introduced in Ho and Hsing (1996) for the extension of such behaviours as previously considered for the Gaussian case. Conditional expectations recalled in Definition A.2.2 are essential to define martingales (see Definition B.6.1).

The idea of this section is to give a flavour of results and underlying techniques but the rigorous proofs should be found in the corresponding literature. Using the weak uniform reduction principle, (Giraitis and Surgailis 1999) established the same result for a causal linear process.

Let

$$X_t = \sum_{s=0}^{\infty} b_s \xi_{t-s},$$

where ξ is independent identically distributed and $b_s = L(s)s^{-(\alpha+1)/2}$.

Theorem 10.5.1 (Causal linear process) *Let $f(x)$ be the density of X_0 and $B_{1-\alpha/2}$ the fractional Brownian motion with Hurst index $H = 1 - \alpha/2$. If there exists constants $\delta, C > 0$ such that*

$$\left|\mathbb{E}\left(e^{iu\xi_0}\right)\right| \le C(1+|u|)^{-\delta},$$

and if $\mathbb{E}|\xi_0|^9 < \infty$ then there exists c_α an explicit constant with

$$n^{\alpha/2-1} F_n(x,t) \longrightarrow c_\alpha f(x) B_{1-\alpha/2}(t)$$

in the Skorohod space $\mathcal{D}[-\infty, +\infty] \times \mathcal{D}[0,1]$.

A main tool is uniform control of the approximation of the empirical process by the partial sums process:

Proposition 10.5.1 (Uniform reduction principle) *There exist $C, \gamma > 0$ such that for $0 < \varepsilon < 1$:*

$$\mathbb{P}\left(\frac{n^{\frac{\alpha}{2}-1}}{L(n)} \sup_{\substack{k \le n \\ x \in \mathbb{R}}} \left|\sum_{t=1}^{k} \left(\mathbb{I}_{\{X_t \le x\}} - F(x) + f(x)X_t\right)\right| \ge \varepsilon\right) \le \frac{C}{n^{\gamma}\varepsilon^3}.$$

Sketch of the proof. Set

$$S_n(x) = \frac{\sqrt{n^\alpha}}{nL(n)} \sum_{t=1}^{n} \left(\mathbb{I}_{\{X_t \le x\}} - F(x) + f(x)X_t\right).$$

Then

$$\mathrm{Var}\,(S_n(y) - S_n(x)) \le \frac{n}{n^{1+\gamma}} \cdot \mu([x,y]).$$

where μ is a finite measure on $\mathbb{R}$. Then a technical chaining argument is used to derive tightness.

Remark 10.5.1 (Ho and Hsing 1996) extend the expansion of the reduction principle:

$$S_{n,p}(x) = \frac{n^{p\alpha/2}}{nL(n)}\left(\sum_{t=1}^{n} \mathbb{I}_{\{X_t \le x\}} - F(x) - \sum_{r=0}^{p}(-1)^r F^{(r)}(x) Y_{n,r}\right),$$

$$Y_{n,r} = \sum_{t=1}^{n} \sum_{1 \le j_1 < \cdots < j_r} \prod_{s=1}^{r} b_{j_s} \xi_{t-j_s}.$$

Proposition 10.5.2 (Uniform reduction principle) *If the density of ξ_0 is $(p+3)$-times differentiable and if $\mathbb{E}|\xi_0|^4 < \infty$, there exist $C, \gamma > 0$ such that for $0 < \varepsilon < 1$:*

$$\mathbb{P}\left(\sup_{x \in \mathbb{R}} |S_{n,p}(x)| \ge \varepsilon\right) \le C n^{-(\alpha \wedge (1-p\alpha)) + \gamma} \varepsilon^{-2-\gamma}.$$

A preliminary view to the technique of proof.

A calculation of the variance of $S_{n,p}(x)$ is first needed.

Set

$$f_t(x) = \mathbb{I}_{\{X_t \le x\}} - F(x) - \sum_{r=0}^{p}(-1)^r F^{(r)}(x) Y_{n,r},$$

write the orthogonal decomposition:

$$f_t(x) - \mathbb{E} f_t(x) = \sum_{s=1}^{\infty} \mathbb{E}(f_t(x)|\mathcal{F}_{t-s}) - \mathbb{E}(f_t(x)|\mathcal{F}_{t-s-1}),$$

where $\mathcal{F}_t$ is the σ-field generated by the $\{\xi_s /\ s \le t\}$. Compute the variance of each term using a Taylor expansion. Note that the $\mathcal{F}_t$ are increasing so that many covariances between terms are zero.

It is possible to generalize the previous method to the case of random fields. (X_t) is a *linear* random field:

$$X_t = \sum_{u \in \mathbb{Z}^d} b_u \xi_{t+u}, \quad \forall t \in \mathbb{Z}^d,$$

where $(\zeta_u)_{u \in \mathbb{Z}^d}$ is an iid random field with zero mean and variance 1, and $b_u = B_0(u/|u|)|u|^{-(d+\alpha)/2}$, for $u \in \mathbb{Z}^d$, $0 < \alpha < d$ and B_0 is a continuous function on the sphere.

Let $A_n = [1, n]^d \cap \mathbb{Z}^d$ and $F_n(x) = \frac{1}{n^d} \sum_{t \in A_n} \mathbb{1}_{\{X_t \leq x\}}$, then:

Theorem 10.5.2 (Doukhan et al. 2005) *If there exist* $\delta, C > 0$ *such that*

$$\left|\mathbb{E}\left(e^{iu\xi_0}\right)\right| \leq C(1 + |u|)^{-\delta},$$

and if $\mathbb{E}|\xi_0|^{2+\delta} < \infty$ *then*

$$n^{\alpha/2}(F_n(x) - F(x)) \longrightarrow c_\alpha f(x) Z,$$

in $\mathcal{D}[-\infty, +\infty]$, *where* Z *is a Gaussian random variable.*

Remark 10.5.2 It is remarkable that the limit distribution is extremely simple in this case, indeed Z does not depend on x.

Recall that the weak dependent case yields much more complicated limit behaviours, typically the Brownian bridge in which the Hölder regularity exponent satisfies $\beta < \frac{1}{2}$.

10.6 More LRD Models

This section provides some directions for the extension of LRD to non-linear models. It contains more bibliographical comments than rigorous statements.

10.6.1 Integer Valued Trawl Models

In Doukhan et al. (2017) we introduce an extension of linear models given by an iid sequence $(\gamma_k)_k$ of copies of a process $\gamma : \mathbb{R} \to \mathbb{R}$. For our purpose, we shall restrict to:

- *Symmetric Poisson Process*: $\gamma(u) = P(u) - P'(u)$ for P, P' independent homogeneous Poisson processes,
- *Symmetric Bernoulli Process*: $\gamma(u) = \mathbb{1}_{\{U \leq u\}} - \mathbb{1}_{\{U' \leq u\}}$ for U, U' independent uniform random variables.

Both processes are centred, and respectively:

$$\begin{cases} \mathrm{Var}\, \gamma(u) & = 2u, \quad \text{or } 2u(1-u), \\ \mathrm{Cov}\,(\gamma(u), \gamma(v)) & = 2(u \wedge v), \quad \text{or } 2(u \wedge v - uv). \end{cases}$$

Consider some non-decreasing sequence $a_k \geq 0$ with $a_k \sim c k^{-\alpha}$ $(k \to \infty)$, for some $1 < \alpha < 2$.

Those models are defined as:

$$X_n = \sum_{j=0}^{\infty} \gamma_{n-j}(a_j). \tag{10.1}$$

The above conditions ensure existence and stationarity of this model.

Moreover

$$\begin{aligned} \text{Cov}\,(X_0, X_k) &\sim c' k^{1-\alpha} \\ \text{Var}\, S_n &\sim c'' n^{3-\alpha}, \end{aligned} \tag{10.2}$$

where we again set $S_n = X_1 + \cdots + X_n$.

Exercise 61 *(Covariances of trawl processes)* Assume that the seed γ is a unit Poisson process (or any other square integrable Lévy process with

$$\text{Cov}\,(\gamma(u), \gamma(v)) = u \wedge v),$$

and also,

$$X_n = \sum_{k=0}^{\infty} \gamma_{n-k}(a_k).$$

Assume moreover that a_j is a non-increasing positive sequence such that

$$\sum_{j=0}^{\infty} a_j < \infty.$$

Then the above process is strictly stationary and in $\mathbb{L}^2$.

Moreover if

$$A_k = \sum_{j=k}^{\infty} a_j, \quad \text{with} \quad \sum_{j=0}^{\infty} A_j = \infty,$$

then:

- Its covariance is $\text{Cov}\,(X_0, X_n) = A_n$.
- Deduce (10.2) for the case of a Poisson seed.
- Prove that an analogue result holds for the case of Bernoulli seeds (which are not Lévy processes).

Hints. The series is normally convergent in $\mathbb{L}^2$ since

$$\|\gamma_{n-k}(a_k)\|_2 = \|\gamma(a_k)\|_2 = \sqrt{a_k}$$

is a square summable series. Thus independence of the sequence (γ_j) allows to conclude.

Stationarity is standard and relies on the fact that the vector valued random variables $(X_{k+1}, \dots, X_{k+\ell})$ are limits in $\mathbb{L}^2$ (and thus in probability) of

$$(X^{(N)}_{k+1}, \dots, X^{(N)}_{k+\ell}) = F^{(N)}_\ell(\xi_{k+\ell}, \dots, \xi_{k-N}),$$

obtained by replacing the above infinite series by series for $0 \le j \le N$; their distribution clearly does not depend on k from the stationarity of inputs (ξ_j).

Now

$$\text{Cov}\,(X_0, X_n) = \sum_{k=0}^{\infty} a_k \wedge a_{k+n} = \sum_{k=0}^{\infty} a_{k+n} = A_n$$

Then (10.2) follows easily for the case of Poisson seeds.

For the case of Bernoulli seeds

$$\begin{aligned}\text{Cov}\,(X_0, X_n) &= \sum_{k=0}^{\infty} (a_k \wedge a_{k+n} - a_k a_{k+n}) \\ &= \sum_{k=0}^{\infty} a_{k+n} - \sum_{k=0}^{\infty} a_k a_{k+n} \\ &= A_n - B_n = A_n(1 + o(1)),\end{aligned}$$

with $0 \le B_n \le A_0 a_n = o(A_n)$ since a_k admits a Riemannian decay rate.[1] Equation (10.2) again follows in this case.

Equation (10.2) implies that, for $H = (3 - \alpha)/2$, the centred process $S_n(t) = n^{-H} S_{[nt]}$ satisfies

$$\lim_{n\to\infty} \text{Cov}\,(S_n(s), S_n(t)) = \text{Cov}\,(B_H(s), B_H(t)).$$

However $S_n(t)$ converges in probability to 0 for each $t > 0$; for this, the formula (4.4) entails $\mathbb{E}S_n^2(t) = \mathcal{O}(n^{\alpha-2}) \to_{n\to\infty} 0$.

In fact the sequence of processes $n^{H-\frac{1}{\alpha}} S_n(t)$ converges to a symmetric Lévy stable process with index α in a certain sense; precisely convergence holds in the M_1-topology on $\mathcal{D}[0, 1]$, see in Jakubowski (1997) which is quite weaker that the usual J_1-topology.

Such unusual behaviours contradict the Dehling definition of LRD since this model is $\mathbb{L}^2$-LRD and it is SRD in the distribution sense of Dehling. Such atypical behaviours may be seen to happen for high traffic models and (Konstantopoulos and Lin 1998) first exhibited such behaviours; note that the latter distributional conver-

[1] Note that the same relation $a_n = o(A_n)$ does not hold when a_k admits a geometric decay rate since in this case tails A_n and the first term a_n of a series admit the same order of magnitude.

gence was proved afterwards in Resnick and Van den Berg (2000). The previous authors worked with the different shot-noise models.

Exercise 62 is a main tool to prove a suitable limit theory for partial sums since it proves their decomposition as a sum of independent random variables and may as well allow the use of the Lindeberg Lemma 2.1.1, to derive alternative Gaussian behaviours (see again in Doukhan et al. 2017).

Exercise 62 *(Decomposition lemma)* Let (X_k) be as in (10.1), setting $S_n = X_k + \cdots + X_n$, then we have

$$S_n = \sum_{s=-\infty}^{n} Z_{s,n} \tag{10.3}$$

$$Z_{s,n} = \sum_{k=1\vee s}^{n} \gamma_s(a_{k-s}).$$

Prove that the random variables $(Z_{s,n})_{s\leq n}$ are independent.

Hint. This just needs a careful observation of X_k: order summed elements wrt to the index s of the involved seed γ_s.

Notice also that the original decomposition in Exercises 91 and 95 again yields a decomposition of the above variables $Z_{s,n}$ as sums of very simple random variables, for the above Poisson case and for Bernoulli distributed inputs. Those exercises allow very precise controls of higher order moments of the partial sums which are a main interest for further ongoing contributions. They yield the following exercise.

Exercise 63 *(Symmetric trawl processes)* Assume that inputs are either symmetric Poisson processes or Bernoulli processes (as in Exercises 94 and 96).

Let $p \geq 2$ be an even integer, then prove that for $1 < \alpha < 2$, there exists a constant $C_{\alpha,p} > 0$ such that $\mathbb{E}S_n^2 \sim C_{\alpha,1}n^{3-\alpha}$ and $\mathbb{E}S_n^{2p} \sim C_{\alpha,p}n^{2p-\alpha}$ as $n \to \infty$ if $p > 1$.

Note that if $m = 2p + 1$ is odd, a symmetry argument entails $\mathbb{E}S_n^m = 0$.

10.6.2 LARCH-Models

As in Sect. 7.2.2 we model the stationary solution of the recursion:

$$X_n = \Big(b_0 + \sum_{j=1}^{\infty} b_j X_{n-j}\Big)\xi_n.$$

This also admits LRD behaviours if the iid sequence (ξ_t) is centred and

$$\mathbb{E}\xi_0^2 \sum_{j=1}^{\infty} b_j^2 < 1,$$

but

$$\sum_{j=1}^{\infty} |b_j| = \infty.$$

More general volatility models

$$X_t = \sigma_t \xi_t, \quad \sigma_t^2 = G\Big(\sum_{j=1}^{\infty} b_j X_{t-j}\Big),$$

extend on ARCH(∞)-models ($G(x) = b_0 + x^2$), and asymmetric ARCH(∞)-models ($G(x) = b_0 + (c + x)^2$).

For deriving LRD properties, again one requires

$$\sum_{j=1}^{\infty} b_j^2 < \infty, \qquad \sum_{j=1}^{\infty} |b_j| = \infty.$$

See (Giraitis et al. 2012) for details.

10.6.3 Randomly Fractional Differences

Philippe et al. (2008) introduced time-varying fractional filters $A(\mathbf{d})$, $B(\mathbf{d})$ defined by

$$A(\mathbf{d})x_t = \sum_{j=0}^{\infty} a_j(t) x_{t-j}, \qquad B(\mathbf{d})x_t = \sum_{j=0}^{\infty} b_j(t) x_{t-j}, \tag{10.4}$$

where $\mathbf{d} = (d_t, t \in \mathbb{Z})$ is a given function of $t \in \mathbb{Z}$.

We also set $a_0(t) = b_0(t) = 1$, and if $j \geq 1$:

$$a_j(t) = \Big(\frac{d_{t-1}}{1}\Big)\Big(\frac{d_{t-2}+1}{2}\Big)\Big(\frac{d_{t-3}+2}{3}\Big)\cdots\Big(\frac{d_{t-j}+j-1}{j}\Big),$$
$$b_j(t) = \Big(\frac{d_{t-1}}{1}\Big)\Big(\frac{d_{t-j}+1}{2}\Big)\Big(\frac{d_{t-j+1}+2}{3}\Big)\cdots\Big(\frac{d_{t-2}+j-1}{j}\Big).$$

If $d_t = d$ is a constant, then $A(\mathbf{d}) = B(\mathbf{d}) = (I - L)^{-d}$ is the usual fractional integration operator ($Lx_t = x_{t-1}$ is the backward shift).

Doukhan et al. (2007a) consider the two following processes, for centred independent identically distributed inputs ϵ_t,

$$X_t^A = \sum_{j=0}^{\infty} a_j(t) \xi_{t-j}, \qquad X_t^B = \sum_{j=0}^{\infty} b_j(t) \xi_{t-j}.$$

If d_t is independent identically distributed and $\mathbb{E}d_t = \bar{d} \in (0, \frac{1}{2})$ then the asymptotic behaviour of partial sums of this process is the same as for ARFIMA$(0, 0, \bar{d}, 0)$ which corresponds to the case of a constant sequence d_t.

If ϵ_t is standard Normal, then

- X_t is Gaussian with a variance $A(t) = \mathbb{E}(X_t^2|\mathcal{D})$, conditionally wrt $\mathcal{D}$, the sigma-algebra generated by $\{d_s/\ s \in \mathbb{Z}\}$,
- $g_k(A) = A^{-2k}\mathbb{E}[h(X)H_k(X;A)]$, where $H_k(x;A) = A^k H_k\Big(\dfrac{x}{A}\Big)$, denotes Hermite polynomials with variance A, for $k = 0, 1, 2, \ldots$.

Then the Gaussian limit theory extends with Hermite coefficients replaced by

$$\beta_k = \mathbb{E}\big(g_k(A(0))Q^k\big)$$

for a random variable Q related to the random coefficients d_t, $\mathbb{E}d_t = \bar{d}$ and d_t admits a finite range, and

$$\mathbb{E}|h(B\xi_t)|^a < \infty, \qquad \text{for some} \quad a > 2.$$

10.6.4 Perturbed Linear Models

Doukhan et al. (2002a) study the empirical process of perturbed linear models:

$$X_t = Y_t + V_t, \quad t \in \mathbb{Z},$$

where (Y_t) is a long-range dependent causal linear process and

$$V_t = H(\xi_t, \xi_{t-1}, \ldots)$$

denotes a short-range dependent perturbation.

Then the perturbation does not modify the behaviour of the empirical process which thus behaves as for linear LRD processes.

10.6.5 Non-linear Bernoulli-Shift Models

Doukhan, Lang, and Surgailis (unpublished manuscript, 2006) study the partial sums process of

$$X_t = H(Y_t; \xi_t, \xi_{t-1}, \ldots), \quad Y_t = \sum_{j=0}^{\infty} b_j \xi_{t-j},$$

where $b_j \sim c_0 j^{d-1}$, with $d \in (0, 1/2)$, and ξ_t independent identically distributed innovations, and H is a function of infinitely many variables.

A main goal of the results was to prove that:

There exists a non-Gaussian process X whose partial sums process converges to a second order Rosenblatt process while the partial sums of X^2 converge to the fractional Brownian motion.

The technique extends that of Ho and Hsing (1996); it is based on a martingale decomposition of the partial sums process.

$$S_n(t) = \sum_{s=1}^{[nt]} (X_s - \mathbb{E}X_s), \quad t \in [0, 1].$$

It is possible to give conditions ensuring that, in law:

$$\begin{aligned} S_n(t) &\sim h'_\infty(0) \sum_{s=1}^{[nt]} Y_s, \\ h_\infty(y) &= \mathbb{E}H(y + Y_t, \xi_t, \xi_{t-1}, \dots). \end{aligned}$$

A similar result holds with a second order U-statistic of ξ which asymptotic is related to the Rosenblatt process.

There exists a constant $c_d \in \mathbb{R}$ such that if $d \in]\frac{1}{2}, 1]$ and if $h'_\infty(0) \neq 0$ then:

$$n^{-d-\frac{1}{2}} S_n(t) \xrightarrow{\mathcal{D}[0,1]}_{n\to\infty} c_d h'_\infty(0) B_d(t),$$

if now $d \in \left]\frac{1}{4}, 1\right]$ and $h'_\infty(0) = 0$ and $h''_\infty(0) \neq 0$ then:

$$n^{-2d} S_n(t) \xrightarrow{\mathcal{D}[0,1]}_{n\to\infty} c_d h''_\infty(0) Z_d^{(2)}(t).$$

Chapter 11
Short-Range Dependence

This chapter introduces some simple ideas. We investigate conditions on time series such that the standard limit theorems obtained for independent identically distributed sequences still hold. After a general introduction to weak-dependence conditions an example states the fact that the most classical *strong-mixing condition* from Rosenblatt (1956) may fail to work, see Andrews (1984).

When dealing with any weak-dependence condition (including strong mixing), additional *decay rates* and *moment conditions* are necessary to ensure CLTs. Decay rates will be essential to derive asymptotic results. Coupling arguments as proposed in Sect. 7.4.2 are widely used for this.

Finally to make clearer the need for decay rates, we explain how CLTs may be proved under such assumptions.

The monograph (Dedecker et al. 2007) is used as the reference for weak-dependence; in this monograph we developed more formal results together with their proofs. We refer a reader to this monograph for more rigorous results.

11.1 Weak-Dependence

When looking for asymptotic independence it seems natural to consider conditions from Doukhan and Louhichi (1999):

Definition 11.1.1 Assume that there exist classes of functions

$$\mathcal{F}, \mathcal{G} : \bigcup_{u \geq 1} \mathbb{R}^u \to \mathbb{R},$$

and a function $\psi : \mathcal{F} \times \mathcal{G} \to \mathbb{R}$ (which depends on f, g and on the numbers of their arguments u, v) and on a sequence $\epsilon_r \downarrow 0$ as $r \uparrow \infty$.

P. Doukhan, *Stochastic Models for Time Series*, Mathématiques et Applications 80,
https://doi.org/10.1007/978-3-319-76938-7_11

Fig. 11.1 Asymptotic independence

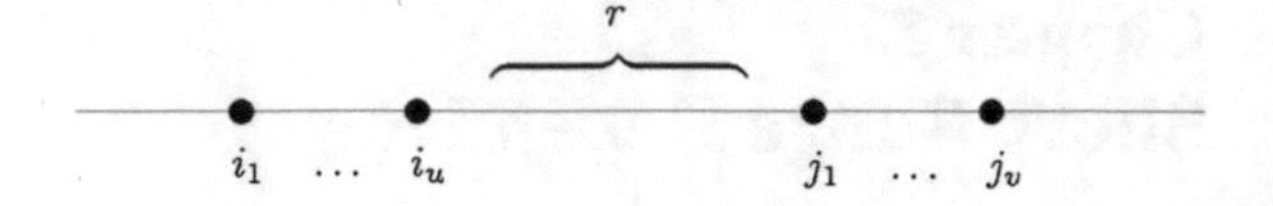

A random process $(X_t)_{t\in\mathbb{Z}}$ is said to be $(\mathcal{F}, \mathcal{G}, \psi, \epsilon)$-weakly dependent in case

$$\left|\text{Cov}\left(f(X_{i_1}, \ldots, X_{i_u}), g(X_{j_1}, \ldots, X_{j_u})\right)\right| \leq \epsilon_r \psi(u, v, f, g) \tag{11.1}$$

for functions f, g belonging respectively to classes $\mathcal{F}, \mathcal{G}$, and

$$i_1 \leq \cdots \leq i_u \leq j_1 - r \leq j_1 \leq \cdots \leq j_v.$$

Epochs are graphically reported in Fig. 11.1.

The following sections are dedicated to examples of these generic notions. Firstly, we explicitly consider strong mixing as well as a simple counter-example; secondly we develop a model-based bootstrap as an example of an application for which weak-dependence notions in Definition 11.4.1 are a reasonable option.

11.2 Strong Mixing

Strong mixing introduced in Rosenblatt (1956) may be seen as a special case of the previous weak-dependence situation. Here

$$\mathcal{F} = \mathcal{G} = \mathbb{L}^\infty, \quad \text{and} \quad \psi(u, v, f, g) = 4\|f\|_\infty \|g\|_\infty, \quad \epsilon_r = \alpha_r.$$

Examples of strongly mixing processes are given in Doukhan (1994). The sup bound of such ϵ_r satisfying the above inequality is denoted α_r; it is also possible to derive:

$$\alpha_r = \sup_{\substack{A \in \sigma(X_i,\ i \leq 0) \\ B \in \sigma(X_j,\ j \geq r)}} |\mathbb{P}(A \cap B) - \mathbb{P}(A)\mathbb{P}(B)|.$$

Indeed the previous inequality extends to (11.1) for non-negative linear combinations of indicator functions. With a density argument it possible to consider arbitrary non-negative functions. A factor 4 appears when one allows functions with values in $[-1, 1]$ since real valued functions are the difference of two non-negative functions. One may refer to Doukhan (1994) for details and examples.

This strong mixing condition does not hold for some models, e.g.

$$X_n = \frac{1}{2}(X_{n-1} + \xi_n), \tag{11.2}$$

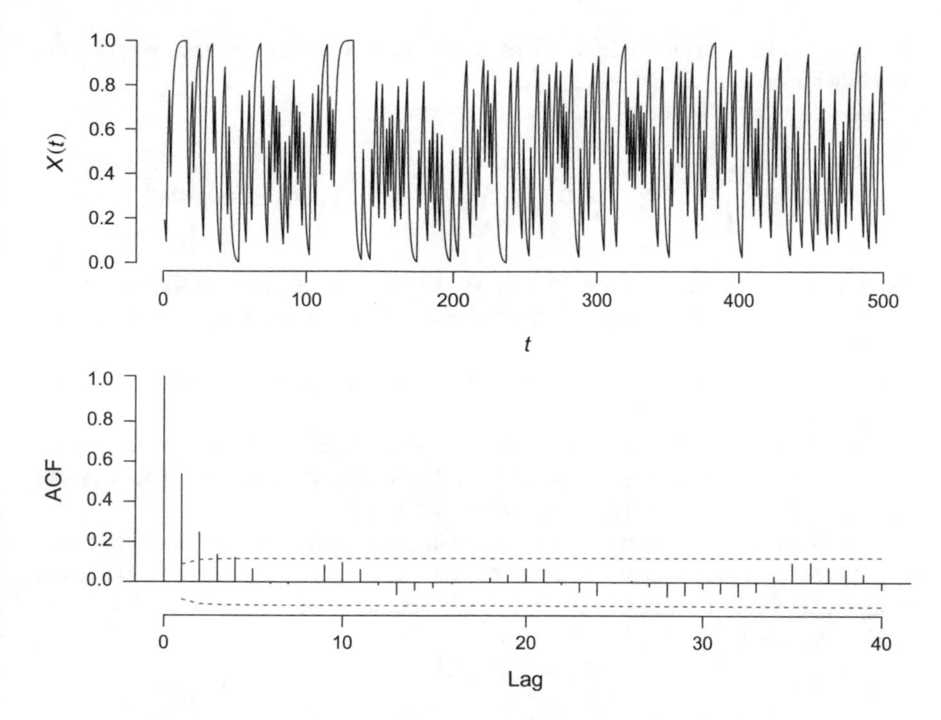

Fig. 11.2 A non-mixing AR(1)-process, and its autocovariances

where the independent identically distributed inputs (ξ_n) admit a Bernoulli distribution with parameter $\frac{1}{2}$.

Remark on the simulation of the model (11.2) that this model admits quite chaotic samples while its covariances admit a fast decay rate, $\text{Cov}\,(X_0, X_t) = 2^{-t}$ (Fig. 11.2).

Here, $X_n = \frac{1}{2}(X_{n-1} + \xi_n)$ and $\xi_n \sim \mathcal{B}(1/2)$.

Proposition 11.2.1 *The stationary solution of equation (11.2) exists and is uniform on the unit interval, moreover it is not strong mixing, more precisely* $\alpha_r \geq \frac{1}{4}$, *for all* $r \geq 1$.

Note. In this case of Eq. (11.2) the process is however weakly dependent under **alternative dependence conditions**, see Example 11.4.1. More precisely $\epsilon_r (= \theta_r) \leq 2^{1-r}$ for $r \in \mathbb{N}$, holds under a dependence assumption for which the considered classes of functions are Lipschitz, see Sect. 11.4 for some more precise statements.

Proof The function $x \mapsto \frac{1}{2}(x + u)$ maps $[0, 1]$ in a subset of $[0, 1]$. This implies that, applying recursively Eq. (11.2), yields

$$X_n = \sum_{k=0}^{p} \frac{1}{2^{k+1}} \xi_{n-k} + \frac{1}{2^{p+1}} X_{n-p}.$$

Hence if we assume that initial values of the model are in the unit interval, the remainder term is $\leq 2^{-1-p} \to_{p\to\infty} 0$.

The stationary solution of the previous equation is

$$X_n = \sum_{k=0}^{\infty} 2^{-1-k} \xi_{n-k} = 0, \xi_n \xi_{n-1} \ldots, \quad \text{in the numeration basis 2.}$$

The expansion of a real number in $x = 0.x_1x_2x_3 \ldots \in [0, 1)$ in basis 2 is in fact unique if one adopts the convention that there does not exist an integer p with $x_k = 1$ for each $k \geq p$.

This restriction does not matter much since it leads to a negligible event, with zero probability.

The marginals of this process are easily proved to be uniformly distributed on [0, 1]; for example choosing an interval with dyadic extremities makes it evident and such intervals generate the Borel sigma field of [0, 1].

Now the previous condition can be written in terms of the sigma-algebra generated by the processes and X_{t-1} is the fractional part of $2X_t$ which implies the inclusion of sigma algebras generated by marginals of such processes.

More precisely

$$X_0 = 0, \xi_0 \xi_{-1} \xi_{-2} \ldots,$$

and

$$X_r = 0, \xi_r \xi_{r-1} \xi_{r-2} \ldots \xi_0 \xi_{-1} \xi_{-2} \ldots$$

the event $A = (X_0 \leq \frac{1}{2})$ can be written as $(\xi_0 = 0)$ and $\mathbb{P}(A) = \frac{1}{2}$.

There exists a measurable function such that

$$X_0 = f_r(X_r) \quad \text{and} \quad A = f_r^{-1}\left(\left[0, \frac{1}{2}\right]\right) \in \sigma(X_r).$$

Namely this function is the r-th iterate of $x \mapsto \text{frac}(2x)$, in terms of the dyadic expansion this function consists simply in suppression of the first r terms in X_r's expansion.

- If $r = 1$ then
$$A = \left(X_1 \in \left[0, \frac{1}{4}\right] \cup \left[\frac{1}{2}, \frac{3}{4}\right]\right).$$

- If $r = 2$ we easily check that
$$A = \left(X_2 \in \left[0, \frac{1}{8}\right] \cup \left[\frac{1}{4}, \frac{3}{8}\right] \cup \left[\frac{1}{2}, \frac{5}{8}\right] \cup \left[\frac{3}{4}, \frac{7}{8}\right]\right).$$

- More generally $A = B$ with $A_r = (X_r \in I_r)$ where I_r is the union of 2^r intervals with dyadic extremities and with the same length 2^{-r-1}.

Thus

$$\alpha_r \geq \sup\{A \in \sigma(X_0), B \in \sigma(X_r)\} \geq \mathbb{P}(A \cap B_r) - \mathbb{P}(A)\mathbb{P}(B_r) = \frac{1}{4}.$$

The previous example proves that strong mixing notions are not enough to consider very reasonable wide classes of statistical models.

11.3 Bootstrapping AR(1)-Models

A main problem for time series is that the exact distributions of many useful functionals are unknown. Such functionals are important since they usually appear as limits (in distribution) of some convergent sequences of functionals

$$G_k = g_k(X_1, \dots, X_k) \xrightarrow{\mathcal{L}}_{k\to\infty} \Gamma.$$

We considered subsampling in Sect. 4.6 as an easy way to proceed. A common way to estimate the quantiles of Γ is due to Efron (1982) and it is known as the bootstrap.

From the previous convergence in distribution the knowledge of quantiles is essential to determine the property of of goodness-of-fit tests (level and power). They also yield asymptotic confidence bands. This is important to be able in simulating many samples

$$\{X_1(\omega_i), \dots, X_k(\omega_i)\}, \quad \text{for } 1 \leq i \leq I.$$

For example, the simple law of large numbers (for independent identically distributed samples) entails the consistency of the empirical quantiles derived from such resampled processes.

We do not intend to provide an abstract theory for the bootstrap but rather to explain how to implement it over a very simple example.

First fit the model

$$X_n = aX_{n-1} + \xi_n. \tag{11.3}$$

Here (ξ_n) is an independent identically distributed and centred sequence with a first order finite moment.

For $|a| < 1$ the Eq. (11.3) admits the solution

$$X_n = \sum_{k=0}^{\infty} a^k \xi_{n-k}.$$

In order to bootstrap we proceed to the following steps.

- The estimator $\widehat{a}_n$ in Exercise 59 is proved to be a.s. convergent by a simple use of the ergodic theorem. Assume that one observes a sample $\{X_1(\omega), \ldots, X_N(\omega)\}$ (which means that $\omega \in \Omega$ is fixed outside of some negligible event) of the stationary solution of the AR(1) process (11.3).

- Then let us use the first n data items $\{X_1(\omega), \ldots, X_n(\omega)\}$ to estimate $\widehat{a}_n$ and from a.s. convergence we may suppose that N is large enough in order that $|\widehat{a}_n| < 1$ for $n \geq N/3$. We only used the first third of the sample to estimate $\widehat{a}_n$ and this allows to estimate residuals $\widehat{\xi}_j = X_j - \widehat{a}_{N/3} X_{j-1}$ for $j > 2N/3$.
  We just omit one third of the data to assume that the random variables $\{\widehat{\xi}_j / j > 2N/3\}$ are almost independent of $\widehat{a}_{N/3}$, in a sense to be precisely set.
- Now assuming that $N = 3n$ we may consider conditionally centred residuals by setting

$$\tilde{\xi}_j = \widehat{\xi}_j - \frac{1}{n} \sum_{k=2n+1}^{N} \widehat{\xi}_k, \qquad 2n < j \leq N.$$

- To the end of resampling statistics we simulate independent identically distributed sequences $(\xi^*_{i,j})_{(i,j) \in \mathbb{N} \times \mathbb{Z}}$ with uniform distribution on the set $\{\tilde{\xi}_j / \ 2n + 1 \leq j \leq 3n\}$.
- This means that we may simply simulate trajectories of the stationary solution of (11.3):

$$X^*_{i,n} = \widehat{a}_n X^*_{i,n-1} + \xi^*_{i,n}, \qquad i \geq 0, \ n \in \mathbb{Z},$$

  which exists since we may choose n large enough for contraction to hold, since $|\widehat{a}_n| < 1$ with a high probability.

As a final remark, the previous stationary solutions of (11.3) are shown to be strongly mixing only in the case when ξ_0's distribution admits an absolutely continuous part, see Doukhan (1994). This is not the case for the resampled process which led e.g. Jens Peter Kreiss and Michael Neumann to simply smooth the discrete distribution ν_* of ξ^*_0 in order to be able to use the necessary asymptotic properties shown under strong mixing in order to prove the consistency of those techniques. They simply convolve ν_* with any probability density to get an absolutely continuous distribution. They just replace ξ^*_0 distribution by $\xi^*_0 + \zeta_0$ for a small random variable ζ_0 independent of ξ^*_0 admitting a density wrt Lebesgue measure (think of $\zeta_0 \sim \mathcal{N}(0, \epsilon^2)$), then the Markov chain obtained is ergodic and strong mixing applies (see (Doukhan 1994) for details) but it is not clear whether this distribution admits a real sense wrt bootstrap.

11.4 Weak-Dependence Conditions

We prove on the simple example of linear processes that an alternative to mixing defined in Definition 11.1.1 is indeed more adapted to such resampling questions.

In the concept of weak-dependence, note that for some processes we are able to get fast speed of decay only for very small classes of functions. It is thus natural to restrict the class of functions f and g which are on some special classes $\mathcal{F}$ and $\mathcal{G}$.

Such simple moving average models will help us to introduce suitable weak-dependence conditions for model-based bootstrap procedures. Weak-dependence conditions also allow to develop a simple asymptotic theory (see Sect. 11.5).

Return to inequality (11.1), the left-hand side of which will be written for simplicity Cov $(\mathbf{f}, \mathbf{g})$ with $\mathbf{f} = f(X_{i_1}, \ldots, X_{i_u})$ and $\mathbf{g} = g(X_{j_1}, \ldots, X_{j_v})$.

Then we consider a simple linear (infinite moving average) model defined through an independent identically distributed sequence with finite first order moments $(\xi_t)_{t\in\mathbb{Z}}$:

$$X_t = \sum_{k\in\mathbb{Z}} a_k \xi_{t-k}.$$

The previous series converge in $\mathbb{L}^1$ in case

$$\|\xi_0\|_1 < \infty, \qquad \text{and} \qquad \sum_{t=-\infty}^{\infty} |a_t| < \infty,$$

and the considered process is then stationary; it corresponds to

$$H((u_t)_{t\in\mathbb{Z}}) = \sum_{t\in\mathbb{Z}} a_t u_{-t}.$$

Then the model is said to be causal in case $a_k = 0$ for $k < 0$ since X_t is measurable with respect to $\mathcal{F}_t = \sigma(\xi_s;\ s \le t)$.

Set

$$X_t^{(p)} = \sum_{|k|\le p} a_k \xi_{t-k}, \qquad \tilde{X}_t^{(p)} = \sum_{0\le k\le p} a_k \xi_{t-k},$$

then it is simple to check that $X_s^{(p)}$ and $X_t^{(p)}$ are independent if $|t-s| > 2p$ and in case $r > 2p$ this also implies that $\mathbf{f}'$ and $\mathbf{g}'$ are independent when setting $\mathbf{f}' = f(X_{i_1}^{(p)}, \ldots, X_{i_u}^{(p)})$ and $\mathbf{g}' = g(X_{j_1}^{(p)}, \ldots, X_{j_v}^{(p)})$ and

$$\text{Cov}\,(\mathbf{f}, \mathbf{g}) = \text{Cov}\,(\mathbf{f}, \mathbf{g} - \mathbf{g}') + \text{Cov}\,(\mathbf{f} - \mathbf{f}', \mathbf{g}').$$

Now if the involved functions are both bounded above by 1 then

$$|\text{Cov}\,(\mathbf{f}, \mathbf{g} - g')| \le 2\mathbb{E}|\mathbf{g} - \mathbf{g}'|, \quad |\text{Cov}\,(\mathbf{f} - \mathbf{f}', \mathbf{g}')| \le 2\mathbb{E}|\mathbf{f} - \mathbf{f}'|.$$

If now those functions are Lipschitz then:

$$|\mathbf{f}-\mathbf{f}'| \le \mathrm{Lip}\, f \sum_{s=1}^{u} |X_{i_s} - X_{i_s}^{(p)}|, \ |\mathbf{g}-\mathbf{g}'| \le \mathrm{Lip}\, g \sum_{s=1}^{v} |X_{j_s} - X_{j_s}^{(p)}|.$$

We also note that, for all t, $\mathbb{E}|X_t - X_t^{(p)}| \le \mathbb{E}|\xi_0| \sum_{k>p} |a_k|$.

Using the bound $|\mathrm{Cov}\,(U, V)| \le 2\|U\|_\infty \mathbb{E}|V|$ yields:

$$|\mathrm{Cov}\,(\mathbf{f}, \mathbf{g}-\mathbf{g}')| \le 2v\mathbb{E}|\xi_0| \sum_{k>p} |a_k|,$$

$$|\mathrm{Cov}\,(\mathbf{f}-\mathbf{f}', \mathbf{g}')| \le 2u\mathbb{E}|\xi_0| \sum_{k>p} |a_k|.$$

In the causal case it is simple to check that $\mathbf{f}$ and $\tilde{\mathbf{g}}$ are independent for $\tilde{\mathbf{g}} = g(\tilde{X}_{j_1}^{(p)}, \ldots, \tilde{X}_{j_v}^{(p)})$.

This implies $\mathrm{Cov}\,(\mathbf{f}, \mathbf{g}) = \mathrm{Cov}\,(\mathbf{f}, \mathbf{g}-\tilde{\mathbf{g}})$ and analogously we obtain:

- $|\mathrm{Cov}\,(\mathbf{f}, \mathbf{g})| \le (u\mathrm{Lip}\, f + v\mathrm{Lip}\, g)\epsilon_r$, if we set

$$\epsilon_r = 2\mathbb{E}|\xi_0| \sum_{|i|>2r} |a_i|,$$

for non-causal linear processes:

$$X_n = \sum_{i=-\infty}^{\infty} a_i \xi_{n-i}.$$

- $|\mathrm{Cov}\,(\mathbf{f}, \mathbf{g})| \le v\mathrm{Lip}\, g \cdot \epsilon_r$, for the causal case, $a_i = 0$ if $i < 0$, with

$$\epsilon_r = 2\mathbb{E}|\xi_0| \sum_{i>r} |a_i|.$$

For the causal case, $a_i = 0$ if $i < 0$:

$$X_n = \sum_{i=0}^{\infty} a_i \xi_{n-i}.$$

Most of the previous models satisfy the conditions as for Bernoulli schemes. Set now:

$$\epsilon_r = 2 \sup_{q>2r} \mathbb{E}\left|H\left((\xi_i)_{i\in\mathbb{Z}}\right) - H\left((\xi_i)_{|i|\le q}\right)\right|,$$

for non-causal Bernouilli shifts $X_t = H(\ldots, \xi_{t+1}, \xi_t, \xi_{t-1}, \ldots)$. And

$$\epsilon_r = 2 \sup_{q>r} \mathbb{E}\left|H\left((\xi_i)_{i\in\mathbb{N}}\right) - H\left((\xi_i)_{0\le i\le q}\right)\right|,$$

for causal Bernouilli shifts $X_t = H(\xi_t, \xi_{t-1}, \ldots)$. The sequence $(\xi_i)_{|i|\le r}$ is obtained by setting 0 for indices with $|i| > r$. Then $\epsilon_r \downarrow 0$ as $r \uparrow \infty$[1], and the following conditions ψ_θ or ψ_η apply according to whether the Bernoulli is causal or not.

Doukhan and Louhichi (1999) write such easy conditions in terms of Lipschitz classes. Some more precise bibliographical comments are given in Remark 8.3.2.

The present chapter is not exhaustive so we will restrict the really general notions in Definition 11.1.1 to a few cases of weak-dependence.

Definition 11.4.1 Set Λ the class of functions $g : \mathbb{R}^v \to \mathbb{R}$ for some integer $v \ge 1$, with $\|g\|_\infty \le 1$ and $\mathrm{Lip}(g) < \infty$ where:

$$\mathrm{Lip}(g) = \sup_{(x_1,\ldots,x_v)\neq(y_1,\ldots,y_v)} \frac{|g(x_1,\ldots,x_v) - g(y_1,\ldots,y_v)|}{|x_1 - y_1| + \cdots + |x_v - y_v|}.$$

Some weak-dependence conditions correspond to $\mathcal{G} = \Lambda$ and, respectively, either $\mathcal{F} = \Lambda$ for the non causal case, or

$$\mathcal{F} = \mathbb{B}_\infty = \{f : \mathbb{R} \to \mathbb{R},\ \text{measurable with } \|f\|_\infty \le 1\}, \quad \text{for the causal case.}$$

Here respectively

$$\begin{aligned}
\psi(u, v, f, g) &= \psi_\eta(u, v, f, g) = u\mathrm{Lip}(f) + v\mathrm{Lip}(g),\\
&= \psi_\theta(u, v, f, g) = v\mathrm{Lip}(g),\\
&= \psi_\kappa(u, v, f, g) = uv\mathrm{Lip}(g)\mathrm{Lip}(g),\\
&= \psi_\lambda(u, v, f, g) = u\mathrm{Lip}(f) + v\mathrm{Lip}(g) + uv\mathrm{Lip}(g).
\end{aligned}$$

Then the process $(X_t)_{t\in\mathbb{Z}}$ is η-weakly dependent (resp. θ, κ or λ) in case the least corresponding sequence ϵ_r given by relation (11.1) converges to 0 as $r \uparrow \infty$; the respective coefficients will be denoted η_r, θ_r, κ_r or λ_r.

Example 11.4.1 (Dependence decay-rates) To derive limit theorems it will be essential to know the decay rates of decorrelation as well as the existence of moments. The following examples aim at filling this important gap.

---

[1]For the special case of the previous linear processes, the present bound ϵ_r is even sharper than those considered above for general Bernoulli shifts.

- Conditions η and θ were checked before to hold for linear causal or non-causal sequences. They also hold analogously for Bernoulli-shifts under assumptions (7.20) if they are $\mathbb{L}^1$-weakly dependent (see Definition 7.4.3). Here respectively

$$\theta_r = 2\delta_r^{(1)}, \quad \text{under a causal condition,}$$
$$\eta_r = 2\delta_{[r/2]}^{(1)}, \quad \text{otherwise.}$$

- Examples of such causal models are Markov stable processes (see Sect. 7.3) satisfy those relations as proved in Theorem 7.3.1. Such Markov models (7.3) indeed satisfy the inequality (7.6). This proves that $\theta_r \le ca^r$ for some constant $c > 0$.
- Non homogeneous Markov chains $X_t = M_t(X_{t-1}, \xi_t)$ extending that in Remark 7.2.1 are easily proved to satisfy such weak dependence conditions $\theta_r \le ca^r$ in case the relations (7.4) and (7.5) are replaced by

$$\sup_t \mathbb{E}\|M_t(u, \xi_0) - Mt(v, \xi_0)\|^p \le a^p \|u - v\|^p,$$
$$\sup_t \mathbb{E}\|M_t(u_0, \xi_0)\|^p < \infty.$$

Such uniform contractive conditions yield weak dependence for general classes of models. A typical situation is provided by a parametric family of Markov equations $X_t = M_\beta(X_{t-1}, \xi_t)$; then select the parameter $\beta = \beta_t$ at each time turns to the above condition. Bardet and Doukhan (2017) suggest $\beta_t = g_t(t/n)$ for some regular periodic family of functions g_t on $[0, 1]$, such that $g_{t+T} = g_t$ with T some known period, see Example 6.6.1. We also derive consistent estimation of those function.
- Linear and Volterra processes are also weakly dependent and tails of coefficients allow to bound ϵ_r in both the causal and the non-causal case.
- In order to consider an explicit example of a chaotic expansion, we consider the LARCH(∞)-models in Sect. 7.2.2. They are solutions of the recursion

$$X_n = \left(b_0 + \sum_{j=1}^{\infty} b_j X_{n-j} \right) \xi_n.$$

The $\mathbb{L}^p$-valued strictly stationary solution of this recursion is

$$X_n = \sum_{k=1}^{\infty} S_n^{(k)},$$

with:

$$S_n^{(k)} = b_0 \xi_n \sum_{l_1=1}^{\infty} \cdots \sum_{l_k=1}^{\infty} b_{l_1} \dots b_{l_k} \xi_{n-l_1} \xi_{n-l_1-l_2} \cdots \xi_{n-(l_1+\cdots+l_k)}.$$

Under the condition

$$B = \|\xi_0\|_p \sum_{l=1}^{\infty} |b_l| < 1,$$

it is simple to derive with the independence of all these products that $\|S_n^{(k)}\|_p \le |b_0|a^k$. Now set $S_n^{(k,L)}$ for the finite sum where each of the indices satisfies $1 \le l_1, \ldots, l_k \le L$ then analogously

$$\|S_n^{(k)} - S_n^{(k,L)}\|_p \le k|b_0|a^{k-1}B_L, \qquad B_L = \|\xi_0\|_p \sum_{l=L+1}^{\infty} |b_l|,$$

here the factor k comes from the fact that in order that only the tail of a series appears, this may occur at any position in the above multiple sums.

Restricting to the case $p = 1$, we now approximate X_n by the following $L \times K$-dependent sequence

$$X_n^{(K,L)} = S_n^{(0)} + S_n^{(1,L)} + \cdots + S_n^{(K,L)}$$

then previous calculations prove that for a constant $C > 0$,

$$\|X_n - X_n^{(K,L)}\|_1 \le C(B_L + a^K).$$

Let $L, K \ge 1$ be such that $LK \le r$ then this implies that wrt ψ_θ,

$$\theta_r \le C \inf_{1 \le L \le r} \left(B_L + a^{\frac{r}{L}} \right).$$

E.g. if $b_l = 0$ for $l > L$ large enough then $\theta_r \le C a^{\frac{r}{L}}$, if B_L h decays geometrically to 0, then $\theta_r \le C e^{-c\sqrt{r}}$, and if $B_L \le cL^{-\beta}$ then $\theta_r \le c'r^{-\beta}$, see Doukhan et al. (2007b).

- Either Gaussian processes or associated random processes (in $\mathbb{L}^2$) are κ-weakly dependent because of Lemma 8.1. Here

$$\kappa_r = \max_{|j-i| \ge r} |\mathrm{Cov}\,(X_i, X_j)|,$$

is the convenient weak-dependence coefficient from inequality (8.1) (in this case absolute values are useless); this inequality also holds for the Gaussian case as proved e.g. in Dedecker et al. (2007).
- Now the function ψ_λ allows to combine both difficulties. For example the sum of one Bernoulli-shift process and of one independent associated process may satisfy such conditions.

Remark 11.4.1 (A comparison) A rigorous comparison of the previous strong mixing conditions and all such weak-dependence is not always possible. α_r and θ_r are obtained in inequality (11.1) as the supremum of covariances

$$|\mathrm{Cov}\,(f(P), g(F))|$$

for functions of past and future where $\|f\|_\infty \le 1$ and where $\|g\|_\infty \le 1$ under mixing or where moreover $\mathrm{Lip}\, g \le 1$ under weak-dependence. Hence

$$\theta_r \le \alpha_r.$$

Various applications of those notions are considered in our monograph (Dedecker et al. 2007). It is however simple to note that such properties are stable through Lipschitz images as an extension of Lemma 7.4.1.

The function

$$g(x_1, \ldots, x_u) = x_1 \times \cdots \times x_u,$$

associated with moments of sums, is more specifically used in the next chapter, it is usually unbounded and non-Lipschitz so that truncations will be needed to derive moment inequalities for partial sums.

The following exercise is a first step to consider the empirical cdf, various generalizations of which may be found in Dedecker et al. (2007).

Exercise 64 *(Heredity for indicators under θ-week dependence)* Let $(X_t)_{t\in\mathbb{Z}}$ be a real valued and θ-weakly dependent process. Assume that there exists a constant $C > 0$ such that $\mathbb{P}(X_i \in [a, b]) \le C(b - a)$ for each $-\infty < a < b < \infty$.
Then:

$$|\mathrm{Cov}\,(g(X_0), g(X_r))| \le (1 + C)\sqrt{\theta_r}.$$

Proof Set g_ϵ the continuous function such that $g_\epsilon(x) = g(x)$ if $x < u$ and $x > u + \epsilon$, and define g_ϵ as affine on $[u, u + \epsilon]$ then $\mathrm{Lip}\, g_\epsilon = 1/\epsilon$:

$$\begin{aligned}
|\mathrm{Cov}\,&(g(X_0), g(X_r))| \\
&\le |\mathrm{Cov}\,(g(X_0), g(X_r) - g_\epsilon(X_r))| + |\mathrm{Cov}\,(g(X_0), g_\epsilon(X_r))| \\
&\qquad\qquad \le C\epsilon + \frac{1}{\epsilon}\theta_r = (1 + C)\sqrt{\theta_r},
\end{aligned}$$

with $\epsilon^2 = \theta_r$ Use the elementary bound $|\mathrm{Cov}\,(U, V) \le 2\|U\|_\infty \mathbb{E}|V|$ in order to conclude.

Exercise 65 *(Heredity under indicators, non causal case)* Extend Exercise 64 under non-causal dependence conditions.

Hint. For non-causal weak-dependences, use

$$|\mathrm{Cov}\,(g(X_0), g(X_r))| \le |\mathrm{Cov}\,(g(X_0), g(X_r) - g_\epsilon(X_r))| \\ +|\mathrm{Cov}\,(g(X_0) - g_\epsilon(X_0)), g_\epsilon(X_r))| \\ + |\mathrm{Cov}\,(g_\epsilon(X_0)), g_\epsilon(X_r))|.$$

The conclusions follow analogously under κ, η and λ-weak-dependences, see (Dedecker et al. 2007) for details.

Exercise 66 *(Heredity, couples)* Let $(U_t)_{t\in\mathbb{Z}}$ and $(V_t)_{t\in\mathbb{Z}}$ be mutually independent η-weakly dependent sequences. Set $W_t = (U_t, V_t)$, prove that the process $(W_t)_{t\in\mathbb{Z}}$ is again η-weakly dependent and moreover that:

$$\eta_{W,r} \le \eta_{U,r} + \eta_{V,r}.$$

Hint. Use Exercise 58 with $X = ((U_{i_1}, \dots, U_{i_u}), (U_{j_1}, \dots, U_{j_v}))$ and $Y = ((V_{i_1}, \dots, V_{i_u}), (V_{j_1}, \dots, V_{j_v}))$. Note that for a function of $f(x, y)$ of two variables, setting $f_x(y) = f(x, y)$ for the partial function yields $\mathrm{Lip}\, f_x \le \mathrm{Lip}\, f$ for each x.

Remark 11.4.2 The above heredity extends to the other weak-dependence conditions, including strong mixing.

Exercise 67 *(Heredity under instantaneous functions)* Consider a sequence $(X_n)_{n\in\mathbb{Z}}$ of $\mathbb{R}^k$-valued random variables. Let $p > 1$.

We assume that there exists some constant $C > 0$ such that

$$\max_{1\le i\le k} \|X_i\|_p \le C.$$

Let h be a function from $\mathbb{R}^k$ to $\mathbb{R}$ such that $h(0) = 0$ and for $x, y \in \mathbb{R}^k$, there exists a in $[1, p[$ and $c > 0$ such that

$$|h(x) - h(y)| \le c|x - y|(|x|^{a-1} + |y|^{a-1})\,.$$

We define the sequence $(Y_n)_{n\in\mathbb{Z}}$ by

$$Y_n = h(X_n), \qquad n \in \mathbb{Z}.$$

Then,

- if $(X_n)_{n\in\mathbb{Z}}$ is η-weak dependent, then $(Y_n)_{n\in\mathbb{Z}}$ is also η-weak dependent, and

$$\eta_{Y,r} = \mathcal{O}\left(\eta_r^{\frac{p-a}{p-1}}\right),$$

- if $(X_n)_{n\in\mathbb{Z}}$ is λ-weak dependent, then $(Y_n)_{n\in\mathbb{Z}}$ also, and

$$\lambda_{Y,r} = \mathcal{O}\left(\lambda_r^{\frac{p-a}{p+a-2}}\right).$$

Remark 11.4.3 Refer to Dedecker et al. (2007) for details. The function $h(x) = x^2$ satisfies the previous assumptions with $a = 2$.

This condition is satisfied by polynomials with degree a. It makes this result useful for spectral estimation, see Sect. 4.4.2.

Proof Let f and g be two real functions as in the above definition.
Denote

$$x^{(M)} = (x \wedge M) \vee (-M), \qquad \text{for} \qquad x \in \mathbb{R}.$$

For $x = (x_1, \ldots, x_k) \in \mathbb{R}^k$, denote $t_M(x) = (x_1^{(M)}, \ldots, x_k^{(M)})$.

Assume that $(\mathbf{i}, \mathbf{j})$ are as in the definition of weak-dependence, and set

$$X_{\mathbf{i}} = (X_{i_1}, \ldots, X_{i_u}), \quad \text{and} \quad X_{\mathbf{j}} = (X_{j_1}, \ldots, X_{j_v}).$$

Define the following functions,

$$F = f \circ h^{\otimes u},\ F^{(M)} = f \circ (h \circ t_M)^{\otimes u},\ G = g \circ h^{\otimes v}, \mathbb{R}^{uk} \to \mathbb{R}$$

and

$$G^{(M)} = g \circ (h \circ t_M)^{\otimes v}, \quad \mathbb{R}^{vk} \to \mathbb{R}.$$

Then:

$$\begin{aligned}
|\mathrm{Cov}\,(F(X_{\mathbf{i}}), G(X_{\mathbf{j}}))| &\le |\mathrm{Cov}\,(F(X_{\mathbf{i}}), G(X_{\mathbf{j}}) - G^{(M)}(X_{\mathbf{j}}))| \\
&\quad + |\mathrm{Cov}\,(F(X_{\mathbf{i}}), G^{(M)}(X_{\mathbf{j}}))| \\
&\le 2\|f\|_\infty \mathbb{E}|G(X_{\mathbf{j}}) - G^{(M)}(X_{\mathbf{j}}))| \\
&\quad + 2\|g\|_\infty \mathbb{E}|F(X_{\mathbf{i}}) - F^{(M)}(X_{\mathbf{i}})| \\
&\quad + |\mathrm{Cov}\,(F^{(M)}(X_{\mathbf{i}}), G^{(M)}(X_{\mathbf{j}}))|.
\end{aligned}$$

We also derive from the assumptions on h and from Markov's inequality that:

$$\begin{aligned}
\mathbb{E}|G(X_{\mathbf{j}}) - G^{(M)}(X_{\mathbf{j}}))| &\le \mathrm{Lip}\, g \sum_{l=1}^{v} \mathbb{E}|h(X_{j_l}) - h(X_{j_l}^{(M)})| \\
&\le 2c\,\mathrm{Lip}\, g \sum_{l=1}^{v} \mathbb{E}\big(|X_{j_l}|^a\, \mathbb{1}_{|X_{j_l}|>M}\big), \\
&\le 2c\, v\, \mathrm{Lip}\, g\, C^p M^{a-p}.
\end{aligned}$$

The same holds for F. Moreover

$$\begin{aligned}\mathrm{Lip}\, F^{(M)} &\le 2cM^{a-1}\mathrm{Lip}\, f, \quad & \mathrm{Lip}\, G^{(M)} &\le 2cM^{a-1}\mathrm{Lip}\, g,\\ \|F^{(M)}\|_\infty &\le \|f\|_\infty, & \|G^{(M)}\|_\infty &\le \|g\|_\infty.\end{aligned}$$

From the definition of weak-dependence of X and the choice of $\mathbf{i}, \mathbf{j}$, setting $A = \left|\mathrm{Cov}\left(F^{(M)}(X_\mathbf{i}), G^{(M)}(X_\mathbf{j})\right)\right|$, we obtain respectively, if $M \ge 1$

$$\begin{aligned}A &= \left|\mathrm{Cov}\left(F^{(M)}(X_\mathbf{i}), G^{(M)}(X_\mathbf{j})\right)\right|\\ &\le 2c(u\mathrm{Lip}\, f\|g\|_\infty + v\mathrm{Lip}\, g\|f\|_\infty)M^{a-1}\eta_r,\\ &\qquad \text{or},\\ &\le 2c(u\mathrm{Lip}\, f\|g\|_\infty + v\mathrm{Lip}\, g\|f\|_\infty)M^{a-1}\lambda_r + 4c^2uv\mathrm{Lip}\, f\mathrm{Lip}\,(g)M^{2a-2}\lambda_r.\end{aligned}$$

Finally, we obtain respectively, if $M \ge 1$:

$$\begin{aligned}|\mathrm{Cov}(F(X_\mathbf{i}), G(X_\mathbf{j}))| &\le 2c(u\mathrm{Lip}\, f\|g\|_\infty + v\mathrm{Lip}\, g\|f\|_\infty)\big(M^{a-1}\eta_r + 2C^pM^{a-p}\big),\\ &\qquad \text{or},\\ &\le c(u\mathrm{Lip}\, f + v\mathrm{Lip}\, g + uv\mathrm{Lip}\, f\mathrm{Lip}\, g)(M^{2a-2}\lambda_r + M^{a-p}).\end{aligned}$$

Now set, either $M = \eta_r^{-\frac{1}{p-1}}$, or $M = \lambda_r^{-\frac{1}{p+a-2}}$, in order to conclude in each case.

11.5 Proving Limit Theorems

There follows a simple way to derive CLTs. The situation chosen is that of stationary and centred processes. Ergodicity indeed allows to recentre such processes.

Moment inequalities, proved in Chap. 12, yield useful controls for $\mathbb{E}|S_n|^p$ and a central limit theorem may be derived by using the following simple dependent Lindeberg inequality.

Lemma 11.5.1 (Dependent Lindeberg (Bardet et al. 2006)) *We set* $f(x) = e^{i<t,x>}$ *for each* $t \in \mathbb{R}^d$.
We consider an integer $k \in \mathbb{N}^*$. *Let* $(X_i)_{1\le i\le k}$ *be* $\mathbb{R}^d$*-valued centred random variables such that:*

$$A_k = \sum_{i=1}^k \mathbb{E}\|X_i\|^{2+\delta} < \infty.$$

Set

$$T(k) = \sum_{j=1}^k \left|\mathrm{Cov}\,(e^{i<t,X_1+\cdots+X_{j-1}>}, e^{i<t,X_j>})\right|.$$

Then

$$\Delta_k \le T(k) + 6\|t\|^{2+\delta}A_k.$$

We denote by $< a, b >$ *the scalar product in* $\mathbb{R}^d$.

Proof Following the proof of Lemma 2.1.1 we only need to reconsider the bound of $\mathbb{E}\delta_j$, for this let a random variable U_j^* be independent of all the other random variables already considered and with the same distribution as U_j. Then we decompose:

$$\delta_j = (g(Z_j + U_j) - g(Z_j + U_j^*)) + (g(Z_j + U_j^*) - g(Z_j + V_j)).$$

The second term admits the bound provided in Lemma 2.1.1, which can be written as stated above since for $f(x) = e^{i<t,x>}$ one easily derives that $\|f^{(p)}\|_\infty = |t|^p$.

Now the first term is the "dependent" one and from the independence of V's and the multiplicative properties of the exponential:

$$\begin{aligned} |\mathbb{E}(g(Z_j + U_j) - g(Z_j + U_j^*))| \\ \leq |\mathbb{E}g(U_1 + \cdots + U_{j-1})(g(U_j) - g(U_j^*))| \\ = |\mathrm{Cov}\,(g(U_1 + \cdots + U_{j-1}), g(U_j))|. \end{aligned}$$

In case the series of covariances is summable we have already remarked that

$$\mathbb{E}S_m^2 \sim \sigma^2 m, \quad \text{for large values of} \quad m.$$

The idea is to compute

$$\Delta_n = \mathbb{E}\left(f\left(\frac{S_n}{\sqrt{n}}\right) - f(\sigma N)\right),$$

for enough functions in the class of $\mathcal{C}^3$-functions.

We need $\Delta_n \to 0$ as $n \to \infty$. For this, the Bernstein blocks technique is sketched. Consider sequences

$$q = q(n) \ll p = p(n) \ll n, \quad \text{as} \quad n \uparrow \infty.$$

Then we decompose

$$\frac{S_n}{\sqrt{n}} = U_1 + \cdots + U_k + V,$$

with

$$k = k(n) = \left[\frac{n}{p(n) + q(n)}\right], \quad \text{and} \quad U_j = \frac{1}{\sqrt{n}} \sum_{i=(j-1)(p+q)+1}^{(j-1)(p+q)+p} X_i.$$

In this case the remainder $\|V\|_2 \to 0$ because

$$V = \frac{1}{\sqrt{n}} \sum_{u \in E} X_u$$

is a sum over some set E with cardinality $m \le q + p = o(n)$.

Indeed

$$\begin{aligned} n\mathrm{Var}\, V &\le \sum_{u,v \in E} |\mathrm{Cov}\,(X_u, X_v)| \\ &= \sum_{u,v \in E} |\mathrm{Cov}\,(X_0, X_{v-u})| \\ &\le m \sum_{j=-\infty}^{\infty} |\mathrm{Cov}\,(X_0, X_j)|. \end{aligned}$$

If X_i and $X_{i'}$ are terms within the sums defining U_j and $U_{j'}$ for $j \ne j'$, then $|i'-i| \ge q$. The variables $U_1, \ldots, U_k$ are thus almost independent and Lemma 11.5.1 may be applied.

To conclude we cite a powerful result adapted to causal cases, see (Rio 2017), its proof is very different:

Theorem 11.5.1 (Dedecker and Rio 2000) *Let $(X_n)_{n\in\mathbb{Z}}$ be an ergodic stationary sequence with $\mathbb{E}X_n = 0$, $\mathbb{E}X_n^2 = 1$. Set $S_n = X_1 + \cdots + X_n$.*

Assume that the random series

$$\sum_{n=0}^{\infty} X_0 \mathbb{E}\Big(X_n \Big| \sigma(X_k/k \le 0)\Big), \quad \textit{converges in } \mathbb{L}^1.$$

Then the sequence $\mathbb{E}(X_0^2 + 2X_0S_n)$ converges to some σ^2.

Moreover[2]*:*

$$\frac{1}{\sqrt{n}} S_{[nt]} \to_{n\to\infty} \sigma W_t, \qquad \textit{in distribution in } \mathcal{D}[0,1].$$

In Dedecker and Doukhan (2003) for the case of θ-weak-dependence and in Dedecker et al. (2007), we derive similar CLTs; assumptions needed to replace such abstract conditions always write in terms of decay rates and moment conditions. We refer the reader to Merlevède et al. (2006) for a complete review of the literature. Previously introduced conditions take into account most of the standard models in statistics.

Exercise 68 Consider a sequence of iid random variables $(R_i)_{i\ge 0}$ (with finite mean) and an independent standard Normal random variable N. Set $X_k = R_k N$:

---

[2]For the Skorohod space, see Definition B.2.2 and the remark following it.

1. Set $\overline{X}_n = (X_1 + \cdots + X_n)/n$ then $\lim_{n\to\infty} \overline{X}_n = \mathbb{E}R_0 \cdot N$ a.s.
2. Deduce that this sequence is not ergodic in case $\mathbb{E}R_0 \neq 0$.
3. If R_k follows Rademacher distribution ($\mathbb{P}(R_k = \pm 1) = \frac{1}{2}$) prove that $\mathrm{Cov}\,(X_k, X_\ell) = 0$ for all $k \neq \ell$ and the sequence is not independent.
4. In this Rademacher case, prove that $\sqrt{n} \cdot \overline{X}_n$ converges in distribution to the product of two independent standard Normal random variables.
5. Prove that the sequence $(X_n)_n$ is not ergodic.

Hints for Exercise 68.

1. The first point follows from the strong law of large numbers.
2. The ergodic theorem (Corollary 9.1.3) does not hold because the limit is non-deterministic, thus we obtained the non-ergodicity.
3. This point is proved in Exercise 2.
4. It follows from the CLT.
5. This sequence is never ergodic since conditionally to N it is ergodic and the tail sigma-field is always the sigma-field generated by N.

Example 11.5.1 Exercise 68 yields an orthogonal stationary sequence of Gaussian random variables such that the law of large numbers holds, but which is not ergodic and which does not satisfy the CLT. This sequence is thus not a Gaussian process.

Remark 11.5.1 The empirical process

$$Z_n(x) = \sqrt{n}(F_n(x) - F(x)), \qquad F_n(x) = \frac{1}{n}\sum_{k=1}^{n} \mathbb{I}_{\{X_k \le x\}},$$

is also of interest and one may consider it as above but in this case heredity of weak-dependence conditions is not ensured directly since the function $u \mapsto \mathbb{I}_{\{u\le x\}}$ is not Lipschitz but concentration conditions as in Lemma 7.4.2 allow to work out the asymptotic properties for such processes. In the remark following Definition B.2.2 we recall a criterion for the convergence of this cumulative distribution. We again refer to Dedecker et al. (2007) for more details.

Exercise 69 *(Subsampling)* Consider a sequence of statistics t_m and a function g. As in (4.9) the index set $E_{m,n}$ admits cardinal N and it is identified to $\{1, \ldots, N\}$. Then $N \sim n - m$ or $N \sim n/m$, respectively for overlapping and non-overlapping schemes and as above,

$$\widehat{K}_n(g) = \frac{1}{N}\sum_{i=1}^{N} g_{i,m}$$

$$g_{i,m} = g(t_m(X_{i+1}, \ldots, X_{i+m})).$$

$$g_{i,m} = g(t_m(X_{(m(i-1)+1}, \ldots, X_{(i+1)m}))$$

Prove that the variance $\text{Var}\,\widehat{K}_n(g) \to_{n\to\infty} 0$, for g a bounded function:

- Under strong mixing this is enough to assume $\lim_r \alpha_r = 0$, and does not rely on the properties of functions t_m.
- Under θ-weak-dependence, use Lipschitz properties of the functions t_m (set $L_m = \text{Lip}\,(t_m))$).
- Derive a consistency result for the case of the empirical mean, as sketched in (4.12).

  Hint. Classically:

$$\text{Var}\,\widehat{K}_n(g) \le \frac{1}{N}\sum_{i=1}^{N} |\text{Cov}\,(g_{0,m}, g_{i,m})|.$$

- In the overlapping and strong mixing case

$$|\text{Cov}\,(g_{0,m}, g_{i,m})| \le \alpha_{r-m+1}, \tag{11.4}$$

  (resp. $\le \alpha_{r/m+1}$) which does not depend on m for this special mixing case. Hence Cesaro's lemma yields the result for this case.

- The cases of weak-dependence are more complex, here $\text{Lip}\, g_m \le L_m \text{Lip}\, g$ and:

$$|\text{Cov}\,(g_{0,m}, g_{i,m})| \le m L_m \text{Lip}\, g \cdot \theta_{r-m+1}, \tag{11.5}$$

  in the overlapping scheme (resp. $\le m L_m \text{Lip}\, g \cdot \theta_{\frac{r}{m}+1}$ in the non-overlapping scheme).
  In the overlapping scheme we obtain for some constant:

$$\text{Var}\,\widehat{K}_n(g) \le \frac{Cm}{n-m}\Big(1 + L_m \text{Lip}\, g \sum_{i=1}^{n-m} |\text{Cov}\,(g_{0,m}, g_{i,m})|\Big).$$

- The normalized empirical means write with

$$t_m(x_1, \ldots, x_m) = \frac{1}{\sqrt{m}}\sum_{i=1}^{m} x_i$$

  then $T_m = \sqrt{m}\cdot\overline{X}_m = t_m(X_1, \ldots, X_m)$ converges to some Gaussian rv as $m \to \infty$, and $L_m = \dfrac{C}{\sqrt{m}}$.

If g is a Lipschitz function, for the above case of means with overlapping scheme, the assumptions

$$\sup_n \frac{1}{\sqrt{m_n}} \sum_{i=1}^{n} \theta_i < \infty, \qquad \lim_{n\to\infty} \frac{m_n}{n} = 0$$

together imply consistency of subsampling.
This holds for instance if

$$\sum_{i=1}^{\infty} \theta_i < \infty, \qquad \lim_{n\to\infty} \frac{m_n}{n} = 0.$$

To derive the above inequality for discontinuous functions $g = \mathbb{I}_{\{\cdot \le u\}}$ one additional step is necessary and it only needs to replace θ_i by $\theta_i^{\frac{1}{3}}$ if marginal distributions admit a bounded density, use Exercise 64. Finally uniform convergence is proved as in (4.12).

Remark 11.5.2 Analogously for subsampling kernel density estimators

$$t_m(X_1, \ldots, X_m) = \sqrt{mh}(f_{m,h}(x) - \mathbb{E} f_{m,h}(x))$$

for a fixed $x \in \mathbb{R}$, then $L_m = \dfrac{C}{h\sqrt{mh}}$ and if f denotes the marginal density of X_0,

$$t_m(x_1, \ldots, x_m) = \frac{1}{\sqrt{mh}} \sum_{i=1}^{m} \left(K\left(\frac{x_i - x}{h}\right) - \int_{\mathbb{R}} K\left(\frac{u - x}{h}\right) f(u)\, du \right).$$

Data-based recentred statistics are considered if two samples $X_1, \ldots, X_n$, and $X'_1, \ldots, X'_n$ are available, two sets $E_{m,n}$, $E'_{m,n'}$ are then built as above and subsampling is provided $\tilde{K}_{n,n'}(g)$ is deduced by replacing $E_{m,n}$ by $E_{m,n} \cup E'_{m,n'}$ and t_m by $\tilde{t}_{2m}$:

$$\tilde{t}_{2m}(x_1, \ldots, x_m, x'_1, \ldots, x'_m) = \frac{1}{\sqrt{mh}} \sum_{i=1}^{m} \left(K\left(\frac{x_i - x}{h}\right) - K\left(\frac{x'_i - x}{h}\right) \right).$$

A divergent sequence of statistics is also considered in Doukhan et al. (2011)

$$t_m(x_1, \ldots, x_m) = \max_{1 \le i \le m} x_i.$$

Higher order moments are considered in Exercise 84.

Chapter 12
Moments and Cumulants

This chapter is devoted to moment methods. The use of moments relies on their importance in deriving asymptotic of several estimators, based on moments and limit distributions.

Cumulants are linked with spectral or multispectral estimation which are main tools of time series analysis.

$$g(\lambda) = \sum_{k=-\infty}^{\infty} \text{Cov}\,(X_0, X_k)e^{-ik\lambda}.$$

Such functions do not characterize the dependence of non-linear processes; indeed we have already examples of orthogonal and non-independent sequences. This motivates the introduction of higher order characteristics.

A multispectral density is defined over $\mathbb{C}^{p-1}$ by

$$g(\lambda_2, \ldots, \lambda_p) = \sum_{k_2=-\infty}^{\infty} \cdots \sum_{k_p=-\infty}^{\infty} \kappa(X_0, X_{k_2}, \ldots, X_{k_p})e^{-i(k_2\lambda_2+\cdots+k_p\lambda_p)}.$$

Remark. The periodogram in Definition 4.4.1 is not enough to deal with non-Gaussian stationary time series. Indeed:

- As stressed in Exercise 2, the covariances are not enough to prove independence. In order to test for stochastic independence, higher-order spectral estimators are thus useful.

- Gaussian laws are characterized by the fact that cumulants with order >2 vanish. This provides hints to test Gaussianness.

P. Doukhan, *Stochastic Models for Time Series*, Mathématiques et Applications 80,
https://doi.org/10.1007/978-3-319-76938-7_12

12.1 Method of Moments

Recall that the method of moments yields limit theorems:

Theorem 12.1.1 (Feller) *Suppose that the sequence of real valued random variables U_n is such that*

$$\mathbb{E}U_n^p \to_{n\to\infty} \mathbb{E}U^p, \qquad \textit{for each integer } p \geq 0.$$

If moreover U admits an analytic Laplace transform around 0.[1] *Then*

$$U_n \stackrel{\mathcal{L}}{\to}_{n\to\infty} U.$$

Hint. Indeed the *analytic continuation theorem* implies that U's distribution is determined by its moments.

Remark 12.1.1 Cumulant/moment inequalities for partial sums of stationary processes are useful.

For the $\sqrt{n}$-limit theorem indeed, denote:

$$Z_n = \frac{1}{\sqrt{n}} \sum_{k=1}^{n} X_k.$$

It was noted in (4.4) that

$$\text{Var } Z_n \to_{n\to\infty} \sum_{k=-\infty}^{\infty} \text{Cov}\,(X_0, X_k),$$

in case the latter series is summable. Such controls of the moments also make more precise the CLT; for example they may allow to derive convergence rates or large deviation principles.

12.1.1 Notations

Let $Y = (Y_1, \ldots, Y_k) \in \mathbb{R}^k$ be a random vector with $\mathbb{E}(|Y_1|^r + \cdots + |Y_k|^r) < \infty$, then we set

$$\phi_Y(t) = \mathbb{E}e^{it\cdot Y} = \mathbb{E}\exp\Big(i\sum_{j=1}^{k} t_j Y_j\Big)$$
$$m_p(Y) = \mathbb{E}Y_1^{p_1} \ldots Y_k^{p_k}.$$

[1] This holds if there exists $\alpha > 0$ with $\mathbb{E}e^{\alpha|U|} < \infty$.

Moreover for $p = (p_1, \ldots, p_k) \in \mathbb{N}^k$, and $t = (t_1, \ldots, t_k) \in \mathbb{R}^k$, we set

$$|p| = p_1 + \cdots + p_k = r,$$
$$p! = p_1! \ldots p_k!,$$
$$t^p = t_1^{p_1} \ldots t_k^{p_k}.$$

In case the previous condition holds for some integer $r \in \mathbb{N}^*$, the function $t \mapsto \log \phi_Y(t)$ admits a Taylor expansion

$$\log \phi_Y(t) = \sum_{|p| \le r} \frac{i^{|p|}}{p!} \kappa_p(Y) t^p + o(|t|^r), \quad \text{as} \quad t \to 0. \tag{12.1}$$

The coefficients $\kappa_p(Y)$ are named cumulants of Y with order $p \in \mathbb{N}^k$ and they exist if $|p| \le r$.

Replace Y by a vector with higher dimension $s = |p|$ with p_1 repetitions for Y_1, …, p_k repetitions for Y_k allows to consider $p = (1, \ldots, 1)$ and we set $\kappa_{(1,\ldots,1)}(Y) = \kappa(Y)$.

If $\mu = \{i_1, \ldots, i_u\} \subset \{1, \ldots, k\}$ set:

$$\kappa_\mu(Y) = \kappa(Y_{i_1}, \ldots, Y_{i_u}), \qquad m_\mu(Y) = m(Y_{i_1}, \ldots, Y_{i_u}).$$

Lenov and Shiryaev (1959)'s formulae,[2] follow from the uniqueness of Taylor expansions (12.1):

$$\kappa(Y) = \sum_{u=1}^{k} (-1)^{u-1}(u-1)! \sum_{\mu_1,\ldots,\mu_u} \prod_{j=1}^{u} m_{\mu_j}(Y). \tag{12.2}$$

$$m(Y) = \sum_{u=1}^{k} \sum_{\mu_1,\ldots,\mu_u} \prod_{j=1}^{u} \kappa_{\mu_j}(Y). \tag{12.3}$$

Previous sums are taken over all the partitions $\mu_1, \ldots, \mu_u$ of the set $\{1, \ldots, k\}$.

Hint for the proofs of (12.2) *and* (12.3). The Taylor expansion of the analytic function $s \mapsto \log(1+s)$ as $t \to 0$ yields[3]

$$\phi_Y(t) = 1 + \sum_{0<|p|\le r} \frac{i^{|p|}}{p!} m_p(Y) t^p + o(|t|^r),$$

[2]These formulae are proved for example in Rosenblatt (1985), pp. 33–34.

[3]The function $s \mapsto \log(1+s)$ is analytic for $|t| < 1$, and the determination of the logarithm is not a problem in the domain $]-\frac{1}{2}, \frac{1}{2}[$ of $\mathbb{C}$.

and

$$\log \phi_Y(t) = \sum_{u=1}^{r} \frac{(-1)^{u-1}}{u} \left(\sum_{0<|p|\le r} \frac{i^{|p|}}{p!} m_p(Y) t^p \right)^u + o(|t|^r),$$

$$= \sum_{u=1}^{r} \frac{(-1)^{u-1}}{u} \sum_{\substack{0<|p|\le r \\ p_1+\cdots+p_u=p}} \frac{(it)^{|p|}}{p!} \prod_{j=1}^{u} m_{p_j}(Y) + o(|t|^r),$$

hence identifying the coefficient corresponding to $p = (1, \ldots, 1)$ for u-tuples such that $p_1 + \cdots + p_u = p$; choose $r = k$ to derive relation (12.2).
Indeed then $|p| = k$, $p! = 1$ and $(it)^p = i^k t^k$.

A combinatoric coefficient $u!$ appears, which corresponds to the number of permutations in a partition.

Use Eq. (A.5) and Exercise 88 to derive:

Exercise 70 If $X \sim \mathcal{N}(0, 1)$ then $\kappa_2(X, X) = 1$ and $\kappa_p(X, \ldots, X) = 0$ for each $p \neq 2$.

Exercise 71 If $X \sim \mathcal{P}(\lambda)$ then $\kappa_p(X, \ldots, X) = \lambda$ for each $p \in \mathbb{N}^*$.

Exercise 72 Consider the case of compound Poisson processes from Example 4.2.2, see also Exercise 92.

12.1.2 Combinatorics of Moments

Recall now some notions from Saulis and Statulevicius (1991).

Definition 12.1.1 Centred moments of the random vector $Y = (Y_1, \ldots, Y_k)$ are defined with $\widehat{\mathbb{E}}\ (Y_1, \ldots, Y_l) = \mathbb{E} Y_1 c(Y_2, \ldots, Y_l)$ where centred random variable $c(Y_2, \ldots, Y_l)$ are recursively identified by setting $c(\xi_1) = \overbrace{\ \xi_1\ } = \xi_1 - \mathbb{E}\xi_1$ and

$$c(\xi_j, \xi_{j-1}, \ldots, \xi_1) = \xi_j \overbrace{c(\xi_{j-1}, \ldots, \xi_1)}$$
$$= \xi_j \left(c(\xi_{j-1}, \ldots, \xi_1) - \mathbb{E} c(\xi_{j-1}, \ldots, \xi_1) \right).$$

Consider $Y_\mu = (Y_j)_{j\in\mu}$ as a p-tuple for $\mu \subset \{1, \ldots, k\}$.

For example $\widehat{\mathbb{E}}\ (\xi) = 0$, $\widehat{\mathbb{E}}\ (\eta, \xi) = \mathrm{Cov}\ (\eta, \xi)$,

$$\widehat{\mathbb{E}}\ (\zeta, \eta, \xi) = \mathbb{E}(\zeta\eta\xi) - \mathbb{E}(\zeta)\mathbb{E}(\eta\xi) - \mathbb{E}(\eta)\mathbb{E}(\zeta\xi) - \mathbb{E}(\xi)\mathbb{E}(\zeta\eta).$$

Centred moments are a way to generalize covariances. They also quantify the independence of the coordinates for a random vector.

The following result explains the nature of cumulants. This provides a representation in terms of centred moments.

Theorem 12.1.2 (Saulis and Statulevicius (1991))

$$\kappa(Y_1, \ldots, Y_k) = \sum_{u=1}^{k} (-1)^{u-1} \sum_{\mu_1,\ldots,\mu_u} N_u(\mu_1, \ldots, \mu_u) \prod_{j=1}^{u} \widehat{\mathbb{E}}\, Y_{\mu_j}$$

sums are over all the partitions $\mu_1, \ldots, \mu_u$ *of the set* $\{1, \ldots, k\}$ *and the integers* $N_u(\mu_1, \ldots, \mu_u) \in \left[0, (u-1)! \wedge \left[\frac{k}{2}\right]!\right]$ *defined for each partition satisfy*

$$N(k, u) = \sum_{\mu_1,\ldots,\mu_u} N_u(\mu_1, \ldots, \mu_u) = \sum_{j=1}^{u-1} C_k^j (u-j)^{k-1},$$

and $\sum_{u=1}^{k} N(k, u) = (k-1)!$

Lemma 12.1.1 is a simple consequence of Theorem 12.1.2.

Lemma 12.1.1 *Let* $Y_1, \ldots, Y_k \in \mathbb{R}$ *be centred random variables. For each* $k \geq 1$ *set* $M_k = 2^{k-1}(k-1)! \max_{1\leq i\leq k} \mathbb{E}|Y_i|^k$ *then*

$$M_k M_l \leq M_{k+l}, \qquad \text{for } k, l \geq 2, \tag{12.4}$$

$$|\kappa(Y_1, \ldots, Y_k)| \leq M_k. \tag{12.5}$$

Remark 12.1.2 This lemma implies:

$$\prod_{i=1}^{u} \left|\kappa(Y_1, \ldots, Y_{p_i})\right| \leq M_{p_1+\cdots+p_u}. \tag{12.6}$$

Proof of Lemma 12.1.1 The first point follows from the inequality $a!\,b! \leq (a+b)!$ also written $\binom{a+b}{b} = C_{a+b}^a \geq 1$ and the second is deduced from Lemma 12.1.2.

Lemma 12.1.2 *For each* $j, p \geq 1$ *and for all the real valued random variables*

$$\|c(\xi_j, \xi_{j-1}, \ldots, \xi_1)\|_p \leq 2^j \max_{1\leq i\leq j} \|\xi_i\|_{pj}^j,$$

with $\|\xi\|_q = (\mathbb{E}|\xi|^q)^{\frac{1}{q}}$.

Proof of Lemma 12.1.2 Jensen's inequality (Proposition A.2.1) leads to

$$\|c(\xi_1)\|_p \leq \|\xi_1\|_p + |\mathbb{E}\xi_1| \leq 2\|\xi_1\|_p.$$

Set $Z_j = c(\xi_j, \xi_{j-1}, \ldots, \xi_1)$ then $Z_j = \xi_j(Z_{j-1} - \mathbb{E}Z_{j-1})$ and from Hölder's inequality (Proposition A.2.2)

$$\|\xi_j Z_{j-1}\|_p^p \leq \|\xi_j\|_{pj}^p \|Z_{j-1}\|_{\frac{pj}{j-1}}^p.$$

Using recursion for the pair $(q, j-1)$ where $q = pj/(j-1)$ the inequalities of Minkowski (Corollary A.2.1) and Hölder (Proposition A.2.2) yield:

$$\begin{aligned}
\|Z_j\|_p &\le \|\xi_j Z_{j-1}\|_p + \|\xi_j\|_p|\mathbb{E}Z_{j-1}| \\
&\le 2\|\xi_j\|_{pj}\|Z_{j-1}\|_q \\
&\le 2^j\|\xi_j\|_{pj} \max_{0\le i<j} \|\xi_i\|_{q(j-1)}^{j-1} \\
&\le 2^j \max_{0\le i\le j} \|\xi_i\|_{pj}^j,
\end{aligned}$$

with $q = p \cdot \frac{j}{j-1}$. Now the relation $q(j-1) = pj$ allows to conclude.

Proof of Lemma 12.1.1 We replace $\max_{j\le J} \|Y_j\|_p$ by $\|Y_0\|_p$ for clarity. Lemma 12.1.2 yields $|\widehat{\mathbb{E}}\ Y_\mu| \le 2^{l-1}\|Y_0\|_l^l$ with $l = \text{Card}\ \mu$.

Indeed write $Z = c(Y_2, \ldots, Y_l)$ and define p through the identity $\frac{1}{p} + \frac{1}{l} = 1$. Then:

$$\left|\widehat{\mathbb{E}}\ (Y_1, \ldots, Y_l)\right| = |\mathbb{E}Y_1 Z| \le \|Y_0\|_l\|Z\|_p \le 2^{l-1}\ \|Y_0\|_l^l,$$

since $p(l-1) = l$. Theorem 12.1.2 implies

$$\begin{aligned}
|\kappa(Y)| &\le \sum_{u=1}^{k} \sum_{\mu_1,\ldots,\mu_u} N_u(\mu_1, \ldots, \mu_u) \prod_{i=1}^{u} 2^{n(\mu_i)-1}\|Y_0\|_{n(\mu_i)}^{n(\mu_i)} \\
&\le \sum_{u=1}^{k} 2^{k-u}\ N(k,u)\|Y_0\|_k^k \\
&\le 2^{k-1}\ \|Y_0\|_k^k \sum_{u=1}^{k} N(k,u) \\
&= 2^{k-1}(k-1)\,!\|Y_0\|_k^k.
\end{aligned}$$

with $n(\mu) = \text{Card}(\mu)$. The previous relation ends the proof.

12.2 Dependence and Cumulants

The following lemmas are essentially proved for sequences of real valued random variables $(X_n)_{n\in\mathbb{Z}}$ in Doukhan and León (1989).

12.2.1 More Dependence Coefficients

Consider a stationary real valued sequence $(X_n)_{n\in\mathbb{Z}}$. Then consider as in Doukhan and Louhichi (1999)

$$c_{X,q}(r) = \max_{1\le l<q} \sup_{\substack{t_1 \le \cdots \le t_q \\ t_{l+1}-t_l \ge r}} \left|\mathrm{Cov}\left(X_{t_1}\dots X_{t_l}, X_{t_{l+1}}\dots X_{t_q}\right)\right| \tag{12.7}$$

Example 12.2.1 Assume that the η-weak-dependence condition (11.1) associated with the functional ψ_η and with the classes of function $\mathcal{F}=\mathcal{G}=\Lambda$ holds.

If $Y_i = h(X_i)$ for some Lipschitz function h bounded by M, we get

$$c_{Y,q}(r) \le M^{q-1}\mathrm{Lip}(h)\theta_r.$$

Setting $\mu_t = \mathbb{E}|X_0|^t$, the following coefficients are also useful

$$c^\star_{X,q}(r) = \max_{1\le l\le q} c_{X,l}(r)\cdot \mu_{q-l}. \tag{12.8}$$

Define

$$\kappa_q(t_2,\dots,t_q) = \kappa_{(1,\dots,1)}(X_0, X_{t_2},\dots,X_{t_q}).$$

The following decomposition explain the way cumulants behave as covariances.

Precisely this proves that cumulants $\kappa_Q(X_{k_1},\dots,X_{k_Q})$ are small if for some index l the lag $k_{l+1}-k_l$ is large. Here $k_1\le\cdots\le k_Q$ and a weak-dependence condition will be assumed.

This is also a natural extension of an important property of cumulants. A cumulant vanishes in case it involves a couple of independent vectors.

Definition 12.2.1 Let $t=(t_1,\dots,t_p)$ be any p-tuple in $\mathbb{Z}^p$, such that $t_1\le\cdots\le t_p$, we set $r(t)=\max_{1\le l<p}(t_{l+1}-t_l)$, the *maximal lag*.
Define the other alternative dependence coefficient:

$$\kappa_p(r) = \max_{\substack{t_1\le\cdots\le t_p \\ r(t_1,\dots,t_p)\ge r}} \left|\kappa_p\left(X_{t_1},\dots,X_{t_p}\right)\right|. \tag{12.9}$$

Lemma 12.2.1 *Suppose $(X_n)_{n\in\mathbb{Z}}$ is a centred and stationary process with finite moments up to order Q.*

Assume that $Q\ge 2$. By using the notation in Lemma 12.1.1 we derive

$$\kappa_{X,Q}(r) \le c_{X,Q}(r) + \sum_{s=2}^{Q-2} M_{Q-s}\left[\frac{Q}{2}\right]^{Q-s+1}\kappa_{X,s}(r).$$

Proof of Lemma 12.2.1 Set

$$X_\eta = \prod_{i \in \eta} X_i,$$

if $\eta \in \mathbb{Z}^p$ (η may include repetitions).

Suppose $k_1 \le \cdots \le k_Q$ are such that

$$k_{l+1} - k_l = r = \max_{1 \le s < p} (k_{s+1} - k_s) \ge 0.$$

Assume that $\mu = \{\mu_1, \ldots, \mu_u\}$ runs over all partitions of $\{1, \ldots, Q\}$.

One of those μ_i, denoted ν_μ, satisfies

$$\nu_\mu^- = [1, l] \cap \nu_\mu \ne \emptyset \quad \text{and} \quad \nu_\mu^+ = [l+1, Q] \cap \nu_\mu \ne \emptyset.$$

From formula (12.3) we obtain with $\eta = \{1, \ldots, l\}$,

$$\kappa(X_{k_1}, \ldots, X_{k_Q}) = \text{Cov}\,(X_{\eta(k)}, X_{\overline{\eta}(k)}) - \sum_u \sum_{\{\mu\}} \kappa_{\nu_\mu(k)} K_{\mu,k}, \tag{12.10}$$

with

$$K_{\mu,k} = \prod_{\mu_i \ne \nu_u} \kappa_{\mu_i(k)}$$

where he previous sum extends to all partitions

$$\mu = \{\mu_1, \ldots, \mu_u\}, \qquad \text{of} \qquad \{1, \ldots, Q\}$$

such that

$$\mu_i \cap \nu \ne \emptyset, \quad \text{for some} \quad i \in [1, u]$$

and

$$\mu_i \cap \overline{\nu} \ne \emptyset.$$

From $r\big(\nu_\mu(k)\big) \ge r(k)$ it is easy to derive

$$|\kappa_{\nu_\mu(k)}| \le \kappa_{X, \text{Card}\,\nu_\mu}(r).$$

This allows to let the size of lags increase.

With Lemma 12.1.1 we obtain

$$|M_\mu| \le M_{Q-\,\text{Card}\,\mu_\nu}$$

as in (12.6).

The following succession of inequalities is easy proved:

$$
\begin{aligned}
\left|\kappa\left(X_{k_1},\ldots,X_{k_Q}\right)\right| &\le C_{X,Q}(r) \\
&\quad + \sum_{u=2}^{[Q/2]} (u-1)! \sum_{\mu_1,\ldots,\mu_u} M_{Q-\operatorname{Card}\nu_\mu} |\kappa_{\nu_\mu(k)}(X)| \\
&\le C_{X,Q}(r) \\
&\quad + \sum_{u=2}^{[Q/2]} (u-1)! \sum_{s=2}^{Q-2} M_{Q-s}\kappa_{X,s}(r) \sum_{\substack{\mu_1,\ldots,\mu_u \\ \operatorname{Card}\nu_\mu = s}} 1 \\
&\le C_{X,Q}(r) \\
&\quad + \sum_{u=2}^{[Q/2]} (u-1)! \sum_{s=2}^{Q-2} (u-1)^{Q-s} M_{Q-s}\kappa_{X,s}(r) \\
&\le C_{X,Q}(r) \\
&\quad + \sum_{s=2}^{Q-2} \frac{1}{Q-s+1}\left[\frac{Q}{2}\right]^{Q-s+1} M_{Q-s}\kappa_{X,s}(r).
\end{aligned}
$$

The inequality

$$\sum_{u=1}^{U}(u-1)^p \le \frac{1}{p+1}U^{p+1},$$

follows from the comparison of a series with an integral.

Remark 12.2.1 Lemma 12.2.1 wites as

$$\kappa_{X,Q}(r) \le c_{X,Q}(r) + \sum_{s=2}^{Q-2} B_{Q,s}\kappa_{X,s}(r).$$

Compare the above recursion to Lemmas 12.2.2 and 12.2.3.

Those are combinatoric versions of cumulants and moments bounds.

Then we derive:

$$\begin{aligned}
\kappa_{X,2}(r) &\le c_{X,2}(r), \\
\kappa_{X,3}(r) &\le c_{X,3}(r), \\
\kappa_{X,4}(r) &\le c_{X,4}(r) + B_{4,2}\kappa_{X,2}(r) \\
&\le c_{X,4}(r) + B_{4,2}c_{X,2}(r), \\
\kappa_{X,5}(r) &\le c_{X,5}(r) + B_{5,3}\kappa_{X,3}(r) + B_{5,2}\kappa_{X,2}(r) \\
&\le c_{X,5}(r) + B_{5,3}c_{X,3}(r) + B_{5,2}c_{X,2}(r), \\
\kappa_{X,6}(r) &\le c_{X,6}(r) + B_{6,4}\kappa_{X,4}(r) + B_{6,3}\kappa_{X,3}(r) + B_{6,2}\kappa_{X,2}(r) \\
&\le c_{X,6}(r) + B_{6,4}\left(c_{X,4}(r) + B_{4,2}c_{X,2}(r)\right) \\
&\quad + B_{6,3}c_{X,3}(r) + B_{6,2}c_{X,2}(r) \\
&\le c_{X,6}(r) + B_{6,4}c_{X,4}(r) + B_{6,3}c_{X,3}(r) \\
&\quad + (B_{6,2} + B_{6,4}B_{4,2})c_{X,2}(r).
\end{aligned}$$

The Lemma 12.2.1 implies the important Corollary 12.2.1 derived from a recursion with the previous inequalities.

Corollary 12.2.1 *For each $Q \ge 2$ there exists a constant $A_Q \ge 0$ only depending on Q and such that*

$$\kappa_{X,Q}(r) \le A_Q \cdot c^*_{X,Q}(r).$$

Remark 12.2.2

- This lemma proves the equivalence between coefficients $c_{X,Q}(r)$ and $\kappa_Q(r)$ up to universal constants. Precise upper bounds follow from Theorem 12.1.2. For this, decompose the sums corresponding to centred moments in two terms among which one explicitly depends on the maximal lag.
  Formula (12.10) implies with $B_{Q,Q} = 1$,

  $$c_{X,Q}(r) \le \sum_{s=2}^{Q} B_{Q,s}\, \kappa_{X,s}(r).$$

  Hence there exists a constant $\widetilde{A}_Q$ with

  $$c_{X,Q}(r) \le \widetilde{A}_Q \kappa^*_{X,Q}(r), \quad \kappa^*_{X,Q}(r) = \max_{2 \le l \le Q} \kappa^*_{X,l}(r)\mu_{Q-l}.$$

  Hence some constants $a_Q, A_Q > 0$ satisfy

  $$a_Q c^*_{X,Q}(r) \le \kappa^*_{X,Q}(r) \le A_Q c^*_{X,Q}(r).$$

These coefficients are equivalent up to constants only depending on Q.

- The previous formula (12.10) implies that a cumulant

$$\kappa(X_{k_1}, \ldots, X_{k_Q}) = \sum_{\alpha,\beta} K_{\alpha,\beta,k} \mathrm{Cov}\,(X_{\alpha(k)}, X_{\beta(k)}),$$

is a linear combination of such covariances with $\alpha \subset \{1, \ldots, l\}$ and $\beta \subset \{l+1, \ldots, Q\}$ for which coefficients $K_{\alpha,\beta,k}$ are polynomials of cumulants. For this replace the Q-tuple $(X_{k_1}, \ldots, X_{k_Q})$ by $(X_i)_{i\in\nu_\mu(k)}$ for each partition μ in formula (12.10) and use recursion.

This representation is useful if one knows the covariances. For a given vector $(X_{k_1}, \ldots, X_{k_q})$, the behaviour of the cumulant is analogous to that of $c_{X,q}(r(k))$. Cumulants admit an advantage with respect to covariances of products: they don't need the precise indices for which the maximal lag occurs.

Example 12.2.2 The constants A_Q are not explicit.
Explicit bounds are derived from the previous proof for small values of Q:

$$\begin{aligned}
\kappa_{X,2}(r) &= c_{X,2}(r) \\
\kappa_{X,3}(r) &= c_{X,3}(r) \\
\kappa_{X,4}(r) &\le c_{X,4}(r) + 3\mu_2 c_{X,2}(r) \\
\kappa_{X,5}(r) &\le c_{X,5}(r) + 10\mu_2 c_{X,3}(r) + 10\mu_3 c_{X,2}(r) \\
\kappa_{X,6}(r) &\le c_{X,6}(r) + 15\mu_2 c_{X,4}(r) + 20\mu_3 c_{X,3}(r)) + 150\mu_4 c_{X,2}(r).
\end{aligned}$$

However the previous heavy combinatorics give an advantage to the rough bounds in Lemma 12.2.1, in order to bound high order cumulants.

12.2.2 *Sums of Cumulants*

The previous bounds yield

Lemma 12.2.2 *Let*

$$\kappa_Q = \sum_{k_2=0}^{\infty} \cdots \sum_{k_Q=0}^{\infty} \left|\kappa\left(X_0, X_{k_2}, \ldots, X_{k_Q}\right)\right|. \tag{12.11}$$

Use the notation (12.8) for each $Q \ge 2$. There exists a constant B_Q such that

$$\kappa_Q \le B_Q \sum_{r=0}^{\infty} (r+1)^{Q-2} C^*_{X,Q}(r).$$

Proof of Lemma 12.2.2 Consider here the partition of the index set

$$E = \{k = (k_2, \ldots, k_Q) \in \mathbb{N}^{Q-1} / \, k_2 \le \cdots \le k_Q\}$$

as $E_r = \{k \in E / \, r(k) = r\}$ for $r \ge 0$ (according to the size of the maximal lag) and denote:

$$\widetilde{\kappa}_Q = \sum_{r=0}^{\infty} \sum_{k \in E_r} \left| \kappa \left(X_0, X_{k_2}, \ldots, X_{k_Q} \right) \right| .$$

Decompose the sums as follows:

$$\kappa_Q \le (Q-1)! \sum_{k_2 \le \cdots \le k_Q} \left| \kappa \left(X_0, X_{k_2}, \ldots, X_{k_Q} \right) \right| = (Q-1)! \, \widetilde{\kappa}_Q .$$

The previous lemma implies

$$\sum_{k \in E_r} \left| \kappa \left(X_0, X_{k_2}, \ldots, X_{k_Q} \right) \right| \le A_Q \mathrm{Card}\, E_r \cdot C^*_{X,Q}(r),$$

for a constant $A_Q > 0$ and the elementary bound

$$\mathrm{Card}\, E_r \le (Q-1)(r+1)^{Q-2},$$

yields the result.

12.2.3 Moments of Sums

Let $(X_n)_{n \in \mathbb{Z}}$ be a stationary and centred sequence, one expects an asymptotic behaviour analogous to the CLT for partial sums

$$\frac{1}{\sqrt{n}} (X_1 + \cdots + X_n) \xrightarrow{\mathcal{L}}_{n \to \infty} \mathcal{N}(0, \sigma^2).$$

The behaviour of moments in $\mathbb{L}^p$-norm is important. It may be used to derive almost-sure behaviours.

The notion of cumulants allows an elementary approach to such expressions.

Lemma 12.2.3 *If the series (12.11) are summable for each $Q \le p$, set $q = [p/2]$ then:*

$$\left|\mathbb{E}\left(\sum_{j=1}^{n} X_j\right)^p\right| \le \sum_{u=1}^{q} n^u \gamma_u, \tag{12.12}$$

$$\textit{with} \qquad \gamma_u = \sum_{v=1}^{2q} \sum_{p_1+\cdots+p_u=p} \frac{p!}{p_1!\cdots p_u!} \kappa_{p_1}\cdots\kappa_{p_u}.$$

Proof Note that $q = p/2$ for p even, and that $q = (p-1)/2$ otherwise.

As in Doukhan and Louhichi (1999) we derive the bound

$$\begin{aligned}
\left|\mathbb{E}(X_1+\cdots+X_n)^p\right| &= \left|\sum_{1\le k_1,\ldots,k_p\le n} \mathbb{E}X_{k_1}\cdots X_{k_p}\right| \\
&\le p! A_{p,n} \\
&= p! \sum_{1\le k_1,\ldots,k_p\le n} \left|\mathbb{E}X_{k_1}\cdots X_{k_p}\right|.
\end{aligned}$$

Let also $\mu = \{i_1, \ldots, i_v\} \subset \{1, \ldots, p\}$ and $k = (k_1, \ldots, k_p)$ set

$$\mu(k) = (k_{i_1}, \ldots, k_{i_v}) \in \mathbb{N}^v. \tag{12.13}$$

To enumerate the terms with their multiplicity it is simpler to consider multi-indices than partitions.

Cumulants and moments are defined analogously.

As in Doukhan and León (1989) with formula (12.3) and partitions $\mu_1, \ldots, \mu_u$ of $\{1, \ldots, p\}$ with exactly $1 \le u \le p$ elements,

$$\begin{aligned}
A_{p,n} &= \sum_{1\le k_1,\ldots,k_p\le n} \sum_{u=1}^{p} \sum_{\mu_1,\ldots,\mu_u} \prod_{j=1}^{u} \kappa_{\mu_j(k)}(X) \\
&= \sum_{u=1}^{p} \sum_{\mu_1,\ldots,\mu_u} \sum_{1\le k_1,\ldots,k_p\le n} \prod_{j=1}^{u} \kappa_{\mu_j(k)}(X) \\
&= \sum_{r=1}^{p} \sum_{p_1+\cdots+p_r=p} \frac{p!}{p_1!\cdots p_r!} \times \qquad (12.14) \\
&\qquad \times \prod_{u=1}^{r} \sum_{1\le k_1,\ldots,k_{p_u}\le n} \kappa_{p_u}(X_{k_1}, \ldots, X_{k_{p_u}}).
\end{aligned}$$

Thus:

$$|A_{p,n}| \le \sum_{u=1}^{q} n^u \sum_{p_1+\cdots+p_u=p} \frac{p!}{p_1!\cdots p_u!} \prod_{j=1}^{u} \kappa_{p_j}. \tag{12.15}$$

Identity (12.14) follows from a change of variable and takes into account the fact that the number of partitions for $\{1, \ldots, p\}$ into u sets with respective cardinalities $p_1, \ldots, p_u$ is a multinomial coefficient.

For $\lambda \in \mathbb{N}$ one may deduce from the stationarity of X that

$$\sum_{1\le k_1,\ldots,k_\lambda \le n} |\kappa_{p_u}(X_{k_1}, \ldots, X_{k_\lambda})| \le n\kappa_\lambda.$$

Cumulants with order 1 always vanish and non zero terms are such that if there exist u indices $p_j \ge 2$ then $u \le q$. Indeed $p_1, \ldots, p_u \ge 2$ thus $2u \le p$. We obtain (12.15).

Remark 12.2.3 If there exists $C > 0$ with $\kappa_s \le C^s$ for each $s \le p$, then due to the multinomial identity the bound (12.15) simply yields

$$\left|\mathbb{E}(X_1 + \cdots + X_n)^p\right| \le p!C^p \sum_{1\le j \le \frac{p}{2}=1}^{q} j^{[\frac{p}{2}]} n^j.$$

12.2.4 Rosenthal's Inequality

As in Doukhan and Louhichi (1999) we derive a Rosenthal inequality involving coefficients $c_{X,l}(r)$.

As before:

$$\left|\mathbb{E}(X_1 + \cdots + X_n)^p\right| \le p!A_{p,n} = p! \sum_{1\le k_1,\ldots,k_p\le n} \left|\mathbb{E}X_{k_1}\cdots X_{k_p}\right|.$$

Each term T_k ($k = (k_1, \ldots, k_p)$) in the sum $A_{p,n}$ admits a maximal lag $r = r(k) = \max_j(k_{j+1} - k_j) < n$,

$$T_k \le c_{X,p}(r) + \left|\mathbb{E}X_{k_1}\cdots X_{k_l}\right| \cdot \left|\mathbb{E}X_{k_{l+1}}\cdots X_{k_p}\right|.$$

Partition the multi-indices k according to the value of $r(k)$ and the smallest index $l = l(k)$ such that $r(k) = k_{l+1} - k_l = r$, for r and l fixed there exists fewer than $n(r+1)^{p-2}$ such multi-indices.

We obtain a main recursion which allows to to extend Rosenthal inequalities (extensions of Lemma 2.2.1 under dependence)

$$A_{p,n} \le (p-1)n \sum_{r=0}^{n-1} (r+1)^{p-2} c_{X,p}(r) + \sum_{l=2}^{p-2} A_{l,n} A_{p-l,n}. \tag{12.16}$$

By using such inequalities (Doukhan and Louhichi 1999) prove a Rosenthal type inequality.

Remark 12.2.4 We make explicit the above recursions for small exponents.

Denote

$$C_{m,n} = \sum_{k=0}^{n-1} (r+1)^{m-2} c_{X,m}(r), \qquad m \ge 2.$$

Iterating the previous relation yields

$$A_{2,n} \le nC_{2,n},$$

$$A_{3,n} \le 2nC_{3,n},$$

$$\begin{aligned} A_{4,n} &\le 3nC_{4,n} + A_{2,n}^2 \\ &\le 3nC_{4,n} + n^2 C_{2,n}^2, \end{aligned}$$

$$\begin{aligned} A_{5,n} &\le 4nC_{5,n} + 2A_{2,n}A_{3,n} \\ &\le 5nC_{5,n} + 4n^2 C_{2,n}C_{3,n}, \end{aligned}$$

$$\begin{aligned} A_{6,n} &\le 5nC_{6,n} + 2A_{2,n}A_{4,n} + A_{3,n}^2 \\ &\le 5nC_{6,n} + 2n^2\left(2C_{3,n}^2 + 3C_{2,n}C_{4,n}\right) + 8n^3 C_{2,n}^3. \end{aligned}$$

We denote (for a fixed q)

$$C_{m,n}^{(q)} = \sum_{k=0}^{n-1} (r+1)^{m-2} c_{X,q}(r).$$

Generally if $p = 2q$ or $p = 2q+1$ we obtain

$$A_{p,n} \le \sum_{j=1}^{q} c_{j,n} n^j,$$

where $c_{j,n}$ is a polynomial with respect to the quantities $C_{i,n}^{(q)}$, for $i \le j$.

Precisely this is a linear combination of expressions

$$\prod_{s=1}^{t} C_{i_s,n}^{(q_s)}, \qquad \text{with} \qquad i_1 + \cdots + i_t = j,\ q_1 + \cdots + q_t = p.$$

Hence for $c_{X,p}(r) = \mathcal{O}(r^{-q})$ one deduces the Marcinkiewicz–Zygmund inequality

$$\left|\mathbb{E}(X_1 + \cdots + X_n)^p\right| = \mathcal{O}(n^q).$$

Rosenthal inequalities yield sharp bounds for centred moments of kernel density estimators or for the empirical process.

12.3 Dependent Kernel Density Estimation

This section describes all the different items related to kernel density estimation under dependence extending Sect. 3.3. Assume that the marginals of the stationary process (X_n) admit a density f.

Let the kernel K be symmetric compactly supported and Lipschitz and suppose we have a window sequence $h_n \downarrow 0$ with $nh_n \to \infty$ and $x \in \mathbb{R}$.

Omitting the additional subindex n we set $U = (U_j)_{j\in\mathbb{Z}}$ with

$$U_j = K\left(\frac{X_j - x}{h_n}\right) - \mathbb{E}K\left(\frac{X_j - x}{h_n}\right).$$

Assume that θ-weak-dependence holds. It is easy to prove that:

$$\operatorname{Lip} h \le \ell 2^{p-1} \cdot \frac{\operatorname{Lip} K}{h_n},$$

if we denote:

$$h(t_1, \ldots, t_l) = \prod_{j=1}^{l} \left\{ K\left(\frac{t_j - x}{h_n}\right) - \mathbb{E}K\left(\frac{X_j - x}{h_n}\right) \right\},$$

in case there exists a constant $M > 0$ such that, for each $n > 0$, the joint density $f_n(x, y)$ of the couple (X_0, X_n) exists and

$$\| f_n(\cdot, \cdot)\|_\infty \le M. \tag{12.17}$$

Exercise 73 *(Sufficient conditions for* (12.17)*)* Assume that (X_n) is a stationary real valued Markov chain with an absolutely continuous Markov transition kernel $P(x, A) = \mathbb{P}(X_1 \in A | X_0 = x)$.

This means that one may write

$$\mathbb{P}(X_1 \in A | X_0 = x) = \int_A p(x, y)\, dy,$$

for a measurable function p. Condition (12.17) holds if $\|f\|_\infty < \infty$ and the transition probabilities admits a transition with a density with $\|p\|_\infty < \infty$.

Integrating the relation (12.17) yields $\|f(\cdot)\|_\infty \le M$ and

$$c_{U,p}(0) \le 2^p f(x) \int_{-\infty}^{\infty} K^2(s) ds.$$

A direct calculation coupled with a weak-dependence inequality yields two distinct controls of $c_{U,p}(r)$ for $r > 0$, hence:

$$c_{U,p}(r) \le 2^{p-1} \left(p \cdot \mathrm{Lip}\, K \frac{\theta_r}{h_n} \right) \wedge (2Mh_n^2),$$

there exists a constant $C > 0$ with

$$C_{p,n} = Ch_n \left(1 + \sum_{k=1}^{n-1} (r+1)^{p-2} \left(h_n \wedge \frac{\theta_r}{h_n^2} \right) \right).$$

Exercise 74 *(Functional AR(1)-model)* Let $X_n = r(X_{n-1}) + \xi_n$ with $\mathrm{Lip}\, r < 1$, then if $\mathbb{E}|\xi_0| < \infty$, and ξ_0 admits a bounded density g wrt Lebesgue measure the bounds in Exercise 73 hold.

Proof Check that Proposition 7.3.2 implies the existence of a stationary distribution and the relation $f(x) = \int_{\mathbb{R}} p(x, y) f(y)\, dy$ implies with $p(x, y) = g(y - r(x))$ that $M = \|g\|_\infty$.

Exercise 75 *(NLARCH(1)-models)* The model as well as ξ_0 are vector valued in $\mathbb{R}^d$. Let

$$X_n = r(X_{n-1}) + s(X_{n-1})\xi_n, \quad \text{with} \quad \mathrm{Lip}\, r + \mathrm{Lip}\, |||s||| \cdot \|\xi_0\|_p < 1$$

($|||\cdot|||$ denotes the operator norm of a $d \times d$-matrix), then if ξ_0 admits a bounded density g wrt Lebesgue measure and $\inf_x |||s^{-1}(x)||| > 0$ the bounds in Exercise 73 hold.

Hint. Proposition 7.3.2 again implies the existence of a stationary distribution and the above proof still holds.

The following elementary inequality is often useful when two different bounds of a quantity are available. In our case, two inequalities appear either from dependence properties, or from an analytic consideration.

Exercise 76

$$u \wedge v \le u^\alpha v^{1-\alpha}, \quad \text{if } u, v \ge 0, \quad 0 \le \alpha \le 1. \tag{12.18}$$

Hint. From the symmetry of u, v's roles assume that $u \le v$ then $u \le 1$ implies $u \wedge v = u = u^\alpha \cdot u^{1-\alpha} \le u^\alpha \cdot v^{1-\alpha}$.

As a simple application of the previous inequalities, for $p = 2$ we obtain the following result.

Proposition 12.3.1 *Assume that $\theta_r \le Cr^{-a}$ for some $a > 3$, then*

$$\lim_{n\to\infty} nh_n \text{Var}\, \widehat{f}(x) = f(x) \int K^2(t)\, dt.$$

Proof First $c_{U,2}(0) \sim h_n f(x) \int K^2(s)\, ds$ and one simply needs to derive that

$$\lim_{n\to\infty} \frac{1}{h_n} \sum_{r=1}^{\infty} c_{U,2}(r) = 0,$$

for some constant and from relation (12.18), we obtain:

$$\frac{1}{h_n} c_{U,2}(r) \le C\left(\frac{\theta_r}{h_n^2} \wedge h_n\right) \le h_n^{1-3\alpha}\theta_r^\alpha.$$

The assumption $a > 3$ implies that there exists some $\alpha < \frac{1}{3}$ such that

$$\sum_{r=1}^{\infty} \theta_r^\alpha < \infty.$$

Hence the dependent part of those variances is indeed negligible and the asymptotic $\mathbb{L}^2$-behaviour of kernel density estimators is the same as under independence.

Exercise 77 Using inequality $\theta/h^2 \wedge h \le \theta^{1/3}$ derived from Exercise 12.18, prove that $C_{p,n} = \mathcal{O}(h_n)$ if

$$\sum_{r=0}^{\infty} (r+1)^{p-2}\theta_r^{\frac{1}{3}} < \infty. \tag{12.19}$$

More generally if $p \ge 2$, from recursion and by using assumption (12.19) and Exercise 77 we get $\left|\mathbb{E}(\widehat{f}(x) - \mathbb{E}\widehat{f}(x))^p\right| \le C(nh_n)^{p-q}$. This bound has order $(nh_n)^{-\frac{p}{2}}$ for even p and $(nh_n)^{-\frac{p-1}{2}}$ if p is odd.

Consider now some even integer $p > 2$. Almost-sure convergence of such estimators also follows from the Markov inequality and the Borel–Cantelli Lemma B.4.1 in case:

$$\sum_{n=1}^{\infty} \frac{1}{(nh_n)^{\frac{p}{2}}} < \infty.$$

Exercise 78 Derive the uniform *a.s.* behavior over a compact interval.

Hint. Use Exercise 15.

Those bounds fit with the underlying CLT:

Theorem 12.3.1 (Bardet et al. (2006)) *Suppose the assumptions in Proposition 12.3.1 hold then:*

$$\sqrt{nh_n}(\widehat{f}(x) - \mathbb{E}\widehat{f}(x)) \to_{n\to\infty} \mathcal{N}\left(0, f(x)\int K^2(t)dt\right).$$

Proof Use Lemma 11.5.1 then arguing as in Proposition 12.3.1 allows a tight control of the dependent terms again (the result is left as an exercise).

Let

$$x_\ell = z_\ell - \mathbb{E}z_\ell, \qquad \text{with} \qquad z_\ell = \frac{1}{\sqrt{nh_n}} K\left(\frac{X_\ell - x}{h_n}\right),$$

$s_1 = 0$ and $s_\ell = x_1 + \cdots + x_\ell$, then we need to prove that

$$\Delta_n = \sum_{k=1}^{n} |\text{Cov}\,(e^{its_{k-1}}, e^{itx_k})| \to_{n\to\infty} 0.$$

This is done by using the following exercises. Use the notation in Exercise 79. First from Exercises 80 and 81 together with inequality (12.18), we derive for each $0 < b < 1$:

$$C_{k,\ell} \le \frac{c}{n}(h \wedge \frac{\theta_{k-\ell}}{h^2}) \le \frac{c}{n} h^{1-3b}\theta_{k-\ell}^b.$$

Now use Exercise 79 to derive:

$$\begin{aligned}
\Delta_n &\le \sum_{k=1}^{n}\sum_{\ell=1}^{k-1} C_{k,\ell} \\
&\le \frac{c}{n} h^{1-3b} \sum_{k=1}^{n}\sum_{\ell=1}^{k-1} \theta_{k-\ell}^b \\
&\le ch^{1-3b} \sum_{j=1}^{n-1} \theta_j^b.
\end{aligned}$$

The last inequality follows from the fact that the identity $j = k - \ell$ occurs for fewer than n couples (k, ℓ).

Now if $\theta_r \le Cr^{-a}$ for some $a > 3$ then

$$\sum_{j=1}^{n-1} \theta_j^b < \infty, \qquad \text{if} \quad ab > 1.$$

There exists such $b < 1/3$ which concludes the proof.

Exercise 79 Set $s_0 = 0$ then prove the decomposition

$$\mathrm{Cov}\,(e^{its_{k-1}}, e^{itx_k}) = \sum_{0 \le \ell < k} C_{k,\ell},$$

with $C_{k,\ell} = \mathrm{Cov}\,(e^{its_\ell} - e^{its_{\ell-1}}, e^{itx_k} - 1)$.

Hint. Remark first that

$$|\mathrm{Cov}\,(e^{its_\ell} - e^{its_{\ell-1}}, e^{itx_k})| = |\mathrm{Cov}\,(e^{its_\ell} - e^{its_{\ell-1}}, e^{itx_k} - 1)|.$$

Then $\mathrm{Cov}\,(e^{its_k}, e^{itx_k}) = \mathrm{Cov}\,(e^{its_k} - 1, e^{itx_k})$, and the decomposition follows from $s_0 = 0$.

The expression is then derived by considering a telescopic sum.

Exercise 80 Assume that the marginal density $f_j(u, v)$ of the random vector (X_0, X_j) satisfies $C = \sup_j \sup_{(u,v)} f_j(u, v) < \infty$ then prove that for some constant $c > 0$

$$C_{k,\ell} \le ct^2 \cdot \frac{h}{n}.$$

Hint. Set $j = k - \ell$ and $g_j(u, v) = f_j(u, v) + f(u)f(v)$ then there exists a constant such that $g_j(u, v) \le C$ for each j, u, v. Then, the relation $|e^{iz} - 1| \le |z|$ entails:

$$\begin{aligned}
C_{k,\ell} &\le t^2 \mathbb{E}|x_{k-1}x_k| + \mathbb{E}|x_{k-1}| \cdot \mathbb{E}|x_k| \\
&= \frac{1}{nh} \int \left| K\left(\frac{u-x}{h}\right) K\left(\frac{v-x}{h}\right) \right| g_j(u, v)du\,dv \\
&= \frac{h^2}{nh} \int |K(y)K(z)| g_j(x - hy, x - hz)du\,dv \\
&\le C \cdot \frac{h}{n} \left(\int |K(y)|dy \right)^2.
\end{aligned}$$

This yields the conclusion of this exercise.

Exercise 81 Assume now that the θ-weak-dependence condition holds (condition associated with $\Psi_\theta(f, g) = v\mathrm{Lip}\, g \|f\|_\infty$) and K is a Lipschitz function.

Then there exists a constant $c > 0$ such that

$$C_{k,\ell} = |\text{Cov}\,(e^{its_\ell} - e^{its_{\ell-1}}, e^{itx_k})| \le ct^2 \cdot \frac{\theta_{k-\ell}}{nh^2}.$$

Hint. Set $f(X_1, \ldots, X_\ell) = e^{its_\ell} - e^{its_{\ell-1}}$ then

$$\|f\|_\infty \le |t|\|x_\ell\|_\infty \le \frac{2\|K\|_\infty}{\sqrt{nh}}$$

and, with

$$g(u) = \exp\left\{it\left(K\left(\frac{u-x}{h}\right) - \mathbb{E}K\left(\frac{X_0 - x}{h}\right)\right)\right\}$$

(and $g(X_k) = e^{itx_k}$), it is simple to check that

$$\text{Lip}\, g \le \frac{|t|\text{Lip}\, K}{h\sqrt{nh}}.$$

Those two bounds entail with the definition of weak-dependence that:

$$C_{k,\ell} \le 2t^2\|K\|_\infty \text{Lip}\, K \cdot \frac{\theta_{k-\ell}}{nh^2}.$$

The proof is complete.

Exercise 82 Extend this whole section to the case of associated processes.
Explain precisely how (12.19) should be modified in this case.

Hint. Use inequality (8.1).

Exercise 83 Extend those results to the case of regression estimators (3.5), both under weakly dependent or under associated frameworks.

Hint. In case Y is a bounded regressor the result does not change too much. Otherwise a truncation technique may be used.

Remark 12.3.1 Standard extensions are possible for the other weak-dependence conditions as well as under strong mixing.

These exercises are left to the reader.

The case of subsampling is analogous:

Exercise 84 Subsampling from Sect. 4.6 may also be considered as in Exercise 69. Higher-order moments may be bounded (see Doukhan et al. 2011) under weak-dependence conditions in order to derive almost-sure convergence from the Borel–Cantelli Lemma B.4.1 in case some even number $p \in 2\mathbb{N}$ satisfies

$$\sum_{n=1}^{\infty} \mathbb{E}\big(\widehat{K}_n(g) - \mathbb{E}\widehat{K}_n(g)\big)^p < \infty.$$

Hints. The proof is as for kernel density estimation based on bounds for coefficients $c_{Z,r}(r)$ in (12.8) with $Z_i = g(t_m(X_{i+1}, \ldots, X_{i+m}))$ in the overlapping scheme.

For $h(x) = \mathbb{1}_{\{x \le z\}}$ analogously to Exercise 64, bounds of

$$c_{Y,r}(r) = \sup_{\mathbf{i},\mathbf{j}} |\mathrm{Cov}\,(h(X_{i_1}) \times \cdots \times h(X_{i_u}), h(X_{j_1}) \times \cdots \times h(X_{j_v}))|,$$

$u + v = p$, $i_1 \le \cdots \le i_u$, $j_1 \le \cdots \le j_v$ with $j_1 - i_u = r$ allow to bound higher order moments.

Set $I_1 = h(X_{i_1})$, $I_1^\epsilon = h_\epsilon(X_{i_1})$, $J_1 = h(X_{j_1}), \ldots$ since those functions are bounded and $\mathrm{Lip}\, h_\epsilon = 1/\epsilon$, using the following inequalities yields e.g. under η-weak-dependence:

$$\begin{aligned}
|\mathrm{Cov}\,(I_1 \cdots I_u, J_1 \cdots J_v))| &\le |\mathrm{Cov}\,(I_1^\epsilon \cdots I_u^\epsilon, J_1^\epsilon \cdots J_v^\epsilon))| \\
&\quad + 2\sum_{s=1}^{u} \mathbb{E}|I_s - I_s^\epsilon| + 2\sum_{t=1}^{u} \mathbb{E}|J_t - J_t^\epsilon| \\
&\le \frac{u+v}{\epsilon}\eta_r + (u+v)\epsilon = 2(u+v)\sqrt{\eta_r}
\end{aligned}$$

with $\epsilon^2 = \eta_r$.

Such bounds do not need this last step under strong mixing.

Erratum to: Non-linear Processes

Erratum to:
Chapter 7 in: P. Doukhan, *Stochastic Models for Time Series*, Mathématiques et Applications 80, https://doi.org/10.1007/978-3-319-76938-7_7

In the original version of the book, the belated correction from author to add the content. **Remark**. The model is also L^2-contractive if $\mathbb{E}Y_0^2 < 1$. If Y_0 admits a Poisson distribution with parameter 1 this means that $a + a^2 < 1$; this is written $0 \leq a \leq (\sqrt{5} - 1)/2 \sim 0.6 < 1$ has to be incorporated at the end of Page 148 in Chapter 7. The erratum chapter and the book have been updated with the change.

The updated online version of this chapter can be found at
https://doi.org/10.1007/978-3-319-76938-7_7

P. Doukhan, *Stochastic Models for Time Series*, Mathématiques et Applications 80,
https://doi.org/10.1007/978-3-319-76938-7_13

Appendix A
Probability and Distributions

This appendix is a short introduction to the basic concepts of probability spaces, such as developed in standard textbooks, refer for example to Kallenberg (1997) and Feller (1968). It is a notational index rather than a real introductory text on probability; it is dedicated to readers with some knowledge of probability theory.

Models of time series are based (here) on random inputs for physical reasons. Thus the appendix recalls some standard facts concerning some useful distributions. One may refer to Feller (1968) for additional examples.

We provide a short introduction to Gaussian distributions with first the standard Normal and then its vector valued extension. Such random variables are needed to define Gaussian processes. Finally γ-distributions are considered; they lead many explicit calculations.

A.1 Notations

For any space E, a sigma-algebra, σ-algebra $\mathcal{E}$ is a subset of $\mathcal{P}(E)$, (the set of subsets of E), such that

- $\emptyset \in \mathcal{E}$,
- $\forall A \in \mathcal{E}: \qquad A^c \in \mathcal{E}$,

  where we denote by $A^c = E \setminus A$ the complementary set of A,
- $\forall A_n \in \mathcal{E}, n = 1, 2, 3 \ldots : \qquad \bigcup_{n \in \mathbb{N}} A_n \in \mathcal{E}$.

A measurable space is any couple $(E, \mathcal{E})$, composed of a set and a σ-algebra on the set E. Elements of $\mathcal{A}$ are called events.

A probability space $(\Omega, \mathcal{A}, \mathbb{P})$ is a measurable space $(\Omega, \mathcal{A})$ equipped with a probability, that is a function $\mathbb{P} : \mathcal{A} \to [0, 1]$ such that:

- $\mathbb{P}(\emptyset) = 0$,
- $\mathbb{P}(\Omega) = 1$,

P. Doukhan, *Stochastic Models for Time Series*, Mathématiques et Applications 80,
https://doi.org/10.1007/978-3-319-76938-7

- $\forall A, B \in \mathcal{A}$:
$$A \cap B = \emptyset \Rightarrow \mathbb{P}(A \cup B) = \mathbb{P}(A) + \mathbb{P}(B),$$
- $\forall A_i \in \mathcal{A}, i = 1, 2, 3 \ldots$:
$$A_1 \subset A_2 \subset \cdots \Rightarrow \lim_{n \to \infty} \mathbb{P}(A_n) = \mathbb{P}(A),$$
where we denote $A = \bigcup_{n=1}^{\infty} A_n$.

Example A.1.1 Examples of measurable spaces $(\Omega, \mathcal{A})$ follow. We recall here that one usually needs two measurable spaces; a space of values or realizations $(E, \mathcal{A})$ and an abstract probability space $(\Omega, \mathcal{A})$ which needs a probability function $\mathbb{P}$.

We list some such simple spaces.

- If Ω is a finite set with n elements then a reasonable choice of sigma-algebra is $\mathcal{A} = \mathcal{P}(\Omega)$ which admits 2^n elements as it may be seen from the fact the application: $A \mapsto \mathbb{I}_A$ defined for $\mathcal{P}(E)$ on the set of functions from E to $\{0, 1\}$ is a bijective function.
- For denumerable finite sets Ω again $\mathcal{A} = \mathcal{P}(\Omega)$ is a suitable framework.
- $\mathbb{R}$ may be equipped with its Borel σ-field, the smallest sigma-algebra containing all the intervals.
- More generally a topological space Ω is measurable with $\mathcal{A}$ the smallest σ-field containing all the open sets. This σ-field is called the Borel σ-field.
- Products of two measurable spaces are still measurable, and here the σ-field is again the smallest containing products $A \times B$ with clear notations.
- Infinite products are again possible; for a family of measurable spaces $(\Omega_i, \mathcal{A}_i)_{i \in I}$ the product $\Omega = \prod_{i \in I} \Omega_i$ and $\mathcal{A}$ is equipped with the smallest σ-field containing all the events $\prod_{i \in I} A_i$ with $A_i \in \mathcal{A}_i$ for each $i \in I$ and $A_i = \Omega_i$ for each $i \notin J$ with $J \subset I$, a finite subset of I.
- Some examples of probability spaces are related to the generation of random variables. They will be considered in Example A.2.3.

The sigma-algebra $\mathcal{A}$ is complete in case $A \in \mathcal{A}$, $\mathbb{P}(A) = 0$ and $B \subset A$ imply $B \in \mathcal{A}$ (roughly speaking, it contains the nullsets).

The σ-fields considered are usually those obtained from a measurable space equipped with some measure (often, probability measures); the completed σ-field is the smallest containing both all the events $A \in \mathcal{A}$ and each set $B \subset A$ for each $A \in \mathcal{A}$ with $\mathbb{P}(A) = 0$.

We also recall the Landau notations used throughout those notes:

Definition A.1.1 For real valued sequences u_n, v_n, the Landau notation $v_n = \mathcal{O}(u_n)$ as $n \to \infty$, means that there is some constant $C > 0$ such that $|v_n| \le C|u_n|$ for all n.

Moreover if u_n, v_n are random, then

- $\mathcal{O}_{\mathbb{P}}(\cdot)$, means that the random variable C is bounded in probability, this can be written as

$$\mathbb{P}\left(\bigcup_{n\geq 1}(C \leq n)\right) = 1.$$

- $\mathcal{O}_{a.s.}(\cdot)$ means $|v_n| \leq C_n |u_n|$ for random variables $C_n = C_n(\omega) > 0$ such that ω-a.s., $\sup_n C_n(\omega) < \infty$.
- $\mathcal{O}_{\mathbb{L}^p}(\cdot)$ means that $C = C(\omega)$ is bounded in $\mathbb{L}^p$, $\mathbb{E}|C|^p < \infty$.
- $v_n = \mathcal{O}(u_n)$ as $n \to \infty$, means that for a sequence $C_n > 0$ such that $\lim_n C_n = 0$, we have $|v_n| \leq C_n |u_n|$ for all n.

A.2 Distributions and Random Variables

Let $X : \Omega \to E$ be an arbitrary function defined on the measurable space $(\Omega, \mathcal{A})$, taking values in another measurable space $(\mathbb{E}, \mathcal{E})$.

We introduce the probabilist notation:

$$(X \in A) = X^{-1}(A) = \{\omega \in \Omega / \ X(\omega) \in A\}, \qquad \text{for all A } \subset E.$$

A random variable $X : \Omega \to E$ is a measurable function between these two measurable sets; this means that $X^{-1}(\mathcal{E}) \subset \mathcal{A}$. In other terms:

$$\forall A \in \mathcal{E} : \qquad (X \in A) \in \mathcal{A}.$$

Note also that $\sigma(X) = X^{-1}(\mathcal{E})$ is the σ-algebra generated by the random variable $X : \Omega \to E$, that mean it is the smallest sub-σ-algebra $\mathcal{F}$ of $\mathcal{A}$ which makes the application $X : (\Omega, \mathcal{F}) \to (E, \mathcal{E})$ measurable.

Also the image distribution or the law of X is the probability distribution defined as

$$P_X(A) = \mathbb{P}(X \in A), \qquad \forall A \in \mathcal{E}.$$

Let E be any topological space, its Borel sigma algebra $\mathcal{E}$ is the smallest sigma-algebra containing all the open sets; it also contains intersections of open sets but also much more complicated sets. In most of the cases $E = \mathbb{R}$ will be endowed with its Borel sigma-algebra, completed when necessary. X's distribution probability is also defined through its cumulative distribution function:

$$F(x) = \mathbb{P}(X \leq x) = P_X((-\infty, x]), \qquad \forall x \in \mathbb{R}.$$

In some cases $E = \mathbb{R}^d$ is a finite dimensional vector space but we shall avoid more complicated situations as much as possible.

The following exercise is useful in the present setting of time series:

Exercise 85 Let $X, Y \in \mathbb{R}^d$ be two random variables. If $\mathbb{E}g(X) = \mathbb{E}g(Y)$ for each Lipschitz function on $\mathbb{R}^d$, then the random vectors X and Y admit the same distribution.

Hint. Approximate indicators of rectangles R by a sequence of Lipschitz functions such that $f(x) = 0$ in case the distance of x to R is more than some arbitrary $\epsilon > 0$. First consider $d = 1$ and then use the tensor product of such Lipschitz functions of a real random variable. The derived function admits a Lipschitz coefficient with order $1/\epsilon$.

For a column vector $v \in \mathbb{R}^d$, set v' the corresponding row vector. We identify such matrix and vector notations in those notes due to the standard duality in an Euclidean space.

Definition A.2.1 For $E = \mathbb{R}^d$ one defines the mean of a random variable $X \in \mathbb{R}^d$:

$$\mathbb{E}X = \int_E x P_X(dx) \in \mathbb{R}^d,$$

in case the integrals converge.[1]

We write $X \in \mathbb{L}^p$ in case $\mathbb{E}\|X\|^p < \infty$ for any norm $\|\cdot\|$ on $\mathbb{R}^d$.

If $p \geq 1$ we shall write $\|X\|_p = (\mathbb{E}\|X\|^p)^{\frac{1}{p}}$, and it is clear that:

$$\|X\|_p = 0 \Longleftrightarrow X = 0, a.s.$$

In case $d = 1$, let $\mathbb{L}^p(\Omega, \mathcal{A}, \mathbb{P})$ be the space of classes of a.s. equal and $\mathbb{L}^p$-integrable random variables. This space is then a Banach space; this needs Corollary A.2.1.

Moreover in case $X \in \mathbb{L}^2$, we define the covariance:

$$\text{Cov}\,(X) = \mathbb{E}XX' - \mathbb{E}X(\mathbb{E}X)'.$$

This is a symmetric positive $n \times n$-matrix. In case $X = (X_1, X_2)$, we also write:

$$\text{Cov}\,(X) = \begin{pmatrix} \text{Var}\, X_1 & \text{Cov}\,(X_1, X_2) \\ \text{Cov}\,(X_1, X_2) & \text{Var}\, X_2 \end{pmatrix}.$$

Note that if $d = 1$, then

$$X \geq 0 \Longrightarrow \mathbb{E}X \geq 0. \tag{A.1}$$

and the notation of the variance can be written as

$$\text{Var}\ X = \text{Cov}\,(X).$$

[1] Or if they can be defined, as is the case for $d = 1$ and $E = \mathbb{R}^+ = [0, +\infty)$. In this case integrals take values in the space $[0, +\infty]$.

An essential result is the following theorem:

Theorem A.2.1 (Markov inequality) *Assume that $V \geq 0$ is a real valued non-negative random variable, then its expectation exists in $\overline{\mathbb{R}} = \mathbb{R} \cup \{\pm\infty\}$ and:*

$$\mathbb{P}(V \geq u) \leq \frac{\mathbb{E}V}{u}, \qquad \forall u > 0.$$

Proof Set $A = (V \geq u)$ then using (A.1) we derive:

$$\mathbb{E}V = \mathbb{E}V\ \mathbb{I}_A + \mathbb{E}V\ \mathbb{I}_{A^c} \geq \mathbb{E}V\ \mathbb{I}_A \geq u\mathbb{P}(A).$$

The result is proved.

Exercise 86 If $\mathbb{E}X_0^2 < \infty$ prove that there exists a function $H : \mathbb{R}^+ \to \mathbb{R}^+$ such that $\lim_{x\to\infty} H(x)/x^2 = \infty$, $\mathbb{E}H(|X_0|) < \infty$.

Hint. For each $k > 0$ choose a non-decreasing sequence $M_k > 0$ such that

$$\mathbb{E}|X_0|^2\ \mathbb{I}_{\{|X_0|\geq M_k\}} \leq \frac{1}{k^2}.$$

Set $H(x) = kx^2$ for $M_k \leq |x| < M_{k+1}$ to conclude.

Proposition A.2.1 (Jensen inequality) *The Jensen inequality holds for each function $g : C \to \mathbb{R}$ convex and continuous on the convex set $C \subset \mathbb{R}^d$.*
If $Z \in C$ a.s. (and if the following expectations are well defined)

$$\mathbb{E}g(Z) \geq g\left(\mathbb{E}Z\right). \tag{A.2}$$

Proof We begin with the case $d = 1$. In this case we assume that $C = (a, b)$ is an interval, then $g : (a, b) \to \mathbb{R}$ is differentiable except possibly on some denumerable set.

At each point of C the left and right derivatives exist (at the extremities, only one of them may be defined).

Moreover, for any $x, y, z \in C$, then if $x < y < z$, one derives:

$$g'(x+) \leq g'(y-) \leq g'(y+) \leq g'(z-),$$

with

$$g'(y\pm) = \lim_{h\to 0^+} \frac{g(y \pm h) - g(y)}{\pm h},$$

then for each $x_0 \in C$ choose any $u \in [g'(x_0-), g'(x_0+)]$.

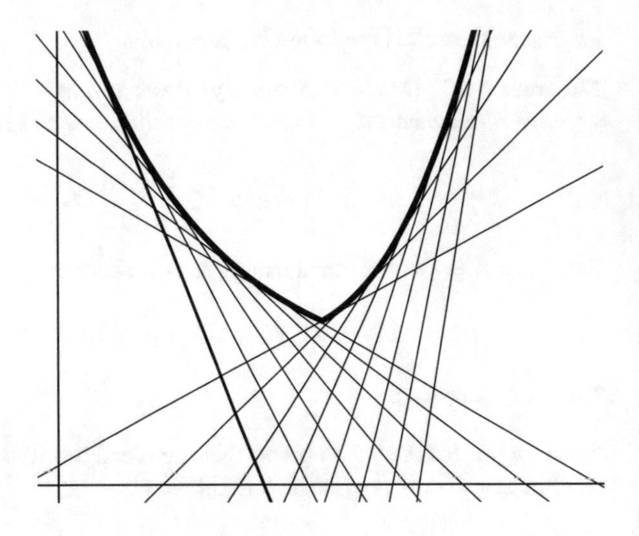

Fig. A.1 Convex function as supremum of affine functions

Then the affine function

$$f(x) = u(x - x_0) + g(x_0)$$

satisfies $f \leq g$ and $f(x_0) = g(x_0)$ by convexity.

Thus:

each convex function g is the upper bound of affine functions $f \leq g$.

From linearity of integrals $f(\mathbb{E}Z) = \mathbb{E}f(Z)$ and $f(\mathbb{E}Z) \leq \mathbb{E}g(Z)$. Now the relation $\sup_f f(\mathbb{E}Z) = g(\mathbb{E}Z)$ allows to conclude.

If now $d \geq 1$ then from the most elementary variant of the Hahn–Banach theorem, the same representation of g holds and the proof is the same, see Fig. A.1.

In the Hilbert case the orthogonal projection provides an elementary way to separate a point from a disjoint closed convex set: take its orthogonal projection y of x then the hyperplane with direction $x^\perp$ and containing the middle of the interval $[x, y]$ is a valuable solution of the Hahn-Banach separation problem.

Remark A.2.1

- This inequality is an equality for each affine function.
- The inequality is strict if g is strictly convex and Z is not a.s. constant. The case of power functions is investigated in Lemma 7.3.1.
- Let $\mathcal{B} \subset \mathcal{A}$ be a sub-σ algebra of $\mathcal{A}$, a conditional variant of this inequality is[2]:

$$\mathbb{E}^{\mathcal{B}} g(Z) \geq g\left(\mathbb{E}^{\mathcal{B}} Z\right). \tag{A.3}$$

[2]For this, a conditional version of the dominated convergence theorem is needed. See Definition A.2.2.

Definition A.2.2 Let $\mathcal{B} \subset \mathcal{A}$ be a sub-σ-algebra of $\mathcal{A}$ on a probability space $(\Omega, \mathcal{A}, \mathbb{P})$ and $X \geq 0$ be a non-negative random variable on this space.
Then $Z = \mathbb{E}^{\mathcal{B}} X$ is the $\mathcal{B}$-measurable random variable defined $\mathcal{B}$-a.s. such that

$$\mathbb{E}(\mathbb{E}^{\mathcal{B}} X)\, \mathbb{I}_B = \mathbb{E} X\, \mathbb{I}_B, \qquad \forall B \in \mathcal{B}.$$

If $\mathbb{E}|X| < \infty$ decomposing $X = X^+ - X^-$ allows to again define properly the conditional expectations. This is a linear non-negative operator on $\mathbb{L}^1(\Omega, \mathcal{A}, \mathbb{P})$.

Remark A.2.2 The equivalent notation is $\mathbb{E}^{\mathcal{B}} X = \mathbb{E}(X|\, \mathcal{B})$ will be indifferently used for clarity.

Remark A.2.3 If $\mathbb{E}X^2 < \infty$ then this definition may be rewritten as

$$\mathbb{E}(\mathbb{E}^{\mathcal{B}} X)Y = \mathbb{E}XY, \qquad \forall Y \in \mathbb{L}^2(\Omega, \mathcal{B}, \mathbb{P}).$$

This operator is also interpreted as the orthogonal projector

$$\mathbb{L}^2(\Omega, \mathcal{A}, \mathbb{P}) \to \mathbb{L}^2(\Omega, \mathcal{B}, \mathbb{P}) \subset \mathbb{L}^2(\Omega, \mathcal{A}, \mathbb{P}).$$

The following standard inequality is also important:

Proposition A.2.2 (Hölder inequality) *Let $X_1 \in \mathbb{L}^{p_1}$, ..., $X_u \in \mathbb{L}^{p_u}$ be real valued random variables, then:*

$$\mathbb{E}|X_1 \cdots X_u| \leq \|X_1\|_{p_1} \cdots \|X_1\|_{p_u}, \quad \text{if} \quad \frac{1}{p_1} + \cdots + \frac{1}{p_u} = 1.$$

Hint. For $z_1, \ldots, z_u > 0$ the convexity of the exponential function implies

$$z_1 \cdots z_u \leq \frac{1}{p_1} z_1^{p_1} + \cdots + \frac{1}{p_u} z_u^{p_u}.$$

Now set $z_j = |X_j| / \|X_j\|_{p_j}$ to conclude.

A standard application of Proposition A.2.2 implies the important idea of sub-linearization, which may be transposed in other settings:

Lemma A.2.1 (Sub-linearization)[3] *Let $p \geq 1$ and $X \in \mathbb{L}^p$ satisfies $\mathbb{E}X = 0$ then*

$$\|X\|_p = \sup\left\{\mathbb{E}XY /\ \|Y\|_q = 1\right\}, \qquad \frac{1}{p} + \frac{1}{q} = 1.$$

Hint. The upper bound follows from Proposition A.2.2. The lower bound follows with the special choice $Y = c \cdot \text{sign}(X)|X|^r$ with $r = p - 1$ (so that $rq = p$), and $c^q = \|X\|_p^{-p}$. Hence this Y satisfies $\|Y\|_q = 1$. Now $\mathbb{E}XY = c\mathbb{E}|X|^p = \|X\|_p^{p-p/q} = \|X\|_p$ yields the lower bound.

[3] See Polya and Szegö (1970).

An immediate consequence is:

Corollary A.2.1 (Minkowski inequality) *Let* $p \geq 1$ *and* $X, Y \in \mathbb{L}^p$ *then*

$$\|X + Y\|_p \leq \|X\|_p + \|Y\|_p.$$

Remark A.2.4 We refer to the beautiful and comprehensive presentation in Polya and Szegö (1970) for the above convexity results. Beyond standard Banach spaces inequalities, their sublinearization technique is a powerful tool.

Definition A.2.3 Let $X \in \mathbb{R}^d$ be a vector valued random variable then its characteristic function is defined as

$$\phi_X(t) = \mathbb{E}e^{it \cdot X}, \qquad \forall t \in \mathbb{R}^d.$$

The Laplace transform of the law of X is:

$$L_X(z) = \mathbb{E}e^{z \cdot X}, \qquad \text{for all } z \in \text{Dom}(L_X) \subset \mathbb{C}^d,$$

($\text{Dom}(L_X)$ is the set of such z such that this expression is well defined).

The generating function of any integer valued random variable X is denoted $g_X(z) = \mathbb{E}z^X$.

Remark A.2.5 First, the characteristic function always exists and $\phi_X(t) = L_X(it)$.

If 0 is interior to the domain of definition of L_X then this function is analytic around 0 as well as ϕ_X.

Exchanging differentiation and integrals is legitimate:

$$\frac{\partial}{\partial t_j}\phi(0) = i \cdot \mathbb{E}X_j.$$

Moreover Fourier integral theory implies that inversion is possible and in this case ϕ_X determines X's distribution.

Simple examples of probability distributions are

- *Discrete random variables*: there exists a finite or denumerable set S such that $\mathbb{P}(X \notin S) = 0$.

  In case the following series are absolutely convergent we denote

  $$\mathbb{E}X = \sum_{x \in S} x \cdot \mathbb{P}(X = x).$$

  In the case when $S \subset \mathbb{Z}$ the generating function $g_X(z) = \mathbb{E}z^X$ will be preferred to the Laplace transform and this function is also defined for $|z| \leq 1$.

Example A.2.1 (*Discrete distributions*)

- The Bernoulli law $b(p)$ with parameter $p \in [0, 1]$ is the law of a random variable with values in $\{0, 1\}$ with

$$\mathbb{P}(X = 1) = p, \quad \text{and} \quad \mathbb{P}(X = 0) = 1 - p.$$

  Here $g_X(z) = pz + q$.

- Binomial law $B(n, p)$ with parameters $n \in \mathbb{N}^*$, $p \in [0, 1]$ is the law of a random variable with values in $\{0, 1, \ldots, n\}$ with

$$\mathbb{P}(X = k) = \frac{n!}{k!(n-k)!} p^k (1-p)^{n-k}.$$

  The origin of this law is that if $X_1, \ldots, X_n \sim b(p)$ are independent identically distributed random variables then

$$X_1 + \cdots + X_n \sim B(n, p).$$

  For this record simply that $g_X(z) = (pz + q)^n$.

- A Poisson distributed random variable $X \sim \mathcal{P}(\lambda)$ with parameter λ takes values in $\mathbb{N}$ and

$$\mathbb{P}(X = k) = \frac{\lambda^k}{k!} e^{-\lambda}.$$

- *Absolutely continuous distributions.*

Definition A.2.4 We assume here that there exists a measurable function $f : E \to \mathbb{R}^+$ such that for each $A \in \mathcal{E}$:

$$P_X(A) = \int_A f(x)\, dx$$

this function is called the density of X distribution.

Remark A.2.6 We also derive that for each function $g : E \to \mathbb{R}$, measurable:

$$\mathbb{E}g(X) = \int_E g(x) f(x)\, dx.$$

The above relation is also the definition of a density.

Example A.2.2 (*Continuous distributions*)

- Uniform $U[0, 1]$-distribution on the unit interval, it admits a density wrt the Lebesgue measure $f(x) = \mathbb{I}_{[0,1]}(x)$.

- Exponential law $\mathcal{E}(\lambda)$ with parameter λ admits the density

$$f(x) = \lambda e^{-\lambda x} \, \mathbb{I}_{\{x \geq 0\}}.$$

- The Normal law $\mathcal{N}(0, 1)$ is the simplest Gaussian law which admits the density

$$f(x) = \frac{1}{\sqrt{2\pi}} \cdot e^{-\frac{x^2}{2}}.$$

More examples of distributions linked with Gaussians as the family of γ-distributions as considered below.
- The Cauchy distribution is defined with

$$f(x) = \frac{1}{\pi} \cdot \frac{1}{1 + x^2}.$$

Clearly the mean of such Cauchy distributed random variables does not exist.

Exercise 87 Let $N \sim \mathcal{P}(\lambda)$, prove that

$$\phi_N(t) = \mathbb{E}e^{itN} = \exp(\lambda(e^{it} - 1)).$$

Hint. Using exponential expansions yields:

$$\phi_N(t) = \sum_{k=0}^{\infty} \frac{(\lambda e^{it})^k}{k!} e^{-\lambda} = e^{\lambda(e^{it}-1)} \sum_{k=0}^{\infty} \frac{(\lambda e^{it})^k}{k!} e^{-\lambda e^{it}} = \exp(\lambda(e^{it} - 1)).$$

A first easy calculation is left to the reader.

Exercise 88 Let $N \sim \mathcal{P}(\lambda)$, prove that its generating function is for each $z \in \mathbb{C}$:

$$g_N(z) = \mathbb{E}z^N = e^{\lambda(z-1)}.$$

Hint. Standard calculations give:

$$g_N(z) = \sum_{k=0}^{\infty} z^k \mathbb{P}(N = k) = \sum_{k=0}^{\infty} \frac{(\lambda z)^k}{k!} e^{-\lambda}.$$

The result follows from the expression of the exponential series.

Exercise 89 Let $N \sim \mathcal{P}(\lambda)$, prove that:

$$\begin{aligned} \operatorname{Var} N &= \lambda, \\ \mathbb{E}(N - \lambda)^3 &= \lambda, \\ \mathbb{E}(N - \lambda)^4 &= \lambda(1 + 3\lambda). \end{aligned}$$

Remark A.2.7 In the case of Gaussian random variables $Z \sim \mathcal{N}(m, \sigma^2)$ for which centred moments $\mathbb{E}|Z - \mathbb{E}Z|^p = \sigma^p \mathbb{E}|\mathcal{N}(0,1)|^p$ admit the order $(\text{Var } Z)^{p/2}$.

Contrary to the Gaussian case, $\mathbb{E}(N - \mathbb{E}N)^p \sim \mathbb{E}N$ as $\mathbb{E}N = \lambda \downarrow 0$, at least for $p = 2, 3, 4$.

Hint for Exercise 89. Set $z = e^{it}$. From Exercise 88 we derive the expression:

$$\psi_N(t) = \mathbb{E}e^{it(N-\lambda)} = e^{\lambda(e^{it}-1-it)}.$$

Now it is simple to see that $i^p \mathbb{E}(N - \lambda)^p = \psi_N^{(p)}(0)$. The first derivatives are determined through a Taylor expansion around $t = 0$,

$$\psi_N(t) = 1 + at + \frac{b}{2}t^2 + \frac{c}{6}t^3 + \frac{d}{24}t^4 + o(t^4), \qquad t \to 0.$$

But as $t \to 0$,

$$\begin{aligned}
\psi_N(t) &= \exp\left(\lambda\left(\frac{(it)^2}{2} + \frac{(it)^3}{6} + \frac{(it)^4}{24}\right) + o(t^4)\right) \\
&= \exp\left(\lambda\left(-\frac{t^2}{2} - i\frac{t^3}{6} + \frac{t^4}{24}\right) + o(t^4)\right) \\
&= 1 + \lambda\left(-\frac{t^2}{2} - i\frac{t^3}{6} + \frac{t^4}{24}\right) \\
&+ \frac{1}{2}\left(\lambda\left(-\frac{t^2}{2} - i\frac{t^3}{6} + \frac{t^4}{24}\right)\right)^2 + o(t^4) \\
&= 1 - \frac{\lambda}{2}t^2 - i\frac{\lambda}{6}t^3 + t^4\left(\frac{1}{24} + \frac{1}{8}\right) + o(t^4).
\end{aligned}$$

Now the results follow from elementary arithmetic.

Exercise 90 Let P be a unit Poisson process then for each $r \geq 1$, and $a_0 > 0$ there exists a constant $C_r > 0$ such that the function defined by $h_r(a) = \mathbb{E}P^r(a)$ satisfies

$$a \leq h_r(a) \leq a + C_r a^2, \qquad \text{if} \qquad 0 \leq a \leq a_0.$$

Hint for Exercise 90. First, notice that the Laplace transform of $P(a)$ can be written as:

$$\phi_a(t) = \mathbb{E}e^{tP(a))} = \exp(a(e^t - 1)) = 1 + a(e^t - 1) + a^2\lambda_a(t),$$

for a function with non-negative analytic expansion such that if $t \geq 0$ then:

$$\lambda_a(t) = \sum_{k=2}^{\infty} \frac{a^{k-2}(e^t - 1)^k}{k!} \leq h(t) = \frac{1}{a_0^2}(e^{a_0(e^t-1)} - 1 - a_0(e^t - 1)).$$

The coefficients of the analytic expansion of λ_a are uniformly bounded. Hence $\phi_a^{(k)}(0) = \mathbb{E}P^k(a) \le a + C_k a^2$ for each integer k if $0 \le a \le a_0$.

For non-integer p consider the integer $k = [r]$, then the Hölder inequality (Proposition A.2.2) with $\alpha = 1/((k+1)-r)$ and $\beta = 1/(r-k)$ (conjugate exponents) gives,

$$\begin{aligned}\mathbb{E}P^r(a) &= \mathbb{E}(P^{k/\alpha}(a)P^{(k+1)/\beta}(a))\\ &\le (\mathbb{E}P^k(a))^{1/\alpha}(\mathbb{E}P^{k+1}(a))^{1/\beta}\\ &\le a + (C_k \vee C_{k+1})a^2,\end{aligned}$$

for all $0 \le a \le a_0$.

The other inequality follows from the inequality $n \le n^r$, valid for each $n \in \mathbb{N}$. Hence,

$$P(a) \le P^r(a),$$

which implies $h_r(a) \ge a$.

The following very standard models of processes are also used to model integer valued GLM time series as in (7.12).

They are useful processes in all areas of probability theory.

Definition A.2.5 (*Poisson processes*) A (homogeneous) unit Poisson process is a process $(P(\lambda))_{\lambda \ge 0}$ such that:

- $P(\lambda) \sim \mathcal{P}(\lambda)$ follows a Poisson distribution with parameter λ,
- It satisfies moreover that $P(\lambda) - P(\mu)$ is independent of the sigma-field $\sigma(P(\nu); \nu \le \mu)$ if $\lambda > \mu \ge 0$.
- The distribution of $P(\lambda) - P(\mu)$ is $\mathcal{P}(\lambda - \mu)$ for $\lambda > \mu \ge 0$.

As a consequence we easily get properties of some related distributions.

Exercise 91 (*Poisson composite distributions*) Let $b_1 > b_2 > \cdots > b_m > b_{m+1} = 0$ and P be a unit Poisson process.

1. Prove that:

$$D = P(b_1) + \cdots + P(b_m) = \sum_{j=1}^{m} j(P(b_j) - P(b_{j+1}))$$

2. There exist independent random variables

$$X_1 \sim \mathcal{P}(b_1 - b_2), \ldots, X_m \sim \mathcal{P}(b_m - b_{m+1})$$

with:

$$D = X_1 + 2X_2 + \cdots + mX_m.$$

3. Derive:

$$\mathbb{E}D = \sum_{j=1}^{m} j(b_j - b_{j+1}), \quad \mathrm{Var}\, D = \sum_{j=1}^{m} j^2(b_j - b_{j+1}).$$

4. Prove that

$$\mathbb{E}(D - \mathbb{E}D)^4 \le 12 \left(\sum_{j=1}^{m} j b_j \right)^2 + 4(1 + 3b_1) \sum_{j=1}^{m} j^3 b_j.$$

Hint. Only the fourth order moment needs clarifications.

We need the last point in Exercise 89.

The Abel transform of series will be useful:

$$\sum_{j=1}^{m} j^2(b_j - b_{j+1}) = \sum_{j=1}^{m} (j^2 - (j-1)^2) b_j \le \sum_{j=1}^{m} (2j - 1) b_j \le 2 \sum_{j=1}^{m} j b_j.$$

For the last inequality, we begin with:

$$\sum_{j=1}^{m} j^4(b_j - b_{j+1}) = \sum_{j=1}^{m} (j^4 - (j-1)^4) b_j$$
$$\le \sum_{j=1}^{m} (4j^3 - 6j^2 + 4j - 1) b_j \le 4 \sum_{j=1}^{m} j^3 b_j.$$

and recall that $h(u) = u(1 + 3u)$; since the series b_j is non-increasing we obtain $h(b_j - b_{j+1}) \le (b_j - b_{j+1})(1 + 3b_1)$.

The result follows from the Rosenthal inequality of order 4 (see Exercise 7).

The successive bounds hold:

$$\begin{aligned} \mathbb{E}(D - \mathbb{E}D)^4 &\le 3 \left(\sum_{j=1}^{m} j^2(b_j - b_{j+1}) \right)^2 + \sum_{j=1}^{m} j^4 h(b_j - b_{j+1}) \\ &\le 3 \left(\sum_{j=1}^{m} j^2(b_j - b_{j+1}) \right)^2 + (1 + 3b_1) \sum_{j=1}^{m} j^4(b_j - b_{j+1}) \\ &\le \frac{1}{2} \left(\sum_{j=1}^{m} j b_j \right)^2 + 4(1 + 3b_1) \sum_{j=1}^{m} j^3 b_j. \end{aligned}$$

The first terms are the square of variances and the last one results from Exercise 89 and they are sums of moments with order 4.

This ends the proof.

Exercise 92 (*Compound Poisson processes*) Let $V_i \geq 0$ be an iid sequence independent of a unit Poisson process. Set

$$N(t) = \sum_{i=1}^{P(t)} V_i.$$

1. Prove that N admits independent and stationary increments.
2. Prove that $\mathbb{E}N(t) = t\mathbb{E}V$ and $\text{Var}\, N(t) = t\mathbb{E}V^2$.
3. Set $L_Z(\lambda) = \mathbb{E}e^{\lambda Z}$ for the Laplace transform of a real valued random variable (in case it is defined), then

$$L_{N(t)} = L_{P(t)} \circ L_V, \qquad L_{N(t)}(\lambda) = \exp\left(t(L_V(\lambda) - 1)\right).$$

If $\mathbb{P}(V_1 \in \mathbb{N}) = 1$ then $\mathbb{P}(N(t) \in \mathbb{N}) = 1$, $\forall t \geq 0$.

Hint.

1. Let $s > t \geq 0$. First $N(s) - N(t) = \sum_{i=P(t)+1}^{P(s)} V_i$ is by nature independent of $N(t)$. Condition with respect to the process P then $N(s) - N(t)$ admits the distribution of the sum of $P(s) - P(t)$ random variables with the same distribution as V_1; from stationarity of P's increments $P(s) - P(t) \sim P(s-t)$, hence N admits independent increments.
2. Condition with respect to P.
3. Condition again with respect to P.
4. From independence of Z and P, we use Exercise 88.

This allows to derive:

$$L_{M(t)}(\lambda) = \mathbb{E}L_{\mathcal{P}}(tZ)(\lambda) = \mathbb{E}e^{tZ(e^\lambda - \lambda)} = L_Z(t(e^\lambda - \lambda)).$$

This ends the proof.

The following other family of processes admit very different properties.

Exercise 93 (*Mixed Poisson process*) Let $Z \geq 0$ be a random variable independent of the unit Poisson process $(P(t))_{t\geq 0}$, prove that $M(t) = P(tZ)$ is again an integer valued process; compute $L_{M(t)}$.

Note that the above process does not have independent increments since all its increments depend on Z.

Remark A.2.8 (*Simulation*) If a cdf F is one-to-one on its image then for each uniform random variable $U \sim U[0, 1]$ the random variable $X = F^{-1}(U)$ admits the cumulative distribution function F.

This is an easy way to simulate real random variables with marginal distribution.

The same relation holds for more general cases when defining:

$$F^{-1}(t) = \inf\{x \in \mathbb{R} |\ F(x) \geq t\}.$$

Simple examples prove that other possibilities are available:

1. Assume that $X \sim b(p)$ then $F(t) = 1$ for $t \geq p$ then one simulates a $b(p)$-distributed random variable by setting $X' = \mathbb{I}_{\{U \leq p\}}$. Other possibilities are $\mathbb{I}_{\{U<p\}}$, $\mathbb{I}_{\{U \geq 1-p\}}$ and $\mathbb{I}_{\{U>1-p\}}$, since $1 - U$ also admits a $U[0, 1]$-distribution.
2. Analogously to Poisson distributed random variables, any integer (or discrete)-valued random variable may be simulated from a uniform one.
   If a random variable admits the discrete support $\{x_0, x_1, \ldots\} \subset \mathbb{R}$, with:

$$\mathbb{P}(X = x_k) = p_k, \qquad \sum_{k=0}^{\infty} p_k = 1,$$

   then with $q_0 = 0$ and $q_k = p_0 + \cdots + p_{k-1}$ for $k \geq 1$, one may define the following random variable with the same distribution as X by setting:

$$Y = \sum_{k=0}^{\infty} x_k \, \mathbb{I}_{\{U \in [q_k, q_k + p_k[\}} \sim X. \tag{A.4}$$

   This principle also allows to define a random process with integer values from a random process with uniform marginal distributions, see Exercise 49 for a hint to this approach. Such models are proved to exist for example in Eq. (11.2).
3. For $\mathcal{E}(\lambda)$-distributions, $F(t) = 1 - e^{-\lambda t}$, so that

$$F^{-1}(t) = -\ln(1 - t)/\lambda;$$

   again simulations of such exponential random variables give

$$X = -\ln(1 - U)/\lambda,$$

   or more accurately

$$X = -\ln(U)/\lambda.$$

Exercise 94 (*Symmetric Bernoulli process*) Let (U, ζ), be independent random variables with $U \sim \mathcal{U}([0, 1])$ and $\mathbb{P}(\zeta = \pm 1) = \frac{1}{2}$.

Set

$$\gamma(x) = \zeta \, \mathbb{I}_{\{U \leq x\}}.$$

1. Prove that $x \mapsto \gamma(x)$ is a càdlàg function on $x \in [0, 1]$.
2. Determine the distribution of $\gamma(x)$, for $0 \leq x \leq 1$.
3. Prove that $(\gamma(b) - \gamma(a))(\gamma(d) - \gamma(c)) = 0$, a.s., if $a \leq b \leq c \leq d$.
4. Compute the mean and the covariance of the process γ.

Hints.

1. For each real number u the function $x \mapsto \mathbb{I}_{\{u \le x\}}$ is cadlag.
2. Note that $(\gamma(x) = 1) = (\zeta = 1) \cap (U = 1)$.
   This implies $\mathbb{P}(\gamma(x) = \pm 1) = x/2$, for $0 \le x \le 1$ and $\mathbb{P}(\gamma(x) = 0) = 1 - x$.
3. Use the fact that the support of the random variable $(\gamma(b) - \gamma(a))$ is $[a, b]$.
4. $\mathbb{E}\gamma(x) = 0$ and by using the above point, $\mathbb{E}\gamma(x)\gamma(y) = x \wedge y$.

Analogously to Exercise 91, we define sums of such symmetric Bernoulli random variables.

Exercise 95 (*Symmetric Bernoulli composite distributions*) Let $b_1 > b_2 > \cdots > b_m > b_{m+1} = 0$ and let γ be the process defined in Exercise 94.

1. Prove that:

$$D = \sum_{j=1}^{m} j(\gamma(b_j) - \gamma(b_{j+1})).$$

2. Prove that, for $p \in \mathbb{N}^*$:

$$\mathbb{E}D^p = \mathbb{E}\zeta^p \sum_{j=1}^{m} j^p(b_j - b_{j+1}).$$

3. Prove that $\mathbb{E}D^p = 0$ for p odd.
   Moreover, if p is even, then prove:

$$\mathbb{E}D^p = p \sum_{j=1}^{m} j^{p-1} b_j + \sum_{j=1}^{m} Q_p(j) b_j,$$

   with $Q_p(\cdot)$ a polynomial with degree $\le p - 2$.
4. Let $b_j = cj^{-\alpha}$, determine equivalents of the previous moments if $\alpha > 0$ and as $m \to \infty$.

Hint.

1. Already proved in Exercise 91.
2. Use point 3 of Exercise 94 to check that all the rectangular terms in this expression vanish.
3. $\mathbb{E}\zeta^p = 0$ is 0 or 1 according to p's parity. Abel transform allows to conclude.
4. If $b_j = cj^{-\alpha}$, then $b_j - b_{j+1} \sim c\alpha j^{-\alpha-1}$.
   For p even

$$\mathbb{E}D^p \sim \frac{p\alpha}{p - \alpha} m^{p-\alpha}$$

   if $p > \alpha$ and it is bounded otherwise.
   An alternative proof use the previous point.

Analogously:

Exercise 96 (*Symmetric Poisson processes*) Extend the bounds of Exercise 91 in case $P(\cdot)$ is replaced by $\zeta P(\cdot)$ as in Exercises 94 and 95.

Example A.2.3 (*Probability spaces*) An example of a probability space is $\Omega = [0, 1]^{\mathbb{Z}}$ endowed with its product σ-algebra.

This is the smallest sigma-algebra containing cylinder events

$$\prod_{n\in\mathbb{Z}} A_n$$

where A_n is a Borel set of $[0, 1]$, such that $A_n \neq [0, 1]$ for only finitely many indices n.

Then a sequence of random variables X_n is defined as the n-th coordinate function $X_n(\omega) = \omega_n$ for all $\omega = (\omega_n)_{n\in\mathbb{Z}}$.

In this case each of the coordinates X_n admits the uniform distribution μ, the Lebesgue measure on $[0, 1]$.

Let now F be the cumulative distribution function of the law ν of a real valued random variable then setting instead $X_n(\omega) = F^{-1}(\omega_n)$ give

$$\mathbb{P}(X_n \in A) = F(A) = \nu(A) = \mathbb{P}(X \in A).$$

One may assign any distribution to these coordinates.

Exercise 97 (*Hoeffding lemma*)

1. Let $Z \geq 0$ be a (*a.s.*-)non-negative random variable then

$$\mathbb{E}Z = \int_0^\infty \mathbb{P}(Z \geq t)\, dt.$$

2. Let $X, Y \in \mathbb{L}^2$ be two real valued random variables

$$\text{Cov}\,(X, Y) = \int_{-\infty}^{\infty}\int_{-\infty}^{\infty}\Big(\mathbb{P}(X\geq s, Y\geq t)-\mathbb{P}(X\geq s)\mathbb{P}(Y\geq t)\Big)dsdt.$$

Hint.

1. Let $z \geq 0$ then

$$z = \int_0^\infty \mathbb{1}_{\{z\geq t\}}\, dt.$$

Set λ the Lebesgue measure on the line. Without any convergence assumption on this integral, the Fubini–Tonnelli theorem applies to the non-negative function $(t, \omega) \mapsto \mathbb{1}_{(Z(\omega)\geq t)}$. This allows to conclude.

2. First for $X, Y \geq 0$ the same trick as above works and

$$\mathbb{E}XY = \int_0^\infty \int_0^\infty \mathbb{P}(X \geq s, Y \geq t)\, ds\, dt.$$

Write $X = X^+ - X^-$, and $Y = Y^+ - Y^-$ for non-negative random variables $X^\pm, Y^\pm$. The formula holds for each of them and

$$\mathbb{P}(X \geq s) = \begin{cases} \mathbb{P}(X^+ \geq s), & \text{if } \ s \geq 0 \\ 1 - \mathbb{P}(X^- > -t), & \text{if } \ s < 0. \end{cases}$$

Now for an arbitrary couple of real valued random variables Cov (X, Y) can be written as a linear combination of four such integrals with respective coefficients ± 1.

A.3 Normal Distribution

A standard Normal random variable is a real valued random variable such that $N \sim \mathcal{N}(0, 1)$ admits the density

$$\varphi(x) = \frac{1}{\sqrt{2\pi}} e^{-\frac{x^2}{2}},$$

with respect to the Lebesgue measure on $\mathbb{R}$.

Exercise 98 The norming constant, yielding $\int_{\mathbb{R}} \varphi(x)\, dx = 1$, in this Normal density is indeed $\sqrt{2\pi}$.

Proof This is checked through the computation of a square as follows:

$$\begin{aligned} \left(\int_{-\infty}^{\infty} e^{-\frac{x^2}{2}} dx \right)^2 &= \int_{-\infty}^{\infty} \int_{-\infty}^{\infty} e^{-\frac{x^2+y^2}{2}} dx\, dy \\ &= \int_0^{2\pi} d\theta \int_0^{\infty} e^{-\frac{r^2}{2}} r\, dr \\ &= 2\pi. \end{aligned}$$

To this aim, use a change in variables with polar coordinates

$$(r, \theta) \mapsto (x, y) = (r \cos\theta, r \sin\theta), \qquad \mathbb{R}^+ \times [0, 2\pi[\to \mathbb{R}^2.$$

This is a bijective change of variable which is a homeomorphism from each open subset $]a, +\infty \times]a, 2\pi[\subset \mathbb{R}^+ \times [0, 2\pi[$ for an arbitrary $a \in]0, 2\pi[$.

It is easy to check that the Jacobian of the previous function

$$J(r, \theta) = \begin{vmatrix} \cos\theta & -r \sin\theta \\ \sin\theta & r \cos\theta \end{vmatrix}$$

is simply r.

Lemma A.3.1 *The characteristic function of this Normal distribution is*

$$\phi_N(s) = \mathbb{E}e^{isN} = e^{-\frac{s^2}{2}}. \tag{A.5}$$

Proof Indeed the Laplace transform $L_N(z) = \mathbb{E}e^{zN}$ is easy to compute in case $z \in \mathbb{R}$:

$$L_N(z) = \mathbb{E}e^{zN} = \frac{1}{\sqrt{2\pi}} \int_{-\infty}^{\infty} e^{zx-\frac{x^2}{2}}\,dx = \frac{1}{\sqrt{2\pi}} \int_{-\infty}^{\infty} e^{\frac{z^2}{2}-\frac{(x-z)^2}{2}}\,dx = e^{\frac{z^2}{2}}$$

with the binomial formula $(x-z)^2 = x^2 - 2zx + z^2$ and after a change in variable $x \mapsto x - z$.

The application $z \mapsto L_N(z)$ is an entire function over $\mathbb{C}$, indeed:

$$\frac{L_N(z+h) - L_N(z)}{h} = \int_{-\infty}^{\infty} e^{zx} \cdot \frac{e^{hx}-1}{h} \varphi(x)\,dx.$$

The Lebesgue dominated convergence theorem proves that $L'_N(z) = \mathbb{E}(Ne^{zN})$. Use $\left|\frac{e^{hx}-1}{h}\right| \le |x|$ and the integrability of $x \mapsto \psi_z(x) = |x|e^{|x\mathbb{R}z|}\varphi(x)$ for it. The latter relation follows from $\lim_{|x|\to\infty} \psi_z(x)e^{x^2/4} = 0$.

The principle of analytic continuation implies that this formula remains valid for each $z \in \mathbb{C}$, and in particular we obtain

$$\phi_N(s) = L_N(is) = e^{-\frac{s^2}{2}}.$$

Equation (A.5) may also be rewritten:

$$\mathbb{E}e^{zN-\frac{z^2}{2}} = 1, \qquad \forall z \in \mathbb{C}. \tag{A.6}$$

From the analyticity of ϕ_N over the whole complex plane $\mathbb{C}$, the distribution of a Normal random variable is given from its characteristic function.

Definition A.3.1 A random variable Y admits the Gaussian law

$$Y \sim \mathcal{N}(m, \sigma^2),$$

if it can be written $Y = m + \sigma N$ for $m, \sigma \in \mathbb{R}$ and for a Normal random variable N.

The density and the characteristic function of such distributions are derived from linear changes in variable:

$$f_Y(y) = \frac{1}{\sigma\sqrt{2\pi}} e^{-\frac{(y-m)^2}{2\sigma^2}}, \qquad \phi_Y(t) = e^{itm} e^{-\frac{1}{2}\sigma^2 t^2}.$$

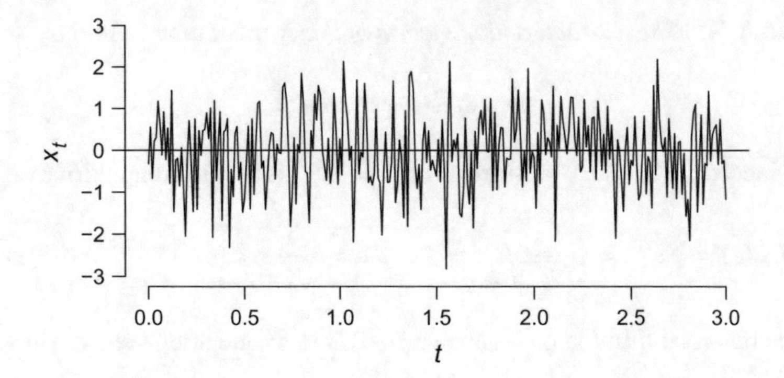

Fig. A.2 Gaussian white noise of variance 1

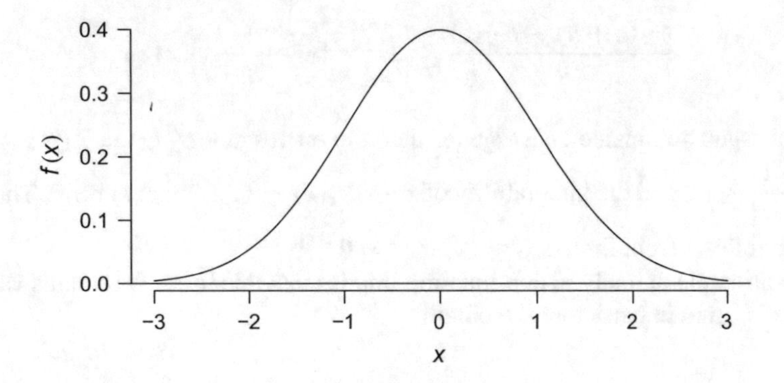

Fig. A.3 Standard normal density

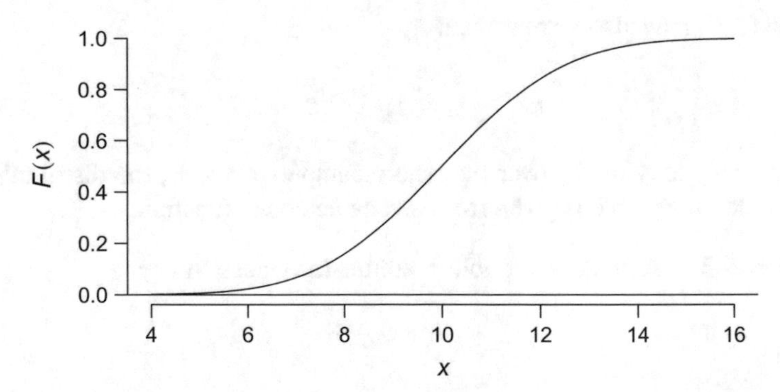

Fig. A.4 Cumulative distribution function of a $\mathcal{N}(10, 2)$

Gaussian samples, Gaussian densities and a Normal distribution function are reproduced in Figs. A.2, A.3, and A.4.

Exercise 99 (*Similarity properties of the Normal law*)

1. An important property is that if random variables $Y_j \sim \mathcal{N}(m_j, \sigma_j^2)$ are independent for $j = 1, 2$, then

$$Y_1 + Y_2 \sim \mathcal{N}(m_1 + m_2, \sigma_1^2 + \sigma_2^2).$$

2. A converse of this result is that if Y_1, Y_2 are independent and have the same distribution μ, if $(Y_1 + Y_2)/\sqrt{2} \sim Y_1 \sim \mu$ admits the same distribution then this distribution μ is centred and Gaussian.

Hints. This property follows from a property of characteristic functions. The characteristic function

$$\kappa(t) = \int_{-\infty}^{\infty} e^{tx} \mu(dx),$$

satisfies

$$\kappa(t) = \kappa^2\left(\frac{t}{\sqrt{2}}\right),$$

from independence. To prove that this characterizes Gaussians, it can be proved that the log-characteristic function is a second degree polynomial.

With this formula, a simple recursion entails that there exists a constant $a \in \mathbb{R}$ such that $\log \kappa(t) = at^2$ for $t = k2^n$ with $k, n \in \mathbb{Z}$. A continuity argument allows to conclude.

A.4 Multivariate Gaussians

Definition A.4.1 A random vector $Y \in \mathbb{R}^k$ is Gaussian if the scalar product $Y \cdot u = Y^t u$ admits a (real valued) Gaussian distribution for each $u \in \mathbb{R}^k$.

We begin with the existence of finite dimensional Gaussian random variables. It is a main step to prove the existence of Gaussian processes.

Lemma A.4.1 *Le Σ be a $k \times k$-symmetric positive matrix and let $m \in \mathbb{R}^d$, then there exists a Gaussian random variable $Y \sim \mathcal{N}_k(m, \Sigma)$.*

Proof of Lemma A.4.1. If Σ is a $k \times k$ symmetric positive matrix, then classically, there exist some orthogonal matrix $P'P = I_d$ and a diagonal matrix with non-negative entries $\lambda_1 \geq \cdots \geq \lambda_k \geq 0$ such that $\Sigma = P'DP$. Then $R = P'\Delta P$ a symmetric positive definite matrix with $R^2 = \Sigma$ when setting Δ the diagonal matrix with entries $\sqrt{\lambda_1} \geq \cdots \geq \sqrt{\lambda_d} \geq 0$.

For $Z = (Z_1, \ldots, Z_k)^t$ independent identically distributed standard Normal random variables and, following the Definition A.4.1, for each $m \in \mathbb{R}^k$:

$$Y = m + RZ \sim \mathcal{N}_k(m, \Sigma).$$

Recall that if Σ is definite then $\lambda_k > 0$ then such a square root is in fact unique; to this aim remark that characteristic spaces coincide.

Some essential features of Gaussian laws follow.

Exercise 100 The law of a Gaussian random variable Y only depends on its mean and on its covariance matrix.

Hint. For

$$u \in \mathbb{R}^k, \qquad \Sigma = \mathbb{E}(Y - \mathbb{E}Y)(Y - \mathbb{E}Y)';$$

we easily check that

$$Y \cdot u \sim \mathcal{N}(\mathbb{E}Y \cdot u, u^t \Sigma u)$$

only depends on u, $\mathbb{E}Y$, and on Σ.

Another way to check this is to compute the characteristic function and to check that it is factorized in case cross covariances vanish. The analyticity of characteristic functions entails they characterize distributions; note that the factorization is equivalent to independence.

Exercise 101 (*Reduction of Gaussian vectors*) Let Y be a Gaussian vector, then prove that $\Sigma = \mathbb{E}(Y - \mathbb{E}Y)(Y - \mathbb{E}Y)^t$, admits a symmetric non-negative square root R such that $\Sigma = R^2$.

Deduce the representation

$$Y = \mathbb{E}Y + RZ,$$

for a random vector Z with iid $\mathcal{N}(0, 1)$-components.

Hint. Σ is non-negative symmetric

$$u'\Sigma u = \mathrm{Var}\,(Y \cdot u) \geq 0, \qquad \forall u \in \mathbb{R}^k.$$

Indeed the above variance is ≥ 0; thus it is diagonalizable in an orthonormal basis thus there exists an orthogonal matrix Ω and a diagonal matrix D with

$$\Sigma = \Omega' D \Omega \quad \text{and} \quad \Omega'\Omega = I_k.$$

Since Σ is non-negative, the matrix D admits non-negative diagonal coefficients (positive if Σ is a definite matrix). The non-negative diagonal matrix Δ with elements the square roots of those of D satisfies $D = \Delta^2$.

Thus

$$R = \Omega^t \Delta \Omega,$$

is a convenient square root (non-negative symmetric) of Σ. This solution may be proved to be unique in case Σ is definite, because eigen-spaces of R and Σ coincide from the fact that those matrices commute.

In this case $Z = R^{-1}(Y - \mathbb{E}Y)$ is a Gaussian vector with orthogonal and Normal $\mathcal{N}(0, 1)$ coordinates. The previous remark proves that these components are independent identically distributed so that $Z \sim \mathcal{N}_k(0, I_k)$.

Exercise 102 (*Density*) Assume that the covariance matrix of Y is invertible, then its density can be written:

$$f_Y(y) = \frac{1}{\sqrt{(2\pi)^k \det \Sigma}} e^{-\frac{1}{2}(y-\mathbb{E}Y)^t \Sigma^{-1}(y-\mathbb{E}Y)}. \tag{A.7}$$

Hint. Use a change in variables. If Σ is invertible then Y admits the suggested density on $\mathbb{R}^k$.

Exercise 103 (*Characteristic function*)

$$\phi_Y(s) = e^{is\cdot\mathbb{E}Y - \frac{1}{2}s^t \Sigma s}.$$

Hint. Even for Σ non-invertible we may write $Y = \mathbb{E}Y + RZ$. For each $s \in \mathbb{R}^k$ we obtain:

$$\begin{aligned}
\phi_Y(s) &= \mathbb{E}e^{is\cdot Y} \\
&= e^{is\cdot\mathbb{E}Y}\mathbb{E}e^{is\cdot RZ} \\
&= e^{it\cdot\mathbb{E}Y}\mathbb{E}e^{iZ\cdot Rs} \\
&= e^{is\cdot\mathbb{E}Y - \frac{1}{2}(Rs)^t(Rs)}
\end{aligned}$$

The expression $\phi_Y(s) = e^{is\cdot\mathbb{E}Y - \frac{1}{2}s^t\Sigma s}$ follows.

Exercise 104 (*Conditioning*) Let $(X, Y) \sim \mathcal{N}_{a+b}(0, \Sigma)$ be a Gaussian vector with covariance matrix written in blocs

$$\begin{pmatrix} I_a & C \\ C' & B \end{pmatrix}$$

for some symmetric positive definite matrix B $(b \times b)$ and a rectangular matrix C with order $a \times b$.

Then:

$$\mathbb{E}(Y|X) = C'X.$$

Hint. Here $Z = Y - C'X$ is orthogonal to X. Hence from the Gaussianity of this vector, they are independent. The result follows.

An important consequence of the previous items is

Exercise 105 For Gaussian vectors, pairwise orthogonality and independence coincide; provide three proofs of the result.

Hint.

1. This results from the fact that pairwise orthogonal Gaussian distributions can be generated from independent Gaussians rvs. Uniqueness implies the conclusion.
2. Alternatively this property may also be derived from the expression of characteristic functions.
3. Finally a proof based upon densities is also straightforward once one knows about the expression of a multivariate Gaussian density.

A.5 γ-Distributions

As an example of the previous sections we introduce another important class of distributions.

Definition A.5.1 The Euler function Γ of the first kind is defined over $]0, +\infty[$ by the relation

$$\Gamma(t) = \int_0^\infty e^{-x} x^{t-1}\, dt.$$

Hints. Let $t \in \mathbb{R}$. The integral $\Gamma(t)$ is that of a positive and continuous function over $]0, +\infty[$.

This is always a convergent integral at infinity but $t > 0$ is necessary to ensure the convergence at the origin.

Integration by parts together with the relation $\frac{d}{dx} x^t = t x^{t-1}$ entails

$$\Gamma(t+1) = \int_0^\infty \frac{d}{dx}\{-e^{-x}\} x^t\, dt = \left[(-e^{-x})x^t\right]_0^\infty - t \int (-e^{-x})x^{t-1}\, dx.$$

Moreover a simple calculation proves that $\Gamma(1) = 1$.

A recursion using the previous identity entails $\Gamma(k) = (k-1)!$ for $k \in \mathbb{N}$:

Lemma A.5.1 *Let $t > 0$ then $\Gamma(t+1) = t\Gamma(t)$ and $\Gamma(k) = (k-1)!$ for each $k \in \mathbb{N}^*$ (with the convention $0! = 1$).*

Definition A.5.2 Set for $b > 0$,

$$c_{a,b} = \frac{b^a}{\Gamma(a)}.$$

For $a, b > 0$, $\gamma(a, b)$ denotes the law with density:

$$f_{a,b}(x) = c_{a,b}\, e^{-bx} x^{a-1}\, \mathbb{I}_{\{x>0\}}.$$

Proof The function $f_{a,b}$ is integrable around infinity in case $b > 0$ and this integral converges at 0 if $a > 0$.

As a density admits the integral 1, we compare both integrals to get:

$$c_{a,b}^{-1} = \int_0^\infty e^{-bx} x^{a-1} dx = b^{-a} \int_0^\infty e^{-y} y^{a-1} dy = b^{-a} \Gamma(a),$$

by using a change of variable $y = bx$. Thus $c_{a,b} = b^a / \Gamma(a)$.

Some simple facts are easily derived:

Lemma A.5.2 *Let $Z \sim \gamma(a, b)$ then for $m > 0$ and $\Re(u) < b$:*

$$\mathbb{E}Z^m = \frac{\Gamma(a+m)}{b^m \Gamma(a)},$$

$$L_{a,b}(u) = \mathbb{E}e^{uZ} = \left(\frac{b}{b-u}\right)^a.$$

Proof

$$\mathbb{E}Z^m = c_{a,b} \int_0^\infty x^m e^{-bx} x^{a-1} dx = \frac{c_{a,b}}{c_{a+m,b}} = \frac{\Gamma(a+m)}{b^m \Gamma(a)}.$$

We compute the Laplace transform $L_{a,b}(u) = \mathbb{E}e^{uZ}$ of Z.
We first assume that $u \in \mathbb{R}$:

$$L_{a,b}(u) = c_{a,b} \int_0^\infty e^{(u-b)x} x^{a-1} \, dx = \frac{c_{a,b}}{c_{a,b-u}} = \left(\frac{b}{b-u}\right)^a.$$

This is an analytic function in case $\Re(u) < b$ since integrals defining $L_{a,b}(u)$ are absolutely convergent because of

$$\left| e^{(u-b)x} x^{a-1} \right| = e^{(\Re u - b)x} x^{a-1}.$$

The same holds for the complex derivative $u e^{(u-b)x} x^{a-1}$. Analytic continuation allows to conclude.

Easy consequences of this lemma follow:

Corollary A.5.1 *Let Z, Z' be two independent random variables with respective distributions $\gamma(a, b)$ and $\gamma(a', b)$, then*

$$Z + Z' \sim \gamma(a + a', b).$$

Proof The previous lemma implies

$$\mathbb{E}e^{u(Z+Z')} = L_{a,b}(u) L_{a',b}(u) = L_{a+a',b}(u)$$

if $\Re u < a \wedge a'$, then the result follows from uniqueness of Laplace transforms in case they are analytic on a domain with a non-empty interior.

We now proceed with an analytic proof of the above result. This proof does not rely on probabilistic concepts of independence or on arguments of complex analysis.

Exercise 106 An alternative proof of Corollary A.5.1.

- Define Euler's function of the second kind for $a, a' > 0$:

$$B(a, a') = \int_0^1 u^{a-1}(1-u)^{a'-1}\,du.$$

(Prove that the above expression is well defined).

- Prove that for $a, a' > 0$:

$$B(a, a') = \frac{\Gamma(a)\Gamma(a')}{\Gamma(a+a')}.$$

- Prove again Corollary A.5.1 without using the notion of Laplace transform and the principle of analytical continuation.

Proof If $a, a' > 0$, the function

$$B(a, a') = \int_0^1 u^{a-1}(1-u)^{a'-1}\,du,$$

is well defined, indeed such integrals converge at origin since $a > 0$ and at point 1, it is due to the fact that $a' > 0$.

Let g be a continuous and bounded function then for such independent $Z \sim \gamma(a, b)$ and $Z' = \gamma(a', b)$ one derives:

$$\begin{aligned}
\mathbb{E}g(Z+Z') &= \int_0^\infty \int_0^\infty g(z+z') f_{a,b}(z) f_{a',b}(z')\,dz dz' \\
&= \int_0^\infty g(u)\,du \int_0^u f_{a,b}(z) f_{a',b}(u-z)\,dz \\
&= c_{a,b}c_{a',b} \int_0^\infty e^{-bu} g(u)\,du \int_0^u z^{a-1}(u-z)^{a'-1}\,dz \\
&= c_{a,b}c_{a',b}\, B(a, a') \int_0^\infty u^{a+a'-1} e^{-bu} g(u)\,du \\
&= \frac{b^{a+a'} B(a, a')}{\Gamma(a)\Gamma(a')} \int_0^\infty u^{a+a'-1} e^{-bu} g(u)\,du
\end{aligned}$$

(with $z = ut$). Then $Z + Z'$ admits a $\gamma(a + a', b)$-distribution.

Now the normalization constant can be written in two different ways which entails:

$$\frac{b^{a+a'} B(a, a')}{\Gamma(a)\Gamma(a')} = \frac{b^{a+a'}}{\Gamma(a+a')},$$

so that $B(a, a') = \dfrac{\Gamma(a)\Gamma(a')}{\Gamma(a + a')}$.

From the above results we obtain:

Exercise 107 The density of the sum S_k of k independent random variables with exponential distribution $\mathcal{E}(\lambda)$ for $\lambda > 0$ is $\gamma(k, \lambda)$.

Hints. For $k = 1$, $S_1 \sim \mathcal{E}(\lambda)$ admits a $\gamma(1, \lambda)$-distribution, then:

$$S_k \sim \gamma(k, \lambda).$$

The addition formula yields the conclusion.

Exercise 108 Define χ^2_k-distribution as the distribution of

$$T_k = N_1^2 + \cdots + N_k^2,$$

for independent and normally distributed $\mathcal{N}(0, 1)$ random variables $N_1, \ldots, N_k$.

Then the law χ^2_k of T_k is $\gamma\Big(\dfrac{k}{2}, \dfrac{1}{2}\Big)$ and $\Gamma\Big(\dfrac{1}{2}\Big) = \sqrt{\pi}$.

Hints. $T_1 = N^2$ is the square of a standard Normal; we compute its density from the expression of $\mathbb{E}g(T_1)$ for each bounded and continuous function $g : \mathbb{R} \to \mathbb{R}$:

$$\begin{aligned}
\mathbb{E}g(T_1) &= \mathbb{E}g(N^2) \\
&= \int_{-\infty}^{\infty} g(x^2)e^{-x^2/2}\frac{dx}{\sqrt{2\pi}} \\
&= 2\int_0^{\infty} g(x^2)e^{-x^2/2}\frac{dx}{\sqrt{2\pi}} \\
&= 2\int_0^{\infty} g(z)\frac{1}{2\sqrt{z}}e^{-z/2}\frac{dz}{\sqrt{2\pi}} \\
&= \int_0^{\infty} g(z)z^{\frac{1}{2}-1}e^{-z/2}\frac{dz}{\sqrt{2\pi}}
\end{aligned}$$

(use the change of variable $z = x^2$ in the above relations).

The density function of T_1's distribution is

$$\frac{1}{\sqrt{2\pi}}z^{\frac{1}{2}-1}e^{-z/2}, \qquad z \geq 0.$$

Up to a constant this is the density $f_{\frac{1}{2},\frac{1}{2}}$ of a $\gamma(\dfrac{1}{2}, \dfrac{1}{2})$-distribution.

Since they are both densities, one infers that $c_{\frac{1}{2},\frac{1}{2}} = 1/\sqrt{2\pi}$ and then $\Gamma\left(\frac{1}{2}\right) = \sqrt{\pi}$. Now the addition formulae allow to conclude for $k > 1$ that

$$T_k \sim \chi_k^2 = \gamma\left(\frac{k}{2}, \frac{1}{2}\right).$$

This completes the proof.

Exercise 109 Let $N \sim \mathcal{N}(0, 1)$ and $m \in \mathbb{N}$. Then

$$\mathbb{E}N^m = \begin{cases} 0, & \text{for } m = 2p+1, \quad \text{an odd number,} \\ \dfrac{(2p)!}{2^p p!}, & \text{for } m = 2p, \quad \text{an even number.} \end{cases}$$

Hints. This follows from Lemma A.5.2 since

$$\mathbb{E}T_1^p = \frac{2^p \Gamma\left(\frac{1}{2}+p\right)}{\Gamma\left(\frac{1}{2}\right)},$$

but this idea needs additional effort. A simpler way to proceed is to use relation (A.5).

Comparing both sides of the expansion of $\mathbb{E}e^{itN} = e^{-t^2/2}$ yields

$$\mathbb{E}e^{itN} = \sum_m \frac{1}{m!}(it)^m \mathbb{E}N^m$$

$$e^{-t^2/2} = \sum_p \frac{1}{p!}\left(-\frac{t^2}{2}\right)^p.$$

Clearly the parity of the characteristic function implies that all odd moments vanish. Now for $m = 2p$, we obtain:

$$\frac{((-t^2)/2)^p}{p!} = \mathbb{E}N^{2p}(-1)^p \frac{t^{2p}}{(2p)!}.$$

The result follows from the above identity.

Appendix B
Convergence and Processes

This appendix is a short introduction to the basic concepts of convergence in a probability space, we refer the reader to Billingsley (1999) for developments and to Jakubowski (1997) for additional extensions.

B.1 Random Processes

Definition B.1.1 A random process is simply a family of random variables $\mathbf{Z} = (Z(t))_{t\in\mathbb{T}}$ with values in a measurable metric complete space E (endowed with its Borel σ-field) for any arbitrary set $\mathbb{T}$. The law of a random process is a distribution on the product space $(E^{\mathbb{T}}, \mathcal{E}^{\otimes\mathbb{T}})$, with $\mathcal{E}^{\otimes\mathbb{T}}$ the σ-algebra generated by cylindric events $\prod_{t\in\mathbb{T}} A_t$ with $A_t \in \mathcal{E}$ for each $t \in \mathbb{T}$ and $A_t = E$ except for finitely many such $t \in \mathbb{T}$.

Remark B.1.1 This σ-algebra $\mathcal{E}^{\otimes\mathbb{T}}$ is the smallest σ-algebra such that $\mathbf{Z}$ is measurable if and only if $Z(t)$ is an E-valued random variable for each $t \in \mathbb{T}$.

Definition B.1.2 Let $Z, \widetilde{Z} : \mathbb{T} \to \mathbb{R}$ be random processes indexed by some arbitrary space $\mathbb{T}$. $\widetilde{Z}$ is a modification of Z in case for all $t \in \mathbb{T}$:

$$\mathbb{P}(Z(t) \neq \widetilde{Z}(t)) = 0.$$

Remark B.1.2 If $\mathbb{T}$ is not denumerable this does not mean that the random variables $\widetilde{\mathbf{Z}} = (\widetilde{Z}(t))_{t\in\mathbb{T}}$ and $\mathbf{Z} = (Z(t))_{t\in\mathbb{T}}$ are a.s. equal as random variables in the product space $\mathbb{R}^{\mathbb{T}}$ equipped with its Borel σ-field $\mathcal{B}(\mathbb{R}^{\mathbb{T}})$. This σ-algebra is again generated by cylindric events, $\prod_{t\in\mathbb{T}} A_t$ with $A_t \in \mathcal{B}(\mathbb{R})$ for each $t \in \mathbb{T}$ and $A_t = \mathbb{R}$ except for finitely many such $t \in \mathbb{T}$. But this is the case under a.s. continuity of both random processes.

P. Doukhan, *Stochastic Models for Time Series*, Mathématiques et Applications 80,
https://doi.org/10.1007/978-3-319-76938-7

A simple example showing the difference between both notions needs non-denumerable sets $\mathbb{T}$. We set $\mathbb{T} = [0, 1]$ for simplicity:

Exercise 110 Set $Z(t) = 0$ for each t and $\widetilde{Z}(t) = \mathbb{I}_{\{U \le t\}}$.
These processes are modifications of each other but

$$\mathbb{P}\Big(\widetilde{Z}(t) = Z(t), \quad \forall t \in [0, 1]\Big) = 0.$$

Use this example to derive that:

Exercise 111 $\mathcal{C}[0, 1] \notin \mathcal{B}(\mathbb{R}^{[0,1]})$.

The Kolmogorov consistency theorem entails the existence of processes on general spaces:

Theorem B.1.1 *One may define a distribution on a product set $E^{\mathbb{T}}$ equipped with the product corresponding σ-algebra in case:*

- *distribution projections exist on each finite subsets $F \subset \mathbb{T}$, denote them P_F, then P_F is a distribution on the measurable set $(E^F, \mathcal{B}(E^F))$;*
- *these finite distributions are coherent in the sense that for $F' \subset F$, the projections satisfy*

$$P_F \circ \pi^{-1}_{F,F'} = P_{F'}$$

where $\pi_{F,F'} : E^F \to E^{F'}$ denotes the projection.

B.2 Convergence in Distribution

We consider a sequence of random variables X_n and a random variable X with values in an arbitrary complete separable metric space (E, d).

Definition B.2.1 The sequence X_n converges in distribution to X, which we denote

$$X_n \to^{\mathcal{L}}_{n\to\infty} X,$$

if

$$\mathbb{E}g(X_n) \to_{n\to\infty} \mathbb{E}g(X), \qquad \forall g \in \mathcal{C}(E).$$

This definition does not depend on the random variables but only on their distribution and $\mathbb{P}_{X_n} \to \mathbb{P}_X$; we really define the convergence of probability measures on a metric space.

Example B.2.1 The first example of a complete metric space used for functional analysis is $\mathcal{C}[0, 1]$ the space of continuous functions $[0, 1] \to \mathbb{R}$, endowed with the norm: $\|f\|_\infty = \sup_{0\le t\le 1} |f(t)|$.

This space is also separable in the sense that there exists a denumerable dense subset of $\mathcal{C}[0, 1]$, e.g. the set of polynomials with rational coefficients.

The following lemma in Chentsov (1956), is usually attributed to the two authors Andrei Kolmogorov and Nikolai N. Chentsov, because of further extensions (see Billingsley 1999 and van der Vaart and Wellner 1998):

Lemma B.2.1 (Chentsov lemma) *If a random process* $Z : [0, 1] \to \mathbb{R}$ *satisfies*

$$\mathbb{E}|Z(t) - Z(s)|^p \leq C|t - s|^a,$$

for some $a > 1$ *then there exists a modification* $\widetilde{Z}$ *of* Z *such that the trajectories of* $\widetilde{Z}$ *are almost surely continuous.*

Moreover the sequence of processes Z_n *is tight in* $\mathcal{C}[0, 1]$*, if for some* $a > 1$ *and for all* $s, t \in [0, 1]$*, the following inequality holds:*

$$\mathbb{E}|Z_n(t) - Z_n(s)|^p \leq C|t - s|^a.$$

Remark B.2.1 In the first point of Lemma B.2.1, the trajectories of $\widetilde{Z}$ are even h-Hölder for each $0 < h < 1 \wedge (a/p)$; Billingsley (1999) and van der Vaart and Wellner (1998) provide more complete statements.

Another important metric space follows.

Definition B.2.2 The Skorohod space $\mathcal{D}[0, 1]$ is the space of the functions $[0, 1] \to \mathbb{R}$, continuous from the right and admitting a limit on the left at each point $t \in [0, 1]$. For short they are called cadlag functions.

Example B.2.2 Such cadlag functions are:

- Continuous functions are cadlag, $\mathcal{C}[0, 1] \subset \mathcal{D}[0, 1]$.
- Indicators are also cadlag, set: $x \mapsto g_t(x) = \mathbb{I}_{\{x \leq t\}}$ for each $t \in [0, 1]$.
- Combinations of previous examples still get the same properties from the classical operative properties of right limits and left continuity.

The metric $d(f, g) = \|f - g\|_\infty = \sup_t |f(t) - g(t)|$ is natural on the space $\mathcal{C}[0, 1]$ of continuous real valued functions on the interval.

Exercise 112 The indicator function $g_{\frac{1}{2}}$ may be *approximated* by a sequence of piecewise affine functions f_n with Lip $f_n = n$ and $f_n(x) = \mathbb{I}_{\{x \leq \frac{1}{2}\}}$ for $|x - \frac{1}{2}| \geq \frac{1}{n}$ but this sequence is not d-Cauchy. Deduce that $\mathcal{D}[0, 1]$ is not separable with the metric d.

Hints. If $\lim_n d(f_n, g_{\frac{1}{2}}) = 0$ then f_n should also have a jump at $\frac{1}{2}$ for large values of n. $\mathcal{D}[0, 1]$ is not separable with the metric d since $d(g_s, g_t) = 1$ if and only if $s \neq t$.

The non-denumerable set $\{g_s/\ s \in [0, 1]\}$ is composed of elements pairwise distant of 1, which allows to conclude.

Remark B.2.2 (*Skorohod metric*) Let $\mathcal{H}$ be the set of monotonic homeomorphisms[4] $\lambda : [0, 1] \to [0, 1]$, then a reasonable metric on $\mathcal{D}[0, 1]$ is

$$\delta(f, g) = \inf_{\lambda \in \mathcal{H}} \left\{ d(f \circ \lambda, g) + \sup_{t \in [0,1]} |\lambda(t) - t| \right\}.$$

This metric makes $\mathcal{D}[0, 1]$ separable but it is not complete.[5] It is simple to prove that $\lim_n \delta(g_t, g_{t+\frac{1}{n}}) = 0$.

Thus $\delta \le d$, and for example the function $f \mapsto \sup_{0 \le t \le 1} f(t)$ is a continuous function on this space $(\mathcal{D}[0, 1], \delta)$.

A criterion for the convergence[6] of the empirical distribution function

$$Z_n(t) = \frac{1}{\sqrt{n}}(F_n(t) - t),$$

of a stationary sequence with uniform marginal distribution is:

- Let $d \in \mathbb{N}^*$. For each d-tuple $t_1, \ldots, t_d \in [0, 1]$, the sequence of random vectors $(Z_n(t_1), \ldots, Z_n(t_d))$ converges in distribution to some Gaussian random variable in $\mathbb{R}^d$.
- There exist constants $a, b, p > 1$ and $C > 0$ such that for each $s, t \in [0, 1]$

$$\mathbb{E}|Z_n(t) - Z_n(s)|^p \le C\left(|t - s|^a + n^{-b}\right)$$

(see e.g. Dedecker et al. 2007).

Billingsley (1999) and Jakubowski (1997) developed the convergence in this space.

Remark B.2.3 Anyway one interesting feature is that the J_1-convergence of two sequences of processes $Z_n^+ = (Z_n^+(t))_{0 \le t \le 1}$ and $Z_n^- = (Z_n^-(t))_{0 \le t \le 1}$. Set $Z_n(t) = Z_n^+(t) + Z_n^-(t)$ and $Z(t) = Z^+(t) + Z^-(t)$. If:

- $Z_n^+ \xrightarrow{\mathcal{L}}_{n\to\infty} Z^+, \quad Z_n^- \xrightarrow{\mathcal{L}}_{n\to\infty} Z^-$, in the J_1-Skorohod topology on $\mathcal{D}[0, 1]$,
- the finite dimensional distributions $(Z_n(t_1), \ldots, Z_n(t_k))$ converge in law to $(Z(t_1), \ldots, Z(t_k))$ for all $k \ge 1$ and $t_1, \ldots, t_k \in [0, 1]$, denoted by $Z_n \to_{fdd} Z^+ + Z^-$,
- the jumps of the limits Z^+ and Z^- are disjoint.

Then $Z_n \xrightarrow{\mathcal{L}}_{n\to\infty} Z^+ + Z^-$, in the J_1-topology.

This property is evident in case the limits are continuous.

---

[4] I.e. bijective continuous functions with a continuous inverse.

[5] See Jakubowski (1997). It is confusing because (Billingsley 1999) modified Skorokhod's metric and proved that this modification is complete and separable.

[6] Note that this result extends of the Chentsov lemma B.2.1.

From now on, we shall restrict to the case $E = \mathbb{R}^d$. In this case,

Lemma B.2.2 (Tightness) *Let X be a rv on $\mathbb{R}^d$. For each $\epsilon > 0$ there exists a compact subset of E such that $\mathbb{P}(X \notin K) < \epsilon$.*

Proof Note that $\Omega = \bigcup_{n=1}^{\infty} A_n$ with $A_n = (|X| \leq n)$. Hence from the sequential continuity of the probability $\mathbb{P}$ there exists n such that $\mathbb{P}(A_n^c) < \epsilon$.

The closed ball with radius n is now a convenient choice $K = B(0, n)$.

Remark B.2.4 This result allows to **restrict to a compact set**. It is easy to prove that the previous convergence holds in case the class of continuous and bounded test functions is replaced by a smaller class of functions.

For example:

- The class of uniformly continuous and bounded functions.[7]
- The class of functions C_b^3 with third order continuous and bounded partial derivatives (see Exercise 113 below).
- The sequence X_n converges in distribution to X if $\phi_{X_n}(t) \to \phi_X(t)$ for each $t \in \mathbb{R}^d$. Indeed, the Stone-Weierstrass theorem asserts the density of trigonometric polynomials on the space $C(K)$ of continuous real valued functions on a compact $K \subset \mathbb{R}^d$, equipped with the uniform norm $\|g\|_K = \sup_{x \in K} |g(x)|$. The Exercise 9 presents the special case $K = [0, 1]$.
- If a sequence of characteristic functions converges uniformly on a neighbourhood of 0 then its limit is also the characteristic function of a law μ (Paul Lévy).

Exercise 113 The convergence in distribution $Z_n \to Z$ of a sequence of real valued random variables holds if $\lim_{n\to\infty} \mathbb{E}g(Z_n) = \mathbb{E}g(Z)$ for each function $g : \mathbb{R} \to \mathbb{R}$ in C_b^3 with third order continuous and bounded partial derivatives.

Hint. From a convolution approximation with a bounded and indefinitely differentiable function with integral 1 ϕ, $f_\epsilon = f \star \phi_\epsilon$ converges uniformly over compact subsets to f as $\epsilon \downarrow 0$, if one sets $\phi_\epsilon(u) = \frac{1}{\epsilon}\phi\left(\frac{u}{\epsilon}\right)$.

Now convolution inherits of ϕ's regularity. Indeed the Lebesgue dominated convergence applies to prove that e.g.

$$f_\epsilon'(u) = \lim_{h\to 0} \frac{1}{h}(f_\epsilon(u+h) - f_\epsilon(u)) = f \star \phi_\epsilon'(u).$$

The result follows.

B.3 Convergence in Probability

From now, on we shall consider **pathwise convergence** only.

[7]The restriction of a continuous function over a compact set is uniformly continuous. Indeed, from the Heine theorem 2.2.1, a continuous over a compact set is uniformly continuous.

Definition B.3.1 The sequence X_n converges in probability to X, which we denote

$$X_n \to^{\mathbb{P}}_{n\to\infty} X,$$

if, for each $\epsilon > 0$:

$$\lim_{n\to\infty} P(|X_n - X| \geq \epsilon) = 0.$$

Lemma B.3.1 *If a real valued sequence of random variables X_n converges in probability to X, then it converges in distribution.*

Proof Assume that convergence in probability holds then from Lemma B.2.2 we may assume that g is uniformly continuous in the definition of convergence in distribution.

Let $\epsilon > 0$, we set $A = (|X_n - X| \geq \epsilon)$. Then:

$$\begin{aligned}|\mathbb{E}(g(X_n) - g(X))| &= \left|\mathbb{E}(g(X_n) - g(X))\, \mathbb{I}_A + \mathbb{E}(g(X_n) - g(X))\, \mathbb{I}_{A^c}\right| \\ &\leq 2\|g\|_\infty \mathbb{P}(A_n) + \sup_{|x-y|<\epsilon} |g(x) - g(y)|.\end{aligned}$$

Uniform continuity of g yields convergence in law.

An alternative proof makes use of Lévy's theorem, see Remark B.2.4 for details.

Definition B.3.2 If $\mathbb{E}|X_n - X|^p \to_{n\to\infty} 0$ we say that the sequence X_n converges to X in $\mathbb{L}^p$.

Remark B.3.1 (*Relations between convergences*)

- Convergence in probability implies convergence in distribution, see Lemma B.3.1.
- Convergence in distribution does not imply convergence in probability.
  A dyadic scheme allows to write $(0, 1]$ as the union of the 2^n disjoint intervals

  $$I_{j,n} =]j2^{-n}, (j+1)2^{-n}], \quad (0 \leq j < 2^n),$$

  with the same measure 2^{-n}.
  It is possible to write $[0, 1] = A_n \bigcup B_n$ where both sets admit the measure $\lambda(A_n) = \lambda(B_n) = \frac{1}{2}$, by setting e.g.

  $$A_n = \bigcup_{j=1}^{2^{n-1}} I_{2j,n}, \qquad B_n = \bigcup_{j=1}^{2^{n-1}} I_{2j-1,n}.$$

  On the probability space $((0, 1], \mathcal{B}((0, 1], \lambda)$, the sequence $X_n = \mathbb{I}_{A_n}$ follows the same Bernoulli distribution $b(\frac{1}{2})$, it converges in distribution to X_0.
  Now the sequence X_n does not converge in probability since

  $$\lambda\left(A_n \cap \left[0, \frac{1}{2}\right]\right) = \frac{1}{4} < \frac{1}{2} = \lambda(A_0).$$

Indeed $\mathbb{P}(X_n < \frac{1}{2}) = \frac{1}{4}$ cannot converge to $\frac{1}{2}$, hence no subsequence of X_n may converge in probability to X_0.

- From the Markov inequality applied to $V = |X_n - X|^p$ it is immediate that $\mathbb{L}^p$-convergence implies convergence in probability.
- However if the random variable Z satisfies $\mathbb{E}|Z|^p = \infty$ and $\mathbb{E}|Z|^q < \infty$ for each $q < p$ then the sequence $X_n = Z/n$ converges to $X = 0$ in probability but not in $\mathbb{L}^p$.

Indeed the Markov inequality implies

$$\mathbb{P}(|X_n| > \epsilon) \leq \frac{\mathbb{E}|Z|^q}{n^q \epsilon^q} \to_{n\to\infty} 0,$$

for each $\epsilon > 0$ in case $q \in (0, p[$.
As an example think of Z with a Cauchy distribution and $p = 1$.

B.4 Almost-Sure Convergence

Definition B.4.1 The sequence X_n converges almost surely to X, which we denote

$$X_n \to^{a.s}_{n\to\infty} X,$$

if there exists an event A with $\mathbb{P}(A) = 0$ such that for each $\omega \notin A$

$$\lim_{n\to\infty} X_n(\omega) = X(\omega).$$

Again the almost-sure (a.s.) convergence implies the convergence in probability.

Definition B.4.2 (*Limit superior*) For a sequence of events $(B_n)_{n\geq 0}$, set

$$\overline{\lim_{n\to\infty}} B_n = \bigcap_{n=0}^{\infty} \bigcup_{k=n}^{\infty} B_k.$$

Remark B.4.1 Note that

$$A_n = \bigcup_{k\geq n} B_k, \quad n = 1, 2, \ldots$$

is a decreasing sequence of events.

Lemma B.4.1 (Borel–Cantelli) *If* $(B_n)_{n\in\mathbb{N}}$ *is a sequence of events such that* $\sum_{n=0}^{\infty}\mathbb{P}(B_n)<\infty$ *then*

$$\mathbb{P}(\overline{\lim_{n\to\infty}} B_n)=0.$$

Exercise 114 If $X_n \to X$ in probability then some subsequence of X_n also converges a.s.

Hint. From $\lim_{n\to\infty}\mathbb{P}(Z_n>1)=0$ with $Z_n=|X_n-X|$, it is possible to extract a subsequence $\phi(m)$ such that $\mathbb{P}(Z_{\phi(m)}>1)\le 1/m^2$:

$$\sum_{m=1}^{\infty}\mathbb{P}(Z_{\phi(m)}>1)<\infty.$$

The result now follows from the Borel–Cantelli lemma B.4.1.

Exercise 115 Exhibit a sequence of random variable converging to 0 in probability but without any a.s. convergent subsequence.

Hint. In Remark B.3.1 we use a dyadic scheme $(I_{j,n})_{0\le j<2^n}$ for $n=1,2,3,\ldots$ hence the sequence $X_n=\mathbb{1}_{A_n}$ does not admit any a.s. convergent subsequence.

B.5 Basic Notations in Statistics

Statistical models are the initial objects in a statistical setting. They are defined from the previous probability framework:

Definition B.5.1 Consider an arbitrary parameter set Θ. Let $(\Omega,\mathcal{A})$ be a measurable space and $(\mathbb{P}_\theta)_{\theta\in\Theta}$ then the triplet $(\Omega,\mathcal{A},(\mathbb{P}_\theta)_{\theta\in\Theta})$ is a statistical model.

A statistic is a measurable function $T:(\Omega,\mathcal{A})\to(E,\mathcal{E})$ on some measurable space $(E,\mathcal{E})$.

Remark B.5.1 The parameter θ is unknown and getting informations concerning it is the aim of statistics. If the value of the parameter is known $\theta=\theta_0$, then the statistical setting turns back to the probability setting and for any event $A\in\mathcal{A}$ and any statistic with values in a vector space $(E,\mathcal{E})$ (say a Banach space for simplicity), then one denotes by $\mathbb{P}_{\theta_0}(A)$ the probability of occurrence of an event and

$$\mathbb{E}_{\theta_0}T=\int_\Omega T(\omega)\mathbb{P}_{\theta_0}(d\omega).$$

Convergence in distribution is also defined in the underlying probability space once the value of the parameter is known.

Remark B.5.2 Parametric settings are associated with $\Theta \subset \mathbb{R}^d$ for some $d \in \mathbb{N}^*$. Real valued statistics are associated with $(E, \mathcal{E}) = (\mathbb{R}, \mathcal{B}(\mathbb{R}))$ but function spaces may also be considered. Finally the parameter θ is often so obvious that it is not even mentioned as an index. Definitely it may be the distribution of a time series, or a marginal probability distribution, or even a probability density of any real parameter or a regression function.

Definition B.5.2 An estimator $\widehat{\theta}$ of a parameter $\theta \in \Theta$ is unbiased in case

$$\mathbb{E}_\theta \widehat{\theta} = \theta, \qquad \forall \theta \in \Theta.$$

More generally

Definition B.5.3 Let $g : \Theta \to E$ be an arbitrary function, then an estimator of $g(\theta)$ is an arbitrary statistic with values in the measurable space $(E, \mathcal{E})$. The estimator of the parameter $g(\theta)$ is an unbiased statistic in case

$$\mathbb{E}_\theta T = g(\theta), \qquad \forall \theta \in \Theta.$$

Definition B.5.4 Let $(T_n)_{n \geq 0}$ be a sequence of statistics with values in a measured metric space $(E, \mathcal{E})$. The various notions of convergence are introduced conditionally with respect to the value of the parameter θ.

Associated convergences are usually called consistences and if the true value of the parameter is θ_0.

- Consistence in probability holds if

$$\lim_{n \to \infty} d(T, g(\theta_0)) = 0, \text{ in } \mathbb{P}_{\theta_0}\text{-probability}, \qquad \forall \theta_0 \in \Theta.$$

- Almost-sure consistence holds if

$$\lim_{n \to \infty} d(T, g(\theta_0)) = 0, \ \mathbb{P}_{\theta_0}\text{-a.s.}, \qquad \forall \theta_0 \in \Theta.$$

- Consistence in $\mathbb{L}^m$ holds if

$$\lim_{n \to \infty} \mathbb{E}_{\theta_0}\big(d^m(T, g(\theta_0))\big) = 0, \qquad \forall \theta_0 \in \Theta.$$

B.6 Basic Notations for Martingales

The notion of martingale is an essential tool to derive limit theorems. Many textbooks consider this topic, we refer the reader to Hall and Heyde (1980) for a thorough presentation, a nice volume on the same topic is Duflo (1996). We simply give some standard basic facts below.

The main attractive feature of martingales is their extremal properties, as e.g. Proposition B.6.1.

It allows to derive strong laws of large numbers such as Theorem B.6.1 without using the Borel–Cantelli lemma B.4.1.

Definition B.6.1 Let $(\Omega, \mathcal{A}, \mathbb{P})$ be a probability space and $\mathbf{F} = (\mathcal{F}_n)_{n\in\mathbb{N}}$ be a filtration (monotonic sequence of sub-σ-fields of $\mathcal{A}$).
If $(X_n)_{n\in\mathbb{N}}$ is an $\mathbf{F}$-adapted sequence of real valued random variables, then:

- $(X_n)_{n\in\mathbb{N}}$ is a *super-martingale* if for each $n \in \mathbb{N}$,

$$\mathbb{E}(X_n \vee 0) < \infty \text{ and } \mathbb{E}^{\mathcal{F}_n} X_{n+1} \geq X_n,$$

- $(X_n)_{n\in\mathbb{N}}$, is a *sub-martingale* if for each $n \in \mathbb{N}$,

$$\mathbb{E}(X_n \wedge 0) > -\infty \text{ and } \mathbb{E}^{\mathcal{F}_n} X_{n+1} \leq X_n,$$

- $(X_n)_{n\in\mathbb{N}}$, is a *martingale* if for each $n \in \mathbb{N}$,

$$\mathbb{E}|X_n| < \infty \text{ and } \mathbb{E}^{\mathcal{F}_n} X_{n+1} = X_n.$$

Lemma B.6.1 *Let $p \geq 1$.*

- *If $(X_n)_{n\in\mathbb{N}}$, is an $\mathbf{F}$-sub-martingale then the sequence $(X_n \vee a)_{n\in\mathbb{Z}}$, is an $\mathbf{F}$-sub-martingale for each $a \in \mathbb{R}$. Moreover the family $(X_n \vee a)_{n\leq N}$ is uniformly integrable for each $N \in \mathbb{Z}$.*
- *If $(X_n)_{n\in\mathbb{N}}$, is an $\mathbf{F}$-martingale and if $\mathbb{E}|X_n|^p < \infty$ for each $n \geq 0$, then $(|X_n|^p)_{n\in\mathbb{Z}}$ is an $\mathbf{F}$-sub-martingale.*

Proposition B.6.1 *Let $(X_n)_{n\in\mathbb{Z}}$ be an $\mathbf{F}$-sub-martingale, then:*

$$c \cdot \mathbb{P}(\sup_n X_n > c) \leq \sup_n \mathbb{E}X_n^+. \tag{B.1}$$

Hence, $\mathbb{P}$-a.s., $\sup_n X_n < \infty$, if $\sup_n \mathbb{E}X_n^+ < \infty$.

Theorem B.6.1 *Let $(X_n)_{n\geq 0}$ be an $\mathbf{F}$-martingale. The following conditions are equivalent:*

1. *the sequence X_n converges in $\mathbb{L}^1$ as $n \to \infty$,*
2. *there exists a random variable $X \in \mathbb{L}^1$ such that $X_n = \mathbb{E}^{\mathcal{F}_n} X$ for each n,*
3. *the sequence X_n is uniformly integrable.*

Then the convergence $X_n \to X$ also holds a.s. and $X_n = \mathbb{E}^{\mathcal{F}_n} X$.

Remark B.6.1 A simple sufficient assumption for the above condition 3. to hold is $\sup_n \mathbb{E}|X_n|^p < \infty$ for some $p > 1$; it also implies the convergence $X_n \to X$ in $\mathbb{L}^p$.

Definition B.6.2 The $\mathbf{F}$-adapted sequence of integrable random variables $(\Delta_n)_{n\geq 1}$ is a sequence of martingale increments if $\mathbb{E}^{\mathcal{F}_n}\Delta_{n+1} = 0$ for each $n \geq 0$.

Let $\mathbf{F} = (\mathcal{F}_n)_{n\geq 0}$ be a filtration indexed by $\mathbb{N}$.

- Any $\mathbf{F}$-martingale $(X_n)_{n\in\mathbb{N}}$ is given by an $\mathbf{F}$-adapted integrable sequence $(\Delta X_n)_{n\in\mathbb{N}^*}$ with $\mathbb{E}^{\mathcal{F}_n}\Delta X_{n+1} = 0$ for all $n \geq 0$, and by a random variable $\mathcal{F}_0$-measurable X_0, through the relation

$$X_n = X_0 + \sum_{k=1}^{n} \Delta X_k.$$

- Conversely $\Delta X_n = X_n - X_{n-1}$ define the increments of a given martingale.

Corollary B.6.1 *Let $(\Delta_n)_{n\geq 1}$ be martingale increments. A martingale (X_n) is defined by $X_0 = 0$, and $X_n = \Delta_1 + \cdots + \Delta_n$, if $n \geq 1$.*

The convergence of the numerical series

$$\sum_{n=1}^{\infty} \mathbb{E}\Delta_n^2$$

implies both the a.s. and the $\mathbb{L}^2$-convergence of the martingale (X_n).

Proof If $k < l$, then $\mathbb{E}\Delta_k\Delta_l = \mathbb{E}\mathbb{E}^{\mathcal{F}_k}\Delta_l = 0$; thus

$$\mathbb{E}X_n^2 = \sum_{k=1}^{n} \mathbb{E}\Delta_k^2.$$

Theorem B.6.1 allows to conclude.

Definition B.6.3 Let $(\Delta_n)_{n\geq 1}$ be $\mathcal{F}$-martingale increments with integrable squares. The compensator of the martingale

$$X_n = \Delta_1 + \cdots + \Delta_n, \quad \forall n \geq 1, \quad \text{and} \quad X_0 = 0,$$

is the process

$$< X >_n = \sum_{j=1}^{n} \mathbb{E}(\Delta_j^2 | \mathcal{F}_{j-1}).$$

Notice that $X_n^2 - < X >_n$ is again an $\mathbf{F}$-martingale.

The following useful Martingale version of the Lindeberg Lemma 2.1.1 is proved in Hall and Heyde (1980), Corollary 3.1, p. 58:

Theorem B.6.2 (Hall and Heyde 1980) *Let $\mathbf{F}_n = (\mathcal{F}_{n,i})_{i\geq 0}$ be a sequence of (nested) filtrations, with $\mathcal{F}_{n+1,i} \subset \mathcal{F}_{n,i}$ for each $i, n \geq 0$.*

Let $(\Delta_{n,i})_{i\geq 0}$ *be a sequence of square integrable* $\mathbf{F}_n$*-martingale increments, define as above a sequence of martingales* $(X_{n,j})_{j\geq 0}$ *as well as their compensators* $(< X_n >_j)_{j\geq 0}$.

Then:

$$\frac{X_{n,n}}{\sqrt{< X_n >_n}} \xrightarrow{\mathcal{L}}_{n\to\infty} \mathcal{N}(0,1),$$

if moreover the following Lindeberg condition holds:

$$\sum_{j=1}^{n} \mathbb{E}\Big(\Delta_{n,j}^2 \, I\!I_{\{|\Delta_{n,j}|\geq\epsilon\}}\Big) \Big| \mathcal{F}_{n,j-1}\Big) \to_{n\to\infty} 0, \qquad \forall \epsilon > 0.$$

Appendix C
R Scripts Used for the Figures

We[8] use the open source software R.

R Core Team (2017). R: A language and environment for statistical computing. R Foundation for Statistical Computing, Vienna, Austria. URL https ://www.R-project.org/.

C.1 Chapter 2

Script C.1 R script producing Figure 2.1

```
normalBinApprox <- function(n, p) {
  ##
  # Accuracy of Gaussian approximation for binomials.
  ##
  # Standardised binomial support
  #
  xk <- (0:n - n*p)/sqrt(n*p*(1 - p))
  #
  # Get the probabilities and adjust the heights
  #
  dist.binom <- dbinom(0:n, n, p)
  delta <- 1/sqrt(n*p*(1 - p))
  result <- dist.binom/delta
  #
  # N(0,1) density function
  #
  x <- seq(-5, 5, 0.01)
  y <- dnorm(x)
  #
  # Plot the hbar chart
  #
  plot(
```

[8]Thanks to Alain Latour, Grenoble.

P. Doukhan, *Stochastic Models for Time Series*, Mathématiques et Applications 80,
https://doi.org/10.1007/978-3-319-76938-7

```
23      xk,
24      result,
25      frame = FALSE,
26      las = 1,
27      type = "h",
28      xlab = expression(italic(x)),
29      ylab = "Density",
30      xlim = c(-4, 4),
31      ylim = c(0, max(y, result))
32  )
33    #
34    # Plot the normal density
35    #
36    lines(x, y)
37    abline(h = 0)
38  }
39  normalBinApprox(10, 3/10)
40  normalBinApprox(100, 3/10)
```

C.2 Chapter 3

Script C.2 R script producing Figure 3.1

```
1  set.seed(101)
2  #
3  # Simulate N tosses of a fair coin
4  #
5  N <- 1000
6  p <- 0.5
7  totn <- 1:N
8  #
9  # Simulate the Bernoulli deviates and estimate p
10 #
11 pn <- cumsum(rbinom(N, 1, p))/totn
12 #
13 # Plot the graphic
14 #
15 plot(
16   pn,
17   type = "l",
18   frame = FALSE,
19   las = 1,
20   ylim = c(0, 1),
21   xlab = expression(italic(n)),
22   ylab = expression(italic(Y/n))
23 )
```

Script C.3 R script producing Figure 3.2

```
1  #
2  data(mtcars)
3  x <- mtcars$mpg
4  #
```

```
5  # Get the empirical cumulative distribution function
6  #
7  rep.val <- ecdf(x)
8  #
9  # and plot it
10 #
11 plot(
12   rep.val,
13   cex = 0.4,
14   main = "",
15   frame = FALSE,
16   las = 1,
17   verticals = TRUE,
18   xlab = expression(italic(x)),
19   ylab = expression(italic(hat(F)(x)))
20 )
21 #
```

Script C.4 R script producing Figure 3.3

```
1  #
2  hist(
3    x,
4    main = "",
5    xlab = expression(italic(x)),
6    freq = FALSE,
7    las = 1,
8  )
9  #
10 # Get the kernel density estimate and plot it on the ↘
   →same graph
11 #
12 d <- density(x)
13 lines(d)
14 #
```

C.3 Chapter 4

Script C.5 R script producing Figure 4.1

```
1  #
2  # Nile flow
3  #
4  # From the "datasets" R package.
5  # R Core Team (2017). R: A language and environment for ↘
   →statistical computing.
6  # R Foundation for Statistical Computing, Vienna, Austria.
7  # URL https://www.R-project.org/
8  #
9  data(Nile)
10 plot(
11   Nile,
12   frame = FALSE,
13   las = 1 ,
14   xlab = "Year",
```

```
15     ylab = expression(italic(X[t])),
16     xlim = c(1860, 1980)
17 )
```

Script C.6 R script producing Figure 4.2

```
1 data(Nile)
2 acf(Nile,frame=FALSE,las=1,ci.type = "ma")
3 pacf(Nile,frame=FALSE,las=1)
```

C.4 Chapter 5

For fractional calculus, we use the R package dvfBM, see Coeurjolly (2009) https://cran.r-project.org/web/packages/dvfBm/dvfBm.pdf.

Script C.7 R script producing Figure 5.1

```
1  set.seed(101)
2  n <- 1024
3  H <- 0.30
4  #
5  fBm.sim <- circFBM(n, H, FALSE)
6  plot(
7     fBm.sim,
8     las = 1,
9     frame = FALSE,
10    xlab = expression(italic(t)),
11    ylab = expression(italic(B[H](t)))
12 )
```

Script C.8 R script producing Figure 5.2

```
1 plot(
2    diff(fBm.sim),
3    las = 1,
4    frame = FALSE,
5    xlab = expression(italic(t)),
6    ylab = expression(italic(nabla*B[H](t)))
7 )
```

Script C.9 R script producing Figure 5.3

```
1  H <- 0.9
2  set.seed(101)
3  fBm.sim <- circFBM(n, H, FALSE)
4  plot(
5     fBm.sim,
6     las = 1,
7     frame = FALSE,
8     xlab = expression(italic(t)),
9     ylab = expression(italic(B[H](t)))
10 )
```

Script C.10 R script producing Figure 5.4

```
plot(
  diff(fBm.sim)*1000,
  las = 1,
  frame = FALSE,
  xlab = expression(italic(t)),
  ylab = expression(italic(nabla*B[H](t) %*% 10 ^ 3))
)
```

Script C.11 R script producing Figure 5.5

```
x <- seq(-6, 6, 0.05)
y0 <- rep(1, length(x))
y1 <- x
y2 <- x ^ 2 - 1
y3 <- x ^ 3 - 3*x
y4 <- x ^ 4 - 6*x ^ 2 + 3
y5 <- x ^ 5 - 10*x ^ 3 + 15*x
y6 <- x ^ 6 - 15*x ^ 4 + 45*x ^ 2 - 15
plot(
  c(-6, 6),
  c(-100, 100),
  las = 1,
  frame = FALSE,
  xlab = expression(italic(x)),
  ylab = expression(italic(H[n](x))),
  type = "n"
)
#abline(h=0,v=0,lwd=1.25)
lines(x, y0, lty = 1)
lines(x, y1, lty = 2)
lines(x, y2, lty = 3)
lines(x, y3, lty = 4)
lines(x, y4, lty = 5)
lines(x, y5, lty = 6)
lines(x, y6, lty = 7, col = "blue")
legend(
  "bottomright",
  title = "Degree",
  legend = paste("n=", 0:6, sep = ""),
  lty = c(1:7),
  col = c(rep("black", 6), "blue"),
  cex = 0.75
)
```

C.5 Chapter 6

Script C.12 R script producing Figure 6.1

```
phi <- 0.6
theta <- 0.7
n <- 1000
```

```
4  set.seed(101)
5  ts.sim <- arima.sim(list(
6    order = c(1, 0, 1),
7    ar = phi,
8    ma = theta),
9    n = n)
10 plot(
11   ts.sim,
12   frame = FALSE,
13   las = 1 ,
14   xlab = "Year",
15   ylab = expression(italic(X[t]))
16 )
```

Script C.13 R script producing Figure 6.2

```
1  acf(ts.sim, frame = FALSE, las = 1)
2  pacf(ts.sim, frame = FALSE, las = 1)
```

Script C.14 R script producing Figure 6.3

```
1  require(arfima)
2  par(mfrow = c(3, 2))
3  set.seed(101)
4  d.all <- c(0.01, seq(0.1, 0.4, 0.1), 0.49)
5  y <- NULL
6  for (d in d.all) {
7    title <- substitute(list(~ italic(d) == a), list(a = d↘
   →))
8    x <- arfima.sim(1000, model = list(
9        phi = .0,
10       dfrac = .3,
11       dint = 0)
12       )
13   y <- cbind(y, x)
14   plot(
15     x,
16     las = 1,
17     frame = FALSE,
18     main = title,
19     xlab = expression(italic(t)),
20     ylab = expression(italic(X(t)))
21   )
22 }
23 colnames(y) <- paste('x', 1:length(d.all), sep = "")
24 par(mfrow = c(1, 1))
```

Script C.15 R script producing Figure 6.4

```
1  par(mfrow = c(3, 2))
2  for (i in 1:length(d.all)) {
3    d <- d.all[i]
4    title <- substitute(list(~ italic(d) == a), list(a = d↘
   →))
5    x <- y[, i]
```

```
6     acf.x <- acf(x, plot = FALSE)
7     plot(
8       acf.x,
9       las = 1,
10      frame = FALSE,
11      main = title,
12      xlab = expression(italic(k)),
13      ylab = expression(italic(r[k]))
14  )
15  }
```

C.6 Chapter 7

Script C.16 R script producing Figure 7.1

```
1  n <- 500
2  b <- 0.75
3  c <- 0.6
4  n.forget <- 20
5  N <- n + n.forget
6  #
7  # We simulate a series of n+n.forget observations.
8  # We eliminate the first ''n.forget'' observations to ↘
   →get rid
9  # of the initial values impact
10 #
11 x <- rep(NA,N)
12 e <- rnorm(N)
13 set.seed(101)
14 x[1] <- rnorm(1)
15 for (i in 2:N) {
16   x[i] <- b*x[i - 1] + e[i - 1] + c*x[i - 1]*e[i - 1]
17 }
18 #
19 # Forget the first values...
20 #
21 x <- x[-(1:n.forget)]
22 plot(
23   x,
24   las = 1,
25   frame = FALSE,
26   xlab = expression(italic(t)),
27   ylab = expression(italic(X(t))),
28   type = "l"
29 )
30 acf(
31   x,
32   frame = FALSE,
33   las = 1,
34   ci.type = "ma",
35   lag.max = 40
36 )
```

Script C.17 R script producing Figure 7.2

```
1  alpha <- 0.5
2  beta <- 0.6
3  gamma <- 0.7
4  n <- 1000
5  n.forget <- 20
6  N <- n + n.forget
7  #
8  # We simulate a series of n+n.forget observations.
9  # We eliminate the first ''n.forget'' observations to ↘
   →get rid
10 # of the impact of the initial values
11 #
12 set.seed(101)
13 xi_t <- rnorm(N)
14 x <- rep(NA, N)
15 x[1] <- abs(alpha)*xi_t[1]
16 sigma2_t <- alpha ^ 2 + beta ^ 2*x[1] ^ 2
17 x[2] <- sqrt(sigma2_t)*xi_t[2]
18 for (t in 3:N) {
19   sigma2_t <-
20     alpha ^ 2 + beta ^ 2*x[t - 1] ^ 2 + gamma ^ 2*x[t - ↘
     →2] ^
21     2
22   x[t] <- sqrt(sigma2_t)*xi_t[t]
23 }
24 t <- 1:N
25 x <- x[-c(1:n.forget)]
26 plot(
27   x,
28   las = 1,
29   frame = FALSE,
30   xlab = expression(italic(t)),
31   ylab = expression(italic(X(t))),
32   type = "l"
33 )
```

Script C.18 R script producing Figure 7.3

```
1  alpha <- 0.5
2  beta <- 0.6
3  gamma <- 0.7
4  n <- 1000
5  n.forget <- 20
6  N <- n+n.forget
7  set.seed(101)
8  xi_t <- rnorm(N)
9  x <- rep(NA, N)
10 x[1] <- abs(alpha)*xi_t[1]
11 sigma2_t <- alpha ^ 2 + beta ^ 2*x[1] ^ 2
12 sigma2_t1 <- sigma2_t
13 x[2] <- sqrt(sigma2_t)*xi_t[2]
14 for (t in 3:N) {
15   sigma2_t <-
```

```
16      alpha ^ 2 + beta ^ 2*x[t - 1] ^ 2 + gamma ^ 2*sigma2↘
        →_t1
17    x[t] <- sqrt(sigma2_t)*xi_t[t]
18    sigma2_t1 <- sigma2_t
19  }
20  t <- 1:N
21  x <- x[-c(1:n.forget)]
22  plot(
23    x,
24    las = 1,
25    frame = FALSE,
26    xlab = expression(italic(t)),
27    ylab = expression(italic(X(t))),
28    type = "l"
29  )
```

Script C.19 R script producing Figure 7.4

```
1  #
2  require(astsa)
3  plot(
4    nyse,
5    main = "",
6    las = 1,
7    frame = FALSE,
8    xlab = expression(italic(k)),
9    ylab = expression(italic(X[k])),
10   ylim = c(-0.2, 0.1),
11   type = "l"
12 )
```

Script C.20 R script producing Figure 7.5

```
1  set.seed(101)
2  beta1 = 0.45
3  n <- 500
4  n.forget <- 20
5  N <- n + n.forget
6  eps <- rbinom(N, 1, 0.95)
7  x <- c(0,rep(NA,N-1))
8  for (i in 2:N)
9    x[i] <- eps[i]*(1 + beta1*x[i - 1])
10 x <- x[-(1:n.forget)]
11 plot(
12   x,
13   las = 1,
14   frame = FALSE,
15   xlab = expression(italic(t)),
16   ylab = expression(italic(X(t))),
17   type = "l"
18 )
19 acf(
20   x,
21   frame = FALSE,
22   las = 1,
```

```
23    ci.type = "ma",
24    lag.max = 40
25  )
```

Script C.21 R script producing Figure 7.6

```
1   set.seed(101)
2   n <- 200
3   n.forget <- 20
4   N <- n + n.forget
5   zeta <- rnorm(N)
6   p <- 0.5
7   x <- c(zeta[1], rep(NA, N-1))
8   for (t in 2:N) {
9     x[t] <- rbinom(1, 1, p)*x[t - 1] + zeta[t]
10  }
11  x <- x[-(1:n.forget)]
12  plot(
13    x,
14    las = 1,
15    frame = FALSE,
16    xlab = expression(italic(t)),
17    ylab = expression(italic(X(t))),
18    type = "l"
19  )
20  acf(
21    x,
22    frame = FALSE,
23    las = 1,
24    ci.type = "ma",
25    lag.max = 40
26  )
27  #
```

Script C.22 R script producing Figure 7.7

```
1   set.seed(101)
2   n <- 200
3   n.forget <- 25
4   N <- n + n.forget
5   zeta <- rpois(N + 1, 2)
6   alpha = 0.5
7   x <- c(zeta[1], rep(NA, N))
8   for (t in 2:(N + 1)) {
9     x[t] <- rbinom(1, x[t - 1], alpha) + zeta[t]
10  }
11  x <- x[-(1:(n.forget + 1))]
12  t <- 1:length(x)
13  plot(
14    x,
15    las = 1,
16    frame = FALSE,
17    xlab = expression(italic(t)),
18    ylab = expression(italic(X(t))),
19    type = "o",
```

```
20    pch = 19,
21    cex = 0.5
22  )
23  acf(
24    x,
25    frame = FALSE,
26    las = 1,
27    ci.type = "ma",
28    lag.max = 40
29  )
30  #
```

Script C.23 R script producing Figure 7.8

```
1  set.seed(101)
2  n <- 150
3  d <- 13
4  lambda0 <- 1
5  gamma_0 <- 2
6  gamma_1 <- 0.5
7  delta_d <- 0.25
8  n.forget <- 20
9  N <- n + n.forget + d
10 x <- c(rep(0, d), rep(NA, N - d))
11 lambda <- c(rep(lambda0, d), rep(NA, N - d))
12
13 for (t in (d + 1):N) {
14   lambda[t] <- gamma_0 + gamma_1*x[t - 1] + delta_d*↘
     →lambda[t - d]
15   x[t] <- rpois(1, lambda[t])
16 }
17 x <- tail(x, n)
18 t <- 1:length(x)
19 plot(
20   x,
21   las = 1,
22   frame = FALSE,
23   xlab = expression(italic(t)),
24   ylab = expression(italic(X(t))),
25   type = "o",
26   pch = 19,
27   cex = 0.5
28 )
29 acf(
30   x,
31   frame = FALSE,
32   las = 1,
33   lag.max = 30
34 )
```

C.7 Chapter 11

Script C.24 R script producing Figure 11.2

```
1  set.seed(101)
2  n <- 500
3  n.forget <- 20
4  N <- n + n.forget
5  x <- c(0, rep(NA, N - 1))
6  xi_n <- rbinom(N, 1, 0.5)
7
8  for (t in 2:N) {
9    x[t] <- (x[t - 1] + xi_n[t])/2
10 }
11 x <- tail(x, n)
12 plot(
13   x,
14   las = 1,
15   frame = FALSE,
16   xlab = expression(italic(t)),
17   ylab = expression(italic(X(t))),
18   type = "l"
19 )
20 acf(
21   x,
22   frame = FALSE,
23   las = 1,
24   ci.type = "ma",
25   lag.max = 40
26 )
```

C.8 Appendix A

Script C.25 R script producing Figure A.1

```
1  dconvex <- function(x) {
2    k1 <- 3.863305 + 0.8709496
3    k2 <- 6.797213
4    (-20*exp(-x))*(x < 2) + (6*(x - 1.25) ^ 2 - 1/10)*(x ↘
   →>= 2)
5  }
6  convex <- function(x) {
7    k1 <- 3.863305 + 0.8709496
8    k2 <- 6.797213
9    ((20*exp(-x) + k1)*(x < 2) + (2*(x - 1.25) ^ 3 - x/10 ↘
   →+ k2) *
10       (x >= 2))
11 }
12 x <- seq(0, 4, 0.01)
13 y <- convex(x)
14 plot(
15   x,
```

```
16    y,
17    xlab = expression(italic(x)),
18    ylab = expression(italic(f(x))),
19    lwd = 3,
20    las = 1,
21    type = "l",
22    xaxt = 'n',
23    yaxt = 'n',
24    xlim = c(0, 4),
25    ylim = c(0, 24),
26    frame = FALSE
27  )
28  abline(h = 0, v = 0)
29  x <- seq(0.25, 3.75, 0.25)
30  for (x0 in x) {
31    y0 <- convex(x0)
32    m <- dconvex(x0)
33    b <- y0 - m*x0
34    if (x0 == 0.25)
35      abline(b, m, lwd = 1.5)
36    abline(b, m, lwd = 0.75)
37  }
```

Script C.26 R script producing Figure A.2

```
1   set.seed(101)
2   mu <- 0
3   sigma <- 1
4   x <- seq(0, 3, 0.01)
5   y <- rnorm(x, mu, sigma)
6   plot(
7     x,
8     y,
9     xlab = expression(italic(t)),
10    ylab = expression(italic(x[t])),
11    ylim = c(-3, 3),
12    las = 1,
13    type = "l",
14    frame = FALSE
15  )
16  abline(h = 0)
```

Script C.27 R script producing Figure A.3

```
1   mu <- 0
2   sigma <- 1
3   x <- seq(mu - 3*sigma, mu + 3*sigma, 0.01)
4   y <- dnorm(x, mu, sigma)
5   plot(
6     x,
7     y,
8     xlab = expression(italic(x)),
9     ylab = expression(italic(f(x))),
10    las = 1,
11    type = "l",
```

```
12    frame = FALSE
13  )
14  abline(h = 0)
```

Script C.28 R script producing Figure A.4

```
1   mu <- 10
2   sigma <- 2
3   x <- seq(mu - 3*sigma, mu + 3*sigma, 0.01)
4   y <- pnorm(x, mu, sigma)
5   plot(
6     x,
7     y,
8     xlab = expression(italic(x)),
9     ylab = expression(italic(F(x))),
10    las = 1,
11    type = "l",
12    frame = FALSE
13  )
14  abline(h = 0)
```

References

Andrews D (1984) Non strong-mixing autoregressive processes. J Appl Probab 21:930–934
Azencott R, Dacunha-Castelle D (1986) Series of irregular observations: forecasting and model building. Applied probability, Springer, New York
Bardet JM, Doukhan P (2017) Non-parametric estimation of time varying AR(1)-processes with local stationarity and periodicity. Preprint arXiv:1705.10140
Bardet JM, Doukhan P, Lang G, Ragache N (2006) Dependent Lindeberg central limit theorem and some applications. ESAIM P & S 12:154–171
Besicovitch AS (1954) Almost periodic functions. Dover, Cambridge
Billingsley P (1999) Convergence of probability measures, 2nd edn. Wiley, New York
Bohr H (1947) Almost periodic functions. Chelsea Publishing, New York
Bradley R, Pruss A (2009) A strictly stationary, n-tuplewise independent counterexample in the central limit theorem. Stoch Process Appl 119(10):3300–3318
Breuer P, Major P (1983) Central limit theorems for non-linear functionals of Gaussian fields. J Multivar Anal 13:425–441
Brockwell PJ, Davis RA (1991) Time series: theory and methods, 2nd edn. Springer series in statistics, Springer, New York
Bulinski AV, Sashkin AP (2007) Advances series in statistical and applied sciences. Limit Theorems for Associated Random Fields and Related Systems, vol 10. World Scientific, Singapore
Chentsov NN (1956) Weak convergence of stochastic processes whose trajectories have no discontinuities of the second kind and the "heuristic" approach to the Kolmogorov-Smirnov tests. Theory Probab Appl 1(1):140–144
Choquet G (1973) Topologie, vol II. Masson, Paris
Cobb GW (1978) The problem of the Nile: conditional solution to a change-point problem. Biometrika 65:243–251
Coeurjolly J-F (2009) dvfBm: discrete variations of a fractional Brownian motion. R package version 1.0
Colomb G (1977) Estimation non paramétrique de la régression par la méthode du noyau: propriété de convergence asymptotiquememt normale indépendante. Annales de la Faculté des sciences Université Clermont. Série Mathématiques 65–5:24–46
Dahlhaus R (2012) Locally stationary processes. Time series analysis: methods and applications, vol 30. Elsevier, Amsterdam
Dedecker J, Doukhan P (2003) A new covariance inequality and applications. Stoch Process Their Appl 106:63–80

P. Doukhan, *Stochastic Models for Time Series*, Mathématiques et Applications 80,
https://doi.org/10.1007/978-3-319-76938-7

Dedecker J, Rio E (2000) On the functional central limit theorem for stationary process. Ann. Inst. H. Poincaré B Probab. Statist. 36(1):1–34
Dedecker J, Doukhan P, Lang G, León JR, Louhichi S, Prieur C (2007) Weak dependence: with examples and applications, vol 190. Lecture Notes in Statistics. Springer, New York
Derriennic Y, Klopotowski A (2000) On bernstein's example of three pairwise independent random variables. Sankhya Ser A 62(3):318–330
Diaconis P, Freedman D (1995) Iterated random functions. SIAM Rev 41(1):45–78
Dobrushin RL, Major P (1979) Non-central limit theorems for non-linear functions of Gaussian fields. Zeitschrift für Wahrscheinlichkeitstheorie und verwande Gebiete 50:27–52
Douc R, Moulines E, Stoffer D (2015) Nonlinear time series: theory, methods, and applications with R examples. CRC Press, Chapman & Hall, Boca Raton, Texts in statistical science
Doukhan P (1988) Formes de Toëplitz associées à une analyse multiéchelle. CRAS, Paris Série 1(306):663–666
Doukhan P (1994) Mixing: properties and examples, vol 85. Lecture notes in statistics. Springer, New York
Doukhan P, Grublyté I, Surgailis D (2016) A nonlinear model for long memory conditional heteroscedasticity. Lith J Math 56(2):164–188
Doukhan P, Jakubowski A, Lopes S, Surgailis D (2017) Discrete-time trawl processes with long memory. Stochastic processes and their applications
Doukhan P, Lang G (2009) Evaluation for moments of a ratio with application to regression estimation. Bernoulli 15(4):1259–1286
Doukhan P, León JR (1989) Cumulants for mixing sequences and applications to empirical spectral density. Probab Math Stat 10(1):11–26
Doukhan P, León JR (1990) Quadratic deviation of projection density estimates. CRAS Paris, Série 1 310(6):425–430
Doukhan P, Louhichi S (1999) A new weak dependence condition and applications to moment inequalities. Stoch Process Appl 84:313–342
Doukhan P, Lang G, Surgailis D, Viano MC (2002a) Functional limit theorem for the empirical process of a class of Bernoulli shifts with long memory. J Theor Probab 18(1):161–186
Doukhan P, Oppenheim G, Taqqu M (2002b) Theory and applications of long-range dependence. Birkhaüser, Boston
Doukhan P, Lang G, Surgailis D (2005) Functional CLTs for short or long memory linear sequences. Ann Inst Henri Poincaré, séries B 38(6):879–896
Doukhan P, Lang G, Surgailis D (2007a) Randomly fractionally integrated processes. Lith Math J 47:3–28
Doukhan P, Madré H, Rosenbaum M (2007b) ARCH-type bilinear weakly dependent models. Statistics 41(1):31–45
Doukhan P, Mayo N, Truquet L (2009) Weak dependence, models and some applications. Metrika 69(2–3):199–225
Doukhan P, Mtibaa N (2016) Weak dependence: an introduction through asymmetric arch models. In: Chaari F, Leskow J, Napolitano A, Sanchez-Ramirez A (eds) Cyclostationarity: theory and methods. Lecture notes in mechanical engineering. Springer, New York
Doukhan P, Prohl S, Robert CY (2011) Extremes of weakly dependent times series, discussion paper. TEST 20(3):447–502
Doukhan P, Sifre J-C (2001) Cours d'Analyse, vol 1. Masson, Paris
Doukhan P, Wintenberger O (2008) Weakly dependent chains with infinite memory. Stoch Process Appl 118:1997–2013
Duflo M (1996) Random iterative models. Springer, New-York
Esary J, Proschan F, Walkup D (1967) Association of random variables with applications. Ann Math Stat 38:1466–1474
Feller W (1968) An introduction to probability theory and its applications, 3rd edn. Wiley series in probability and mathematical statistics, Wiley, New York

Fortuin CM, Kasteleyn PW, Ginibre J (1971) Correlation inequalities on some partially ordered sets. Commun Math Phys 22–2:89–103
Giraitis L, Surgailis D (1999) Central limit theorem for the empirical process of a linear sequence with long memory. J Stat Plan Inference 100:81–93
Giraitis L, Koul H, Surgailis D (2012) Large sample inference for long memory processes. Imperial College Press, London
Hall P, Heyde CC (1980) Martingale limit theory and its application. Academic Press, London
Ho H-C, Hsing T (1996) On the asymptotic expansion of the empirical process of long memory moving averages. Ann Stat 24:992–1024
Hurst HE (1951) Long-term storage capacity of reservoirs. Trans Am Soc Civil Eng 116:770–799
Jakubowski A (1997) A non-Skorohod topology on the Skorohod space. Electron J Probab 2:1–21
Kallenberg O (1997) Foundations of modern probability. Springer, New York
Kazmin JA (1969) Appell polynomial series expansions. Math Notes Acad Sci USSR 5(5):304–311
Kedem B, Fokianos K (2002) Regression models for time series analysis. Wiley, Hoboken
Konstantopoulos T, Lin S-J (1998) Macroscopic models for long-range dependent network traffic. Queueing Syst 28:215–243
Leonov VP, Shiryaev AN (1959) On a method of calculation of semi-invariants. Theory Probab. Appl. 4:319–329
Lorentz GG (1966) Approximation of functions. Chelsea Publishing Company, AMS providence
Major P (1981) Multiple Ito-Wiener integrals, vol 849. Lecture notes in mathematics. Springer, Berlin
Marcus M-B, Pisier G (1981) Random fourier series with applications to harmonic analysis (AM-101). Princeton University Press, Princeton, Annals of mathematical studies
Massart P (2007) In: Picard J (ed) Concentration inequalities and model selection; Ecole d'Eté de Probabilités de Saint-Flour XXXIII-2003. Lecture notes in mathematics, vol 1896. Springer, New York
Merlevède F, Peligrad M, Utev S (2006) Recent advances in invariance principles for stationary sequences. Probab Surv 3:1–36
Newman CM (1984) Asymptotic independence and limit theorems for positively and negatively dependent random variables. In: Tong YL (ed) Inequalities in Statistics and Probability, vol 5. IMS Lecture Notes-Monograph Series. Elsevier, Amsterdam, pp 127–140
Nourdin I, Peccati G, Podolskij M (2011) Quantitative Breuer-major theorems. Stoch Process Appl 121(4):793–812
Paraschakisa K, Dahlhaus R (2012) Frequency and phase estimation in time series with quasi periodic components. J Time Ser Anal 33:13–31
Peligrad M, Utev S (2006) Invariance principle for stochastic processes with short memory. Monograph Series High Dimensional Probability, Vol. 51, IMS Lecture Notes, pp. 18-32
PetrovVV (1975) Limit theorems of probability theory. Sequences of independent random variables. Oxford studies in probability. Oxford University Press, Oxford
Philippe A, Surgailis D, Viano MC (2008) Time-varying fractionally integrated processes with nonstationary long memory. Theory Probab Appl 52(4):651–673
Politis DN (2003) Adaptive bandwidth choice. J Nonparametric. Stat 15:517–533
Polya G, Szegö G (1970) Problems and theorems in analysis, 4th edn. Classics in mathematics, Springer, New York
Priestley ME, Chao MT (1972) Nonparametric function fitting. J R Stat Soc Ser B 34:385–392
Resnick S, Van den Berg E (2000) Weak convergence of high-speed traffic models. J Appl Probab 37:375–397
Rio E (2017) Asymptotic theory of weakly dependent random processes (a first version appeared in French in, (2000) Number 80 in probability theory and stochastic modelling. Springer, New-York
Rosenblatt M (1991) Stochastic curve estimation. NSF-CBMS regional conference series in probability and statistics, vol 3
Rosenblatt M (1956) A central limit theorem and a strong mixing condition. Proc Natl Acad Sci U S A 42:43–47

Rosenblatt M (1961) Independence and dependence. Proceedings of the fourth Berkeley symposium on mathematical statistics and probability 2:431–443
Rosenblatt M (1985) Stationary processes and random fields. Birkhäuser, Boston
Sansone G (1959) Orthogonal functions. Intersciences, New York
Saulis L, Statulevicius VA (1991) Limit theorems for large deviations. Birkhäuser, Boston
Shumway RH, Stoffer DS (2011) Time series analysis and its applications: with R examples, 3rd edn. Springer texts in statistics, Springer, New York
Slepian D (1972) On the symmetrized Kronecker power of a matrix and extensions of Mehler's formula for Hermite polynomials. SIAM J Math Anal 3:606–616
Soulier P (2001) Moment bounds and central limit theorem for functions of Gaussian vectors. Stat Probab Lett 54(2):193–203
Straumann D (2005) Estimation in conditionally heteroscedastic time series models, vol 181. Lecture notes in statistics, Springer, Berlin
Szegö G (1959) Orthogonal polynomials, vol 23. Colloquium Publication, American Mathematical Society, New York
Szewcsak Z (2012) Relative stability in strictly stationary random sequences. Stoch Process Appl 122(8):2811–2829
Tsybakov AB (2004) Introduction à l'estimation non-paramétrique. Springer, Heidelberg
van der Vaart A (1998) Asymptotic statistics. Cambridge series in statistical and probabilistic mathematics. Cambridge University Press, Cambridge
van der Vaart A, Wellner JA (1998) Weak convergence and empirical processes. Springer series in statistics, Springer, New York
Wu WB, Rosenblatt M (2005) Nonlinear system theory: another look at dependence. Proc Natl Acad Sci USA 102(40):14150–14154

Index

Symbols
$(\Omega, \mathcal{A})$, measurable space, 247
$(\Omega, \mathcal{A}, \mathbb{P})$, probability space, 247
B-function, Euler, 272
$B(n, p)$, binomial distribution, 5, 12, 14
$B_H(\cdot)$, fBm, 74
$D(a, r) = \{z \in \mathbb{C}/ |z - a| < r\}$ open disk, 109
$D(a, r) = \{zin\mathbb{C}/ |z - a| < r\}$ open disk, 107
$D(a, r) = \{z \in \mathbb{C}/ |z - a| < r\}$ open disk, 113
H-self-similarity, 77
$W(\cdot)$, Bm, 60, 75
Δ-method, 139
Γ-function, Euler, 270
Λ, set of Lipschitz functions, 171
$B30D \cdot B30D_\infty$, 276
η weak-dependence, 213, 231
γ-distribution, 270
κ weak-dependence, 213
λ weak-dependence, 213
$\mathbb{C}$, complex numbers, 3
$\mathbb{E}$, expectation, 250
$\mathbb{N}$, non-negative integers, 3
$\mathbb{P}$, probability, 3
$\mathbb{R}$, real numbers, 3
$B(n, p)$, binomial distribution, 255
$\Re$, real part of a complex number, 271
$\mathbb{E}^{\mathcal{B}}X, \mathbb{E}(X|\mathcal{B})$, conditional expectation, 253
$|||\cdot|||$, operator norm, 241
$\to^{\mathcal{L}}_{n\to\infty}$, convergence in distribution, 276
$\vee$, maximum, 3
$\wedge$, minimum, 3
$b(p)$, Bernoulli distribution, 255, 261
$\mathcal{C}[0, 1]$, space of continuous functions, 276
$\mathcal{N}(0, 1)$, Normal-distribution, 256
$\mathcal{O}(\cdot)$, Landau notation, 248
$\mathcal{O}_{\mathbb{P}}(\cdot)$, 249
$\mathcal{O}_{\mathbb{L}^p}(\cdot)$, 249
$\mathcal{O}_{a.s.}(\cdot)$, 249
$\mathcal{P}(\lambda)$, Poisson, 255
$\wr(\cdot)$, Landau notation, 248, 257
σ-algebra, σ-field, 247–249
θ weak-dependence, 213
$\to_{fdd}$, finite dimensional convergence, 278
$\to^{\mathbb{L}^p}_{n\to\infty}$, convergence in $\mathbb{L}^p$, 280
$\to^{\mathbb{P}}_{n\to\infty}$, convergence in probability, 280
$\to^{a.s}_{n\to\infty}$, almost-sure convergence, 281
$b(p)$, Bernoulli distribution, 207
m-dependent, 149
$\mathcal{C}^k_b(I)$, $\mathcal{C}^k_b$, smooth functions on an interval, 9
Var , variance, 250

A
Appell polynomials, 117
Associated, 167
Atom, 13
Autocovariance, 58
Autoregression, 129
  integer valued, 147

B
Bernoulli scheme, 155
Bernstein blocks technique, 220
Bootstrap, 69, 206, 209
Borel–Cantelli lemma, 282
Borel sigma-algebra, 249
Brownian motion, 60, 75, 94, 150, 191

P. Doukhan, *Stochastic Models for Time Series*, Mathématiques et Applications 80,
https://doi.org/10.1007/978-3-319-76938-7

C
Cadlag or càdlàg: continue à droite, limite à gauche, left-continuous with limit on the right, 277
Causal, 102, 107, 155
Chain, infinite memory, 130
Chaos, 116
discrete, 115
Gaussian, 73
Hermite, 117
Characteristic function, 254, 265
Complete, 248
Concentration condition, 165
Contrast, 30, 31
Convergence
J_1, 173, 200, 278
M_1, 200
$\mathbb{L}^p$, 280
almost-sure, a.s., 281
in $\mathbb{L}^p$, 44
in distribution, 191, 209, 276
in probability, 279
finite dimensional, fdd, 173, 278
uniform in $\mathbb{L}^p$, 44
Coupling, 162
Cov, covariance, 250
Covariance, 5, 50, 53, 58, 62, 250
Covariogram, 65, 111
Cumulant, 225, 227, 228, 230, 235

D
Decorrelation, 162, 163
Density, 4, 30, 68, 240, 255, 264
multispectral, 225
spectral, 60, 61, 66
Dependence
long-range, LRD, 60, 189, 198, 203
short-range, SRD, 60, 186, 191
Dependence coefficient
α_r, 206, 210, 216
η_r, 213
κ_r, 213
$\kappa_{X,q}(r)$, 231
λ_r, 213
θ_r, 213, 216
$c_{X,q}(r)$, 231
Distribution
continuous, 13
cumulative, 118, 166, 222
image, 249

E
Empirical
cumulative distribution, 28, 88
mean, 27
process, 222
Ergodic, 27, 178
Estimation, 283
consistent, 104, 139, 140, 185, 186, 283
contrast, 30
empirical, 27, 32, 64, 122, 140, 185
empirical covariance, 58
histogram, 31
kernel, 33, 38, 66, 240
kernel regression, 38
least squares, LSE, 31
maximum likelihood, MLE, 30
Nadaraya–Watson, 39, 40, 44, 47
orthogonal projection, 32
semi-parametric, 45
wavelet, 32
Whittle, 67, 136
spectral, 66, 67, 225
Euler
Γ function, 270
B function, 272
Event, 4–6, 177, 208, 247
Expectation, 251
conditional, 253

F
Filtration, nested, 286
Formula
diagram, 92
Hoeffding, 263
Leibniz, 83
Mehler, 86
Fourth moment method, 79, 96
Fractional
Brownian motion, fBm, 74
filter, 202
integration, 202

G
Gaussian
chaos, 78
family, 73
process, 73
vector, 267
Generating function, 254

H
Heat equation, 91
Hermite
  expansion, 79
  rank, 82, 88, 191
Homeomorphism, 278

I
Identifiable, 31
Independent, 3–6
  pairwise, 4
Inequality
  Bennett, 24
  Bernstein, 24
  exponential, 22
  Hoeffding, 22
  Hölder, 230, 253
  Hopf maximal, 181
  Jensen, 135, 251, 252
  Marcinkiewicz–Zygmund, 17, 240
  Markov, 251
  Minkowski, 254
  moment, 17, 216
  Rosenthal, 17
Iterative random model, 129, 150

K
Kernel, 33, 38, 45, 52, 66, 67, 104, 164
  Dirichlet, 66
  order p, 33
  Markov, 129, 240

L
Landau notation, 63, 248
Laplace transform, 254, 265
Law, 249
  γ, 270
  χ^2_k, 273
  $\gamma(a, b)$, 270
  Bernoulli, 20, 145, 207, 255
  binomial, 255
  Cauchy, 256
  exponential, 256, 261, 273
  Gaussian, 222, 225, 247, 256, 265
  Normal, 256
  Poisson, 255, 258
  Rademacher, 22, 222
  uniform, 255, 261
  uniform on the interval, $U[0, 1]$, 255
Lebesgue measure, λ, 84
Leverage, 141
Limit superior, lim sup, $\overline{\lim}$, 281
Lindeberg, 9, 11
  dependent, 219
Long-range, periodic, 113

M
Markov chain, 128
Markov chain, stable, 128
Martingale, 284
  convergence, 284
  sub-, 284
  super-, 284
Mean, 102, 104, 147, 186, 250
Measurable space, 247
Metric
  Skorohod, 278
Model
  AR-ARCH, 131, 132
  ARCH asymmetric, 141
  ARCH(2), 132
  ARFIMA$(0, d, 0)$, 108
  ARFIMA(p, d, q), 112
  ARMA(p, q), 105
  bilinear, 122
  branching, 144
  GARCH, 150, 152
  GARCH(1,1), 133
  generalized linear, GLM, 149, 150, 154
  INAR, 147, 169
  INARCH, 7
  INMA(m), integer moving average, 149
  LARCH(∞), 127, 169, 201, 214
  memory, 120, 128
  non-linear AR, 68, 132, 154, 169
  selection, 31
  switching, 144
  tvAR(1), time varying AR(1), 120
Moment, 126, 133, 137, 138, 140, 151, 166, 205, 225
  method, 137, 145

O
Operator
  backward, 103
  shift, 103
  Steutel–van Harn, 147
  thinning, 147
Orlicz norm, 13

P
Periodic, 52

almost, 52
Periodogram, 65, 225
Polynomial
Appell, 116
Hermite, 80
Jackson, 67
Jacobi, 85
Legendre, 85
orthogonal, 84
Tchebichev, 85
Probability, 247
space, 247
Process
compound Poisson, 57, 260
linear, 101
locally stationary, 113
periodic, 114
Poisson, 112, 129, 150, 258
symmetric Bernoulli, 198, 261
symmetric Poisson, 198, 263
mixed Poisson, 260

R
Random
iterative system, 150
measure, 55, 57
process, 275
variable, 249
Range, 186
Regression, non-parametric, 38
Regression, random design, 38
Resampling, 69, 210

S
Separable, 277
Simulation, 260
Skorohod space, 188, 277
Spectral representation, 57
Stationarity
local, 59, 117, 120
second order, 49
strict, 49, 50, 127, 130, 152, 156, 172, 183, 189
weak, 49, 50, 57, 59, 60, 62, 65, 187
Statistic, 282
Statistical model, 282
Stirling formula, 109
Stochastic volatility model, 132
Strong mixing, 140, 206, 210
Sublinearization, 253
Subsampling, 69, 209, 222, 245
Symmetric
definite, 268
non-negative, 268

T
Theorem
central limit, 12
ergodic, 177
Hahn–Banach, 252
Heine, 22
Herglotz, 54
Kolmogorov, 74, 276
Weierstrass, 20
Trawl, 199

U
Unbiased, 28, 30, 32, 64, 88, 283

V
Volterra expansion, 115, 116, 120, 214

W
Weak-dependence, 163, 167, 172, 205, 206, 211, 222, 231
Wold decomposition, 65

Y
Yule–Walker equation, 107, 136

Printed in the United States
By Bookmasters